# Development and Prospective Applications of Nanoscience and Nanotechnology

***Volume 1: Nanomaterials and their Fascinating Attributes***

**Editor**

**Sher Bahadar Khan**

**Co-Editors**

**Abdullah M. Asiri**

*Center of Excellence for Advanced Materials Research (CEAMR)*
*and Chemistry Department*
*Faculty of Science*
*King Abdulaziz University*
*Jeddah 21589*
*Saudi Arabia*

**&**

**Kalsoom Akhtar**

*Division of Nano Sciences and Department of Chemistry*
*Ewha Womans University*
*Seoul 120-750*
*Korea*

# CONTENTS

# FOREWORD

The field of nanotechnology has emerged just a decade ago, scientists and pundits were questioning if nanotechnology would ever lead to the development of "real" products with "real" value to consumers. Was nanotechnology just another scientist's dream that would not provide any practical value to society?

Today, not only one ecan buy tennis rackets, bicycles, clothes, and band-aids that incorporate nanomaterials, but it is clear that such products provide greater value to the general public than their respective conventional products. Nanotechnology is no longer just a marketing ploy or a fancy advertisement scheme used by Apple. Gone also are the days of naysayers wondering how nanotechnology is different than the fundamental sciences of chemistry, physics, and biology. What continues to separate nanotechnology from the study of fundamental atomic and molecular interactions that a traditional researcher in those fields may accomplish is an emphasis on *new properties of materials* gained when controlling structures at the nanoscale - the atomic and molecular level. It is this continual emphasis on the control of structures at the nanometer level that has led to *significantly changed properties and products* that are advancing numerous fields. This is what separates nanotechnology from other traditional science fields (such as chemistry, physics, and biology).

What also continues to provide a bright future for nanotechnology is the selection of difficult contemporary problems that only nanotechnology (it appears) is providing solutions for. This book will highlight some of these persistent contemporary problems in which nanotechnology is finding solutions, such as building novel sensors, alternative energy sources, and improved semiconductors. This book eloquently defines current problems we have with limited sources of carbon-based energy and provides nanotechnology solutions to develop novel solar cells, among other solutions. It also provides solutions unimaginable without nanomaterials to build more precise sensors to detect levels of contaminants previously not possible. New routes of nanoparticle synthesis are also covered in this book to ensure the safe fabrication of high purity, inexpensive nanomaterials – issues critical for the continued commercial success of nanotechnology. Issues of toxicity are also discussed providing valuable insights into how we should choose nanoparticles with low toxicity for particular energy, sensor, or semi-conductor applications.

In closing, though, as we advance into the next generation of nanotechnology research we should not forget the past. It has been our emphasis of nanoscale events which has led us to where we are today. For example, the theory of atomic matter was suggested as early as 370 BC by the Greek philosopher Democratis; this could certainly be our first indication of the desire to understand nanoscale events to create materials with tailorable properties. Since then many scientists have followed this train of thought, and, of course, in 1959 Richard Feynman predicted the possibility of maneuvering matter atom by atom.

It has been our continual emphasis on creating materials with unique properties and the selection of the world's most pressing problems that has allowed nanotechnology to continue to grow. We must not forget this and continue to address our most significant scientific problems. We must look at our materials and determine ways to improve their properties by understanding and controlling nanoscale assembly. The book reminds us all of the problems that we have today in energy, sensors, semi-conductors and many other applications and how nanotechnology is one of the few fields providing any type of answers. Enjoy reading this book and let the discoveries continue……

***Thomas J. Webster***
Chair and Professor of Chemical Engineering
Northeastern University, Boston, MA 02115, USA

# PREFACE

Nano represent very small thing which has diverse applications and high potency that guiding to industrial and technological development.

Materials of nano size show remarkable physical and chemical properties due to several factors including the rise in surface area compared to volume which takes place as particles get smaller.

Nanotechnology finds its applications in many areas such as novel foodstuffs, medical equipments, chemical coatings, human health testing kits, security system's sensor, water purification items for manned space craft, displays for hand-held computer games, and cinema screens. Nanotechnology is likely to impart an impact on almost every industry. Recently nanotechnology has become known as the science of every tiny thing, more particularly, 'nano'. Nanotechnology is a catch-all expression for resources and tools that function at the nanoscale.

Nanotechnology is a diverse science that has spread its applications in fields such as colloidal science, device physics and supra molecular chemistry. Being a subject of key interest, it was thought out to present recent development in nanotechnology in the form of a book to the scientific community.

There are six chapters in this book. In the first chapter, the recent development of nanotechnology in many applications such as energy/power, medicine/healthcare, water purification, biotechnology, electronics, sporting goods, environmental issues, defense/security, and textile/fabrics is summarized. Chapter two reviews the latest developments in the area of nanocomposites with reference to nanoclays, nanofibers and nanotubes. The classification, characterization and applications of various types of nanocomposites have been discussed in aerospace industry. It also highlighted the applications of nanocomposites in relaunch vehicles, space ladders, ablatives, internal motor casing and nano modified carbon/carbon composites. In third chapter, use and potential applications of nano materials, organic semiconductor complexes and composites sensors, organic field effect transistors, thin-film field effect transistors, temperature sensitive field-effect transistor, solar cells, and light-emitting diodes have been summarized. Fourth chapter describe nanotechnology, nanomaterials, nanocomposites, their importance, synthesis and characterization of nanocomposites. Fifth chapter describe synthesis of nanoparticles of various compositions, their exciting properties and size, and morphologies for a wide range of applications. Chapter six deals with DFT study of nanoparticles using run type of Frequencies in ADF-Molecule interface.

We deem that this book will convey the savor of the nanotechnology properly and illustrate that study on nanomaterials is emerging to be of key interest. We intently anticipate that this book will be beneficial for students, teachers and practitioners.

***Sher Bahadar Khan &***
***Abdullah M. Asiri***
Center of Excellence for Advanced Materials Research
Chemistry Department
Faculty of Science
King Abdulaziz University
Jeddah 21589
P.O. Box 80203
Saudi Arabia

***Kalsoom Akhtar***
Division of Nano Sciences
Department of Chemistry
Ewha Womans University
Seoul 120-750
Korea

# CONTRIBUTORS


| | |
|---|---|
| **Kishwar Khan** | Center for Sustainable Energy Systems (CSES), ANU College of Engineering and Computer Science (CECS), Australian National University (ANU), Canberra ACT 0200, Australia |
| **Sarish Rehman** | College of Engineering, Peking University, Beijing; Institute of Photonics, University of Eastern Finland (UEF), Department of Physics and Mathematics, P.O. Box 111, FI-80101, Joensuu, Finland |
| **Zaffar Mohammad Khan** | Department of Aeronautics and Astronautics, Institute of Space Technology, Islamabad Highway, Islamabad, Pakistan |
| **Mohammad Rizwan Saleem** | Department of Aeronautics and Astronautics, Institute of Space Technology, ISI, Islamabad Highway, Islamabad, Pakistan and Institute of Photones, University of Eastern Finland, (UEF), Department of Physics and Mathematics, FI-80101, Joensuu, Finland |
| **Kh. S. Karimov** | GIK Institute of Engineering Sciences and Technology, Topi, KK, Pakistan and Physical Technical Institute of Academy of Sciences, Rudaki Ave.33, Dushanbe, 734025, Tajikistan |
| **Muhammad Abid** | Interdisciplinary Research Center, COMSATS Institute of Information Technology, Wah Cantt, Pakistan |
| **Sher Bahadar Khan** | Center of Excellence for Advanced Materials Research (CEAMR) and Chemistry Department, Faculty of Science, King Abdulaziz University, P. O. Box 80203, Jeddah 21589, Saudi Arabia; Center of Excellence for Advanced Materials Research, King Abdulaziz University, Jeddah 21589, P.O. Box 80203, Saudi Arabia; Chemistry Department, Faculty of Science, King Abdulaziz University, P. O. Box 80203, Jeddah 21589, Saudi Arabia |
| **Kalsoom Akhtar** | Division of Nano Sciences and Department of Chemistry, Ewha Womans University, Seoul 120-750, Korea |
| **Abdullah M. Asiri** | Center of Excellence for Advanced Materials Research, King Abdulaziz University, Jeddah 21589, P.O. Box 80203, Saudi Arabia; Chemistry Department, Faculty of Science, King Abdulaziz University, P. O. Box 80203, Jeddah 21589, Saudi Arabia; Center of Excellence for Advanced Materials Research (CEAMR) and Chemistry Department, Faculty of Science, King Abdulaziz University, P. O. Box 80203, Jeddah 21589, Saudi Arabia |
| **Shahida B. Niazi** | Institute of Chemical Sciences, Bahauddin Zakariya University, Multan, Pakistan |
| **Abdul Majid** | Physics Department, University of Gujrat, Pakistan |

# Development and Prospective Applications of Nanoscience and Nanotechnology

## *Volume 1: Nanomaterials and their Fascinating Attributes*

**CHAPTER 1**

# Wonders of Nanotechnology

**Kishwar Khan[1,*] and Sarish Rehman[2]**

*[1]Center for Sustainable Energy Systems (CSES), ANU College of Engineering and Computer Science (CECS), Australian National University (ANU), Canberra ACT 0200, Australia and [2] Department of Materials Science and Engineering, College of Engineering, Peking University, Beijing 100871, China*

**Abstract**: The Sector of Nano-Technology, at present, has evolved with a gigantic breakthrough in the World of Technology. Keeping in view the application of Nano-Technology on items such as of range from clothing to *etc*., funding and investment are immensely encompassed for research in materials throughout the world. Nanotechnology is being well-thought-out as another industrial revolution and is expected to be our socio-economic life for the upcoming 40-50 years. Nanotechnology deals with the materials at small level such as atoms and molecules are referred to be as Nano-Technology. Nanotechnology is the science and technology at the level of such small material sizes as that of atoms and molecules. These materials are called to be as "Nanomaterials". Their structures are keenly researched and these materials are used in almost all types of industrial products to improve their quality and performance with large value additions. In order to perceive the small sizes of atoms and molecules, it can be imagined that these sizes are 800,000 times smaller in size as compared to the thickness of human hair or of a paper sheet.

The materials used at that small sizes or scale, what is termed as "nano scale", are termed as nanomaterials and have very exotic properties and improve the performance of the products, thereby adding great value to the similar industrial products which donot use nanoscale materials and which bring much lower prices. These industrial products could be textiles, medicines, pharmaceuticals, sports goods, auto or electronics products, computers, communication equipment, solar energy, clean drinking water, materials of construction industry, oil and gas exploration, defense and security equipment, *etc*.

This is because of the fact that all materials are made of atoms and molecules and the control and use of materials at that small sizes make the products of great quality and performance. In this chapter, it has been endeavored to summarize the recent development in nanotechnology for numerous applications in energy/power, medicine/healthcare, water purification, biotechnology, electronics, sporting goods, environmental issues, defense/security, and textile/fabrics.

**Keywords:** Applications, clean drinking water, communication equipment, computers, electronics products, energy, industrial products, nano hazards, nanomaterials, nanoparticles, nanotchnology, photovoltaic, solar cells, solar energy.

***Corresponding author Kishwar Khan:** Center for Sustainable Energy Systems (CSES), ANU College of Engineering and Computer Science (CECS), Australian National University (ANU), Canberra ACT 0200, Australia; E-mail: kknano@hotmail.com

**Sher Bahadar Khan, Abdullah M. Asiri & Kalsoom Akhtar (Eds)**

# 1. INTRODUCTION

As from elucidate of Professor Richard Feynman, in 1959, gave the first illuminating talk on nano technology, which was entitles as:

There is relative quantity of space at the bottom. Normally, Materials Sciences and Materials Technology is referring by ceramics or crystalline materials, glasses or non-crystalline materials, polymers or harsh chain molecular materials and metals or cohesively-bonded materials. All these materials are widely applied in a changed trend towards improvement of human life. Materials world is boom developed with the latest modern technologies, and, so as to be visible having its progressive outputs. Nanotechnology is one of those fresh and delicate technologies which have been fashion waves in this modern vintage [1-3].

However, Nanotechnology composed of the logical study of Material Physics and Material Chemistry. Nanotechnology is highly alert a fresh study and field into prominence, so consequently, it is an interesting but early field of study under constant development, tender to precise extend liberty of research progress [3].

## 1.1. What is Nanotechnology?

Nanotechnology is the study of small atoms having sizes from 1-100 nanometers as in shown in Fig. (**1**). As from the scientists, dangers of nanotechnology possibly exist in how these small atoms may interact with the surrounding, and more specifically, with the human physical body. Big quantity is being stressed to integrated nanoparticles into obtained that are already being developed in the public eye. When this asset is compared with the proportional deficient research into nanotech health problems, various scientists uneasily become sorrowful [4].

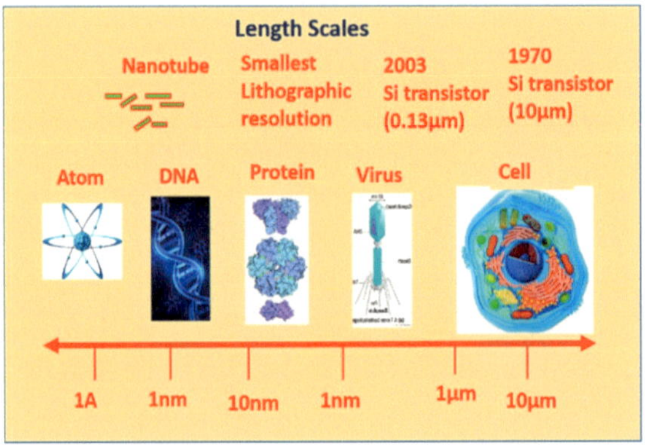


**Figure 1:** Length scales of nanotechnology.

Professor George Robillard, Director (BIOMADE), University of Groningen, Netherlands, expressed Nanotechnology in a better way 'The basic of nanotechnology depend of systems in the size range of nanometers. For instance, a drug-delivery system is nanotechnology. The Molecular assembly in big functional complexes is of high affair. A complex that can produce a protein to a specific area in the body is one of its example [1, 4].

According to the field experts, the elements changing upon at the nanoscale, and exists very differently than their larger complement.

For example the graphite, as its properties are well known and it holds specific position in toxicology guidepost.

Nobel laureate, Richard Smalley, Rice University, has acquired about carbon nanotubes (CNT) and fullerenes (buckyballs) — nanoparticles of carbon — that are lawfully treated as graphite, even, they behave in means dissimilar graphite thus fashioning the arrangement a potentially harmful one [2, 5, 6].

The Worldwide research progress is making nano products with the aim to increasing health care and advancement in other research areas, some of these Nano-Products have registered in the market, more are towards market, and, others are still under progress than to become a fact. These inventions possess noted potential, chart as shown in Fig. (**2**). Although, miner quarries are arise about their long term status and the dangerous–benefit characteristics of their utilization.

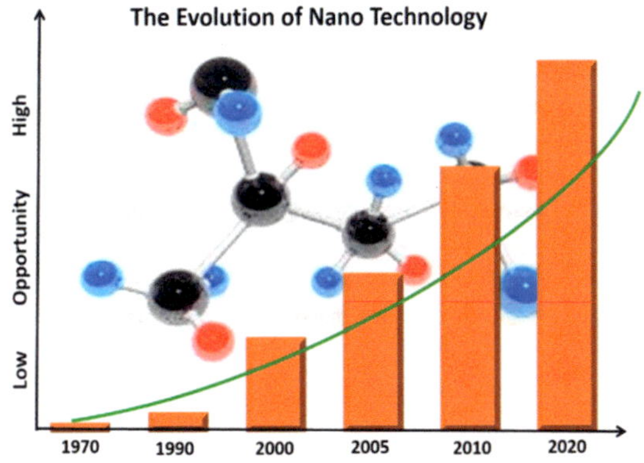


**Figure 2:** The evaluation of nanotechnology.

In the meantime, in 2000, when US President Bill Clinton publicized the establishment of the United State National Nanotechnology Initiative (NNI),

Europe, Japan, & other Asian countries likely to be China, Korea, Iran, India, Taiwan and Nuclear power Pakistan have wonder with efficient funding in national nanotechnology programs.

European Commission, funding about 24% of the publicly financed research in the European Union, and the Euro-Union's fifteen (15) member countries will spend about $180 million (200 million euros) on nano science nanotechnology in 2002 [5].

Nanotechnology covers a diverse range of subject matter and industry areas as shown in Fig. (**3**). It has illimitable present applications and many other applications that are still to be known [3].

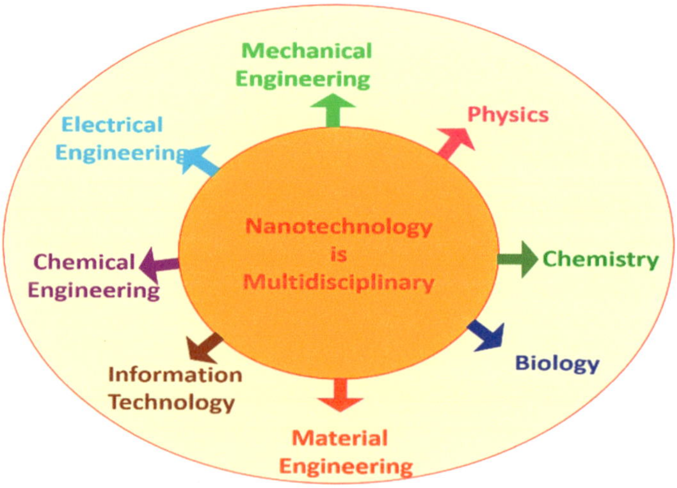


**Figure 3:** Subjects that cover nanotechnology.

Nanotechnology can be expressed as an enabling technology that grows many of our present technologies. Nanotechnology positively mixed current industries by improving their outputs and the quality of their processes in addition to the strength and other in progress features of materials which are applicable in current products [3].

## 1.2. World Investment in Nanotechnology

Fig. (**4**) shows, world Investment nanotechnology is so important that all countries, including the neighboring countries are rapidly and heavily investing billions of dollars in the R&D and on the applications of nanotechnology, detail chart of some countries is shown in Fig. (**5**).

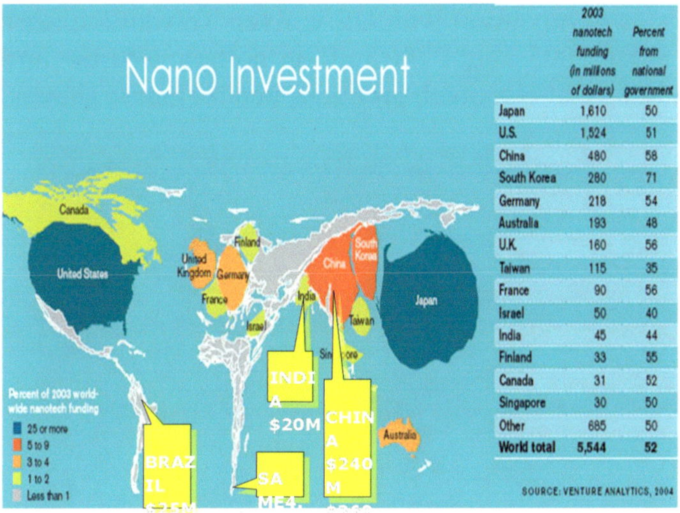


**Figure 4:** Nanotechnology investment.

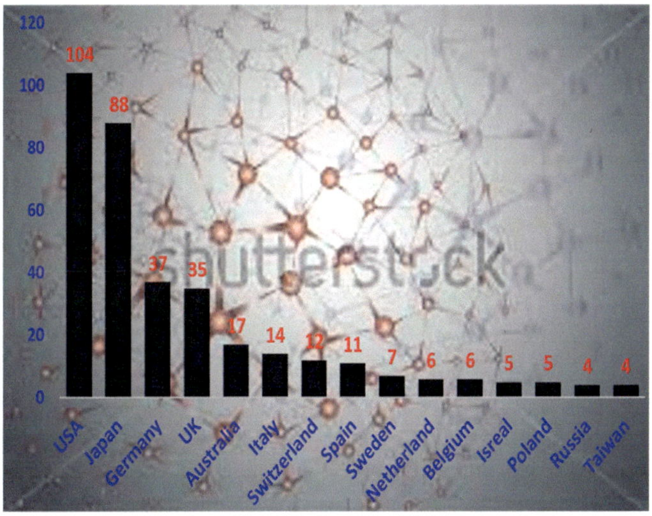


**Figure 5:** Countries investing chart.

To mention briefly, the USA has invested 8.3 billion dollars from 2001-2008 with \$1.5 billion allocation for 2009 for nanotechnology. Russia invested \$7.5 billion to develop nanotechnology for 5 years period since 2007 of which \$5 billion have already been released.China and EU are also investing billions of dollars annually in Nanotechnology in competition with USA. Even Taiwan industry has invested on a Nano- park since 2005 about 700 million dollars with the aim of about 20 billion dollars marketing of Nanotech products, detail chart is shown in Fig. (**6**).

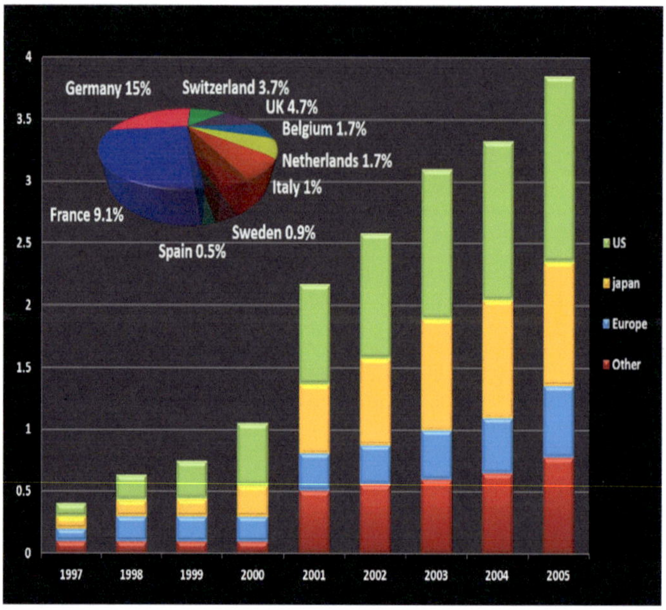


**Figure 6:** Nanotechnology products chart.

## 1.3. World Nanotechnology and Neighboring Countries

Among the neighboring countries, India has invested 250 million dollars to establish three Nanotech Institutes in Bangalore and invested 300 million dollars for a Nano-city in Chandrigarh with promise of billion dollar investment in the next few years. This investment is by a business man who initiated the famous Hot Mail e-mail system. Already in India (Sept 2008) there are 50 Institutes working in R&D of nanotechnology and about 30 industries are involved for making consumer products based on Nanotechnology. Iran is following Nanotechnology with so much seriousness that the President of Iran has already appointed a scientist as "Advisor on Nanotechnology" in the President's office who coordinates nanotechnology programs of Ministries of health, industries, oil and gas *etc* on National - level. Already Iran has signed contracts for commercial plants of Clean–drinking water production, commercial plant for Oil and Gas cleaning and commercial scale production of high tech electronics products for use in defense and consumer industry.

Nanotechnology is so highly commercial and business oriented that over 2500 companies are involved in business in the world over, with about 700 in USA alone. This makes it obvious the commercial importance of the nanotechnology, as shown in Fig. (**7**); as a company will not venture investment unless there is profit insight. The business community estimates that by 2015 there will be a business of about $ 1- 2 trillion in nanotechnologies. In terms of Business it is a huge market.

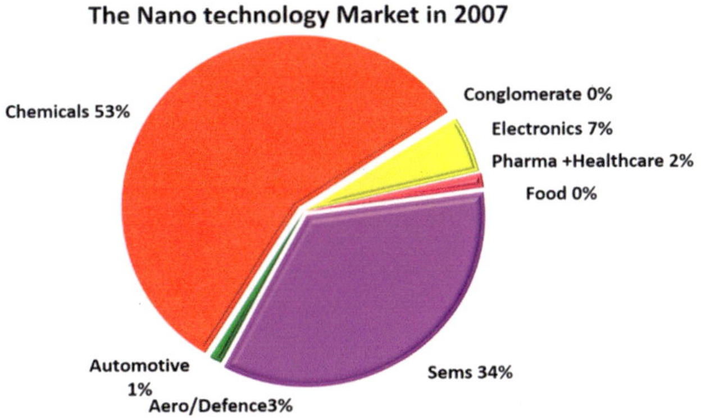


**Figure 7:** Nanotechnology market in 2007.

## 1.4. World Wide Nanotechnology: Universities and Companies

Recently various industrial companies and academic institutes have focused their interest and investment in nanotechnology. The following charts in Fig. (**8**), representing the country wise summery of nanotechnology usage. Biggest part of nano market is the segment of ball sports equipment having 23.7% of the total market value. Japan, France, UK, Germany and USA are the major markets for sporting tolls where USA itself commands 40.6% share of the arcade.

## 1.5. Support of Political Leaders for Nanotechnology

The political leadership of these countries is so supportive that special funds and public resources are allocated to pursue nanotechnology. The US presidents have been allocating billions of dollars for nanotechnology initiatives through the Act of Parliament in order to give nanotechnology the importance which is due to it. Even the Presidents of the neighboring developing countries of India and Iran have laid special emphasis on the development on Nanotech for their countries. There are about 300 laboratories on nanotech in US universities and centers alone. This again shows the extent of seriousness of the US nanotech. Not only the US political leadership including President *Obama* is keen on nanotechnology even the president of the developing countries of India and Iran have been making open statements on the need of Nanotechnology for development of their countries. The former President of India, *Dr. Abul Kalam* made the statement that "nanotech is one of the two technologies which will take India by 2015 to the level of advanced countries. European countries are striving hard to compete with US and Japan in Efforts to support Nanotechnology.

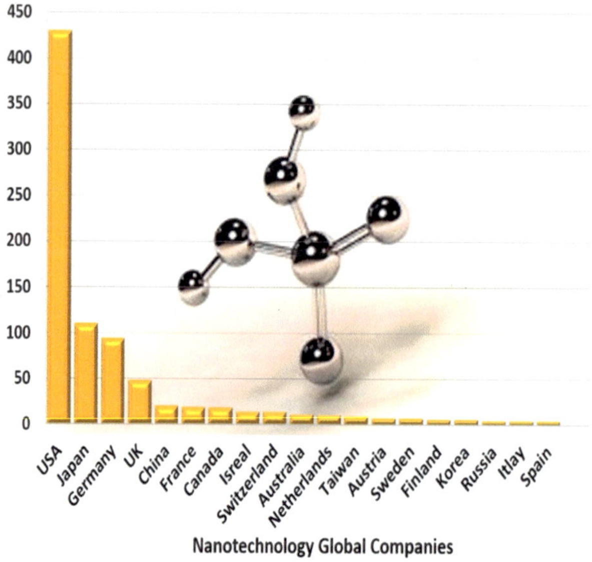


**Figure 8:** Nanotechnology global companies.

## 1.6. Why Nano will Change the Properties of Materials?

When materials are decreased to Nano-Scale Magnitudes, their major properties often get changed vividly. Getting started approximately at 100 nanometers and below, materials become capable to overcome the size barrier at which quantization of energy is quite significant for the electrons. At Nano-scale level materials properties are different compared macro-Scale. Let us try to explain through an example *i.e.*, opaque substances convert to transparent ones (copper); inert materials act as catalysts (platinum); stable materials shot as ignitable (aluminum); solids turn into liquids at room temperature (gold); and insulators become conductors (silicon). Afresh of concepts evolved with Fresh Tools. It has been driven out that the mechanical rules which administrate and manage the Nano-World, are pretty different from our routine, Macro-World Know-How. Embodying the extreme small forces, which act at Nano-Scale, has also steered to further development and progression of instruments.

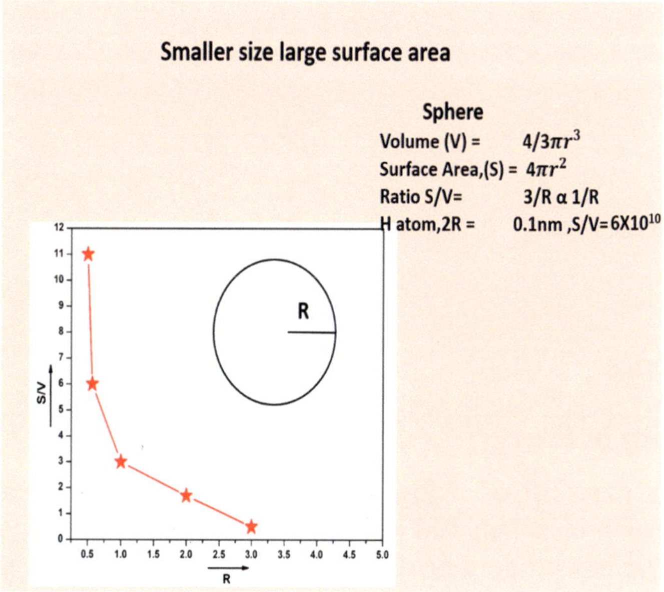


**Figure 9:** Smaller size large surface area relation.

A shown in Fig. (**9**), one more vital trait of Nano-Materials is to discuss their surface area. When they're matched with the exact value of material in substantial form, materials at nano scale level comparatively expose high surface area compared to macro-scale. This lead to high chemical reactivity of material materials in comparison to bulk at which they are inert, become reactive as soon as they are produced in their Nano-Scale Form), and shake their strength or electrical properties.

## 2. APPLICATIONS IN VARIOUS FIELDS

Talking of uses in various fields, the study of Nanotechnology should speed up us to made synthetic product advance, light in weight, with more strength, safe and sound as shown in Fig. (7).

Mentioned below are some zones in which nanotechnology can have remarkable magnitudes:

a. Energy

b. Medicine

c. Water purification

d. Biotechnology

e. Electronics

f. Sports

g. Environmental Issues

h. Defense

i. Textile/Fabrics

## 2.1. Nanotechnology in Energy

Nanotechnology is the skill of using materials and structures that expose one supercritical dimension less then 100 nm that leads to advance applications [7, 8]. At nano level the various properties of materials such as conductivity, mechanical properties optical and magnetic properties are quite different from macro-scale.

Therefore, nanotechnology offers unlimited possibilities to explore latest methods in a resourceful manner to conserve energy and so address energy shortage problems.

Nanotechnology presents tremendous applications in various energy field like, supercapacitor, Lithium-Ion batteries, solar cells, energy saving light bulbs, ductile electronic (mobile phone, tablets), green and safe fuel cells to authorize the advancement of high power car [9].

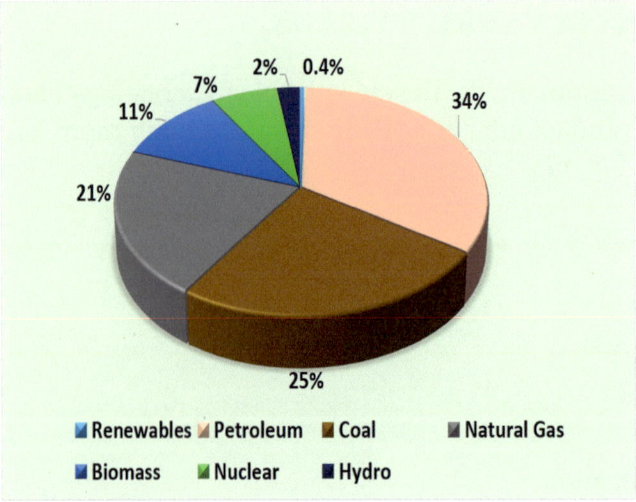


**Figure 10:** Shows EIA World Energy Sources, 2006 [9].

At present, gauss oil reserves are approximately one thousand (1000) billion barrels [10] which is enough to meet the energy requirements for at least only three (03) years. Majorly saying, there exist two solutions to this problem. The first change is to overcome the rate of energy expenditure by increasing the energy efficiency. By doing so, we can decline the equal of sixty four (64) million barrels of oil per day [11].

The second solution may be to find renewable energy resources and techniques, to make the base balance with the existing sources.

Nanotechnology greatly affects the energy from various aspects like production, conservation, transmission and usage *etc*. From Fig. (**10**), we easily summarized, that at present, the best energy sources are natural gas petroleum, and coal tar. It is shown in the above list that the current oil resources are not enough to fulfill the increasing energy demands, it is noticeable to express that coal is available in huge amount but, it is badly linked with $CO_2$ emissions thus making even this energy source altogether unattractive. Biomass Energy Systems have great struggle with food reservoirs [8]. Increase in Living Values and Population Escalation have led to Increase in Consumption of Global Energy. The Usage of Energy on Worldwide Basis is estimated to increase about 40% in the next 20 years, Fig. (**11**), and double of this much by 2050 [12].

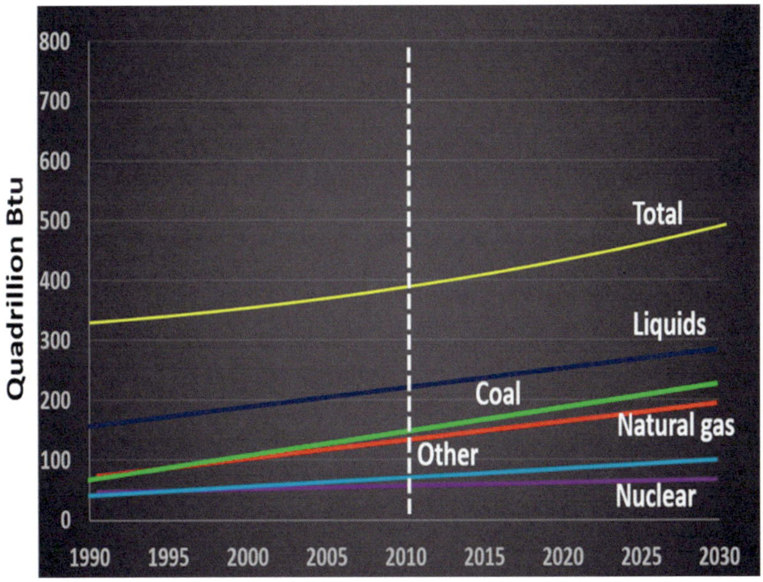


**Figure 11:** Total Energy use up on worldwide basis in quadrillion BTUs (1990-2050) (according to U. S Energy Information center) [13].

Nanoparticles and nano manufacturing activity can have many impacts on energy transmission and conduction system development. These techniques may be in use for the next upcoming legion years.

In a flash speaking, Nanotechnologies may have more and more capable use of transportation fuels.

This, maybe, shall slow down the increase in demand for long distance dispatch of liquid fuels.

Nanotechnology also made construction materials, due to nano materials they are may be stronger and are of lesser volume by comparison with materials of construction being used in past, that will decrease the footsteps required for the maintenance of sewage and electrical communication lines [14]. In this part, there's are detail of some methods in that nanotechnology will quicken the progress of safer, most high power energy sources and usage, that will lessen the need for long-distance transmission of electricity, petroleum distillate fuel, and natural gas [15]. Shortly talking, the possible solution to fulfill the energy needs comfortably is to enhance the efficiency of existing sources and the rate of their consumption. One of the most promising technologies is the Solar Photo-Voltaic Cells in order to fulfill the need of quick deployment at cheap price per watt [8]. Nanotechnology can be helpful to get high power solar cell using multiple practices, like to use, quantum dots that can decrease the gap between conduction and valance band and lead high efficiency. Using non-metallic or crystalline photonic nano-layer systems that have anti-reflective affect can also give high efficiency. By controlling the structure of all types of solar cell using plasma aided method can also lead to high productivity and efficiency [8].

## 2.2. Sources of Energy Generation

### *2.2.1. Photovoltaic Cells*

Photovoltaic Cells (PV) cells transform the visible light (sunshine) into electrical energy surface two useful outputs, low power and high power. Disorganization & low efficiency of conventional Photovoltaic cells cause a loss of 70% of radiant energy to the cell because the entering photons have energy equal to the band gap energy of silicon element. So, nanotechnology can be applied to prepare solar cells with reasonable price and high efficiency same as that of current advance technology. For *e.g*, organic inorganic semiconductor nanorods surrounded within polymer matrix to form elastic and economical solar cells. Narrow superimposed solar cells potentially are cost-effective and economical for their Low materials, low processing temperature and integrated cell detachment. Dye sensitized solar (DSC) cells are generally synthesized by doping titanium dioxide ($T_iO_2$) nanoparticles with dye molecules. That can be as a good upset towards reducing manufacturing costs and can create energy as from low beam light. Spain and UAE made 'GAMA SOLAR POWER PLANT' that are working 24 hours day/night since 2013, and generate 20 MW electricity per day, that energy is enough for 25000 homes for domestic purpose. The European Union has invested in nanotechnology research in a bid to dramatically ramp up the efficiency of solar cells, as shown in Fig. (**12**) below.

### *2.2.2. Fuel Cells*

A fuel cell can be defined as transformation device that transforms electrical energy present in the fuel to chemical energy in the existence of an electrolyte. The important challenge for a fuel cell is storing hydrogen and sluggish oxygen reduction reaction that are catalyzed by Pt based catalyst that are quite expensive, unstable and possess poor methanol crossover affect [8]. Another challenge for the fuel cell is the measurement of the temperature difference inside a fuel cell in an efficient and economical manner. Thus, nanotechnology also play a significant role here *e.g*, complex hydrides such as $LiBH_4$ and nonporous metal-organic composites are suitable for hydrogen storage in solid state fuel tanks [8]. With the help of nanoscience we can use non precious metal/metal oxide (M/MOx) base electrocatalyst water degradation. The temperature of fuel cell membranes can be control *via* the synthesis of polymer functionalization of organic- inorganic nanocomposites [8]. Nanotechnology can solve the problem of hydrogen feeding in the fuel cell by replacing the hydrocarbon with nanocatalysts. So using this type of technology it is possible to have high power density micro-fuel cell instead of currently existing heavy batteries [8].

### *2.2.3. Wind Energy*

This type of energy is less than 0.9 % of energy requirements throughout the globe. The need of wind energy is increasing in comparison to the other sources of energy. But this root has chief problems like low efficiency and trustworthiness of the turbines and hurdles in generation of high power at low cost. So nanosciences also have tremendous influence here on the efficiency of turbines through synthesis of carbon derived nanocomposites. Which are utile to make light-weight and better mechanical strength rotor blades. Nanoscale Varnishes for bearings may be helpful towards repair the effectiveness of wind turbines.

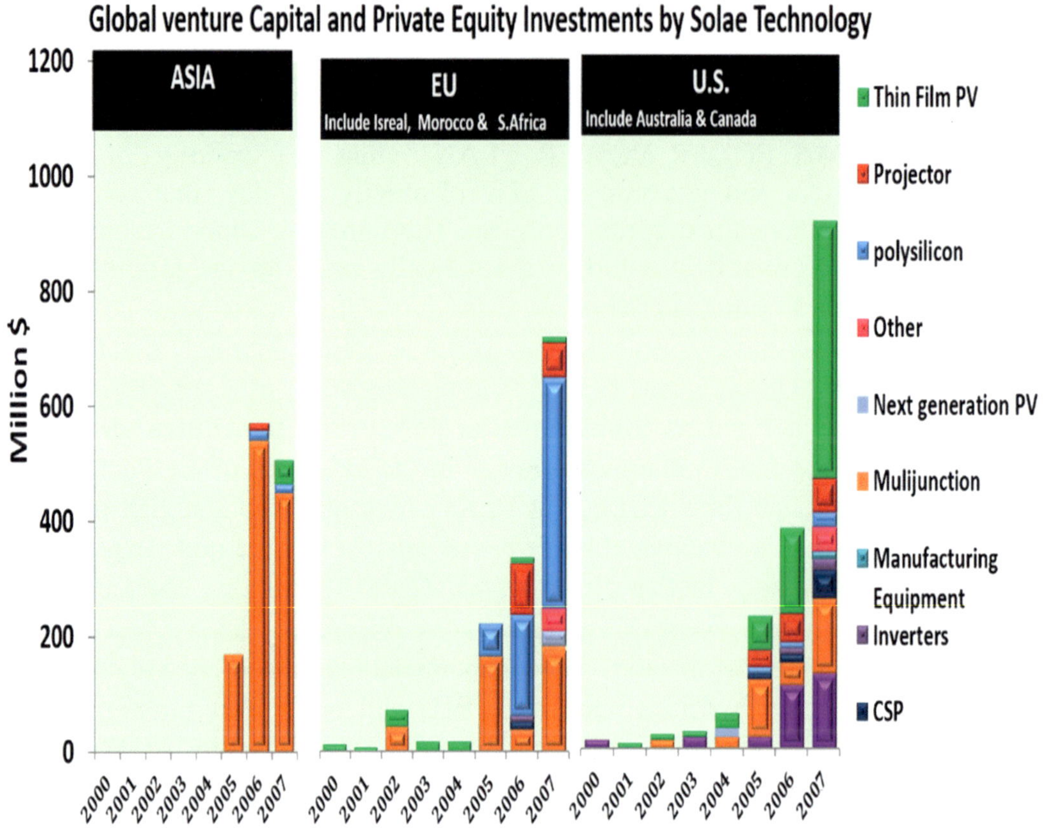


**Figure 12:** Global venture Capital and Private equity investments by Solae Technology during 2000-2007.

## *2.2.4. Fossil Fuel*

At present usage of fossil fuel have exceeded than 85% of the total energy needs worldwide. With passage of time as existing oil resources are getting reduced, some new ways to explore increase oil capture from already available resources and innovative reservoirs of oil for instance shale has been introduced [8]. So, Nanotechnology is tremendously influencing the fossil fuel industry for increase proficiency and production rate. For instance: silicate nanoparticle mixture is using nowadays to control the viscosity of oil during production. Nonporous materials are used for state of impurities in oil deposits and to increase the production rate too. Furthermore nanotechnology has been practiced to use nano-lubricants in order to decline fuel wearing of drill probes which are used in expedition [8].

### *2.2.5. Others*

Other types of technologies using nanoscience also exhibits a significant effect on energy production using alternate reservoirs like geo thermal and coal firepower plants [8]. Like

1. For the separation of carbon dioxide in coal power plants nanostructured membranes are applied.

2. In gas turbines plasma coated membranes are used as thermal barrier layers.

3. Nano particulate varnishing materials are utilized as non-sticking ceramic coating in heat exchangers of coal power plants to decrease coagulation and caking [8].

## 2.3. Sources of Energy Storage

### *2.3.1. Li-Ion Batteries*

LIBs which are the focus of research for high energy production nowadays possess somewhat high power densities but main problem with LIBs is their extreme low/high operating temperature that affect their rate capabilities. Presently, there's no any available electrode material and electrolyte which can deliver best efficiency of high stability, excellent capacity, high operating voltage and better cycle life. Various nanocomposites like lithium titanate have been discovered recently as anode material for LIBs. On the same way, electrolytes that are composed of nanoparticles has been introduced to increase the power capacity of the LIBs. These nanomaterials are capable to operate at safer voltage range, high temperature and high stress whenever need to apply [8].

Nanophosphate lithium ions batteries, as shown in Fig. (**13**). (Nanophosphate which has been used in the recovery of kinetic energy (KERS). Derived fromA123 System's, USA).

### *2.3.2. Supercapacitors*

A supercapacitor has two electrodes enclosed within an electrolyte which is isolated by a separating membrane. The main purpose of the supercapacitor is charge transferring between two electrodes. The exposed surface area of the electrode is directly proportional to the amount of energy produced. Henceforward, the usage of Nanomaterials like 3D carbon derived aerogel, graphene carbon from biomass and carbon nanotubes (CNTs) can favorably increase the enactment of a Supercapacitor [8], as shown in Fig. (**14**).

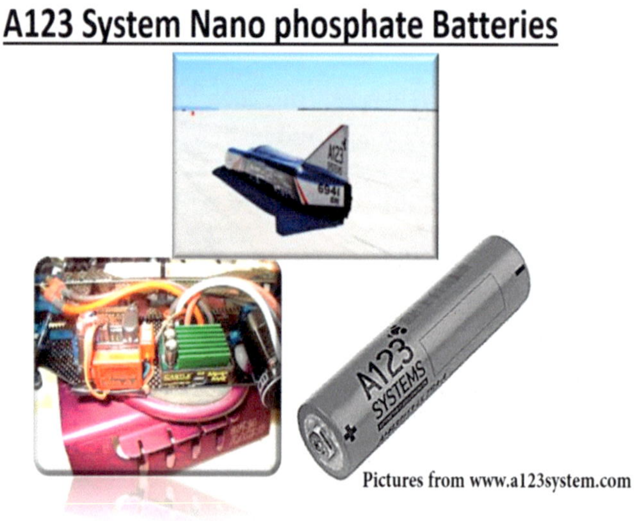


**Figure 13:** A123 System Nano phosphate Batteries.

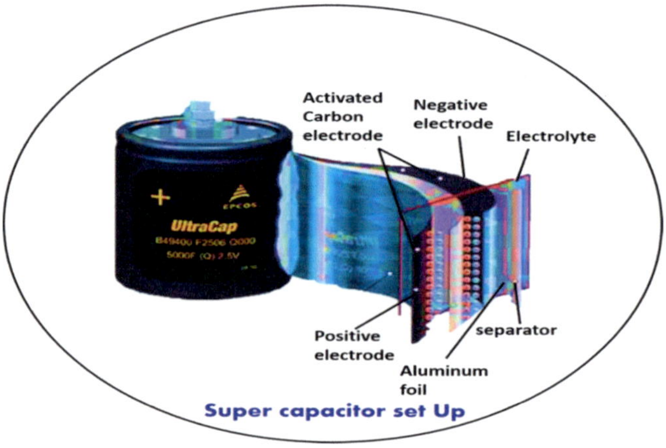


**Figure 14:** Super capacitor set up.

### *2.3.3. Hydrogen Storage*

Hydrogen energy is one of the most powerful sources of energy. The main problem is the transportation of the hydrogen energy which consumes one quarter of the global energy production [16], but still it is the focus of researcher to make it suitable for practical application as it is the green source of energy. One of the ultimate solution for the safe and better transportation of hydrogen storage is to use nanomaterials as hydrogen storage materials, because the tank use for the transportation are dangerous due to high pressure, presence of liquid hydrogen safety issues and high cost. Alternate solution are the morphology control synthesis of nanomaterials exposing active surface area, hybridization of the organic/inorganic composites and exchange of cation/anions as better hydrogen storage materials. Nano scaling has been proof to be very promising to overcome the problem of mass diffusion [17].

## 2.4. Energy Distribution

Present observations show that the energy prices are continuously getting step up. So energy and power section is searching for the alternate solution for minimizing transportation expenses to make the ease of availability at low cost. Nanomaterials like conducting aluminum based or carbon derived composites possess 6-10 times high capacity than aluminum or conductor steel wire which, at present is in common use in daily life [8].

Magneto-Resistive nanosensors made on magnetic nanolayers and electronic tools possess a great importance towards self-calibration and self-diagnosis assisting to analyze the infrastructure. That will be useful in decreasing the augmenting availability and flexibility [8].

## 2.5. Energy Utilization

Utilization of nanotechnology has a great influence to increase the proficiency of energy dissipation in various sections of industry [8].

### *2.5.1. Transportation/Carriage/Cartage Industry*

Nano-catalysts also bring very tremendous revolution in nanotechnology for influence on energy utilization in transport industry. Presently, nanolubricants minimize friction, high power LIBs and low weight nanomaterials do not intermediate on strength.

In 2006, in field trials by Oxonica, nanoparticles manufactured by using oxide of cerium that initiate ignition between diesel (fuel) and air have been proven to economically augment fuel efficiency to 10%.

Boric acid based nanolubricants are capable to decline frictional losses up to 80% to increase the durability of the components [8]. Nanostructured Metal Matrix (MMC) and Polymer Composites are very helpful to decline the overall weight of vehicles in transportation sector. Apart from miniaturization/shrinking of tolls *via* use of nanosensors will further decrease the total weight of the electric vehicles.

### 2.5.2. Building Industry

The upkeep of nanotechnology to the construct industry is through empowering the production and practice of energy saving LEDs centered and based on organic/inorganic semiconductor nanomaterials. Quantum Dots are useful towards increasing the yield and energy efficiency of the LEDs by reducing the smattering & seeped. Various LED especially organic once are the focus of nanotechnology nowadays by optimization of the layer thickness, applying doped materials and field carrier materials [8]. Reinforced polymers/CNTs nanocomposites that ultralight in weigh and highly stable can boost construction materials. Large amount of energy is consumed in overall thermal system of buildings throughout the globe [8].

Silicon Oxide-based micro/mesoporous Aerogels, and, nanoporous polymer foams are nanoporous materials possessing microspore size less than 5 nm and have a significant potential towards reducing these costs.

### 2.5.3. Manufacturing Sector/Industry

Towards energy efficiency of industrial processes nanotechnology have vast applications by using better materials of insulation as discussed above.

High surface area of nanostructured catalysts can boost the production rate. These catalysts can enable us to synthesize, manufacture and blend materials in new vigorously promising ways. Nanoscale powders have the capability to decrease annealing temperatures in high temperature synthesis processes for example, ceramics because of high expose surface area and ability to amalgamate. Nanotechnology can also be widely used to make micro reactors which have remarkable potential to increase Energy Efficiency in the Manufacturing Sector, where heat and mass can be organized in a more enhanced manner [8]. 3. Nanotechnology in Medical Sciences and Health-Care Nanotechnology has great and huge potential when discussing its applications in the sector of medical and biological sciences. Nanotechnologies have saved so many lives by discovering advance technologies to cure so many diseases [18].

## 3. NANOTECHNOLOGY IN MEDICAL SCIENCES AND HEALTH-CARE

Nanotechnology has great and huge potential when discussing its applications in the sector of medical and biological sciences. Nanotechnologies have saved so many lives by discovering advance technologies to cure so many diseases [18].

## 3.1. Use of Nanotechnology in Drug Delivery Systems

### *3.1.1. Nanobots*

Nanobots are robots (mechanical devices) that are few nanometer in size and make some specific function. These nanobots can efficiently deliver drug. As we know that drugs can affect the whole body before it can reach to the targeted site. Nanotechnology has made it possible to target the drug to an exact location that is disease-suffer area. Nanotechnology will deliver the drug to effective site and will abruptly decrease the possibility of side-effects. Fig. (**15**), show the devise that use Nanobots to measure the blood sugar level [19, 20].

**Figure 15:** Nanobots to monitor Blood Contents. (Remarkable Nano-Bots).

### *3.1.2. Use of Nano-Technology in Drug Delivery Procedure*

Drug carriers possess 5 to10 atoms thick outer and 50- 100 nm wide the inner side drug-filled cell. The drug carriers release a beat sense when they receive the signs of the disease, the inner wall, melts to release the drug at the respective site. Pharmaceuticals expert Bio Sante Aston Vicki, expresses: "When drugs are put into Nanostructures, their solubility quite gets enhanced" [20, 21].

### *3.1.3. Advantages of Using Nanobots for Drug Delivery*

The most prominent advantages of use of nanobots is quantity and timings of releasing drug that can be controlled by the electrical pulse [21]. Elan Pharma, a big Firm, is the latest technology of nanobots for drug release stated by Merck's Emend and Wyeth's Rapamune [22].

## 3.2. Disease Diagnosis and Prevention

### *3.2.1. Diagnosis and Imaging*

Nanobiotech researcher and experts have introduced a microchips coated with human being molecules. It structure has been control to accurately emit and release instinct signal when the chip intellect symbols of a disease. Specific sensitive nanobots can be insert inside the blood to check the contents of the blood and symptom of the diseases, as shown in Fig. (**16**). Their additional use includes checking the blood sugar level. The principal importance of using these types' nanobots includes their cheap production rates and easy movability [20, 21].

### *3.2.2. Quantum Dots*

Quantum Dots (QDs) are stated to be as the nanomaterials that spark very brilliantly when irradiated with ultraviolet (UV) light. QDs can attach exactly to the surface of the molecules they want with helping of coating special nanomaterial on their surface. QDs make bright tumors area by attaching to the proteins which are unique to cancer cells [23].

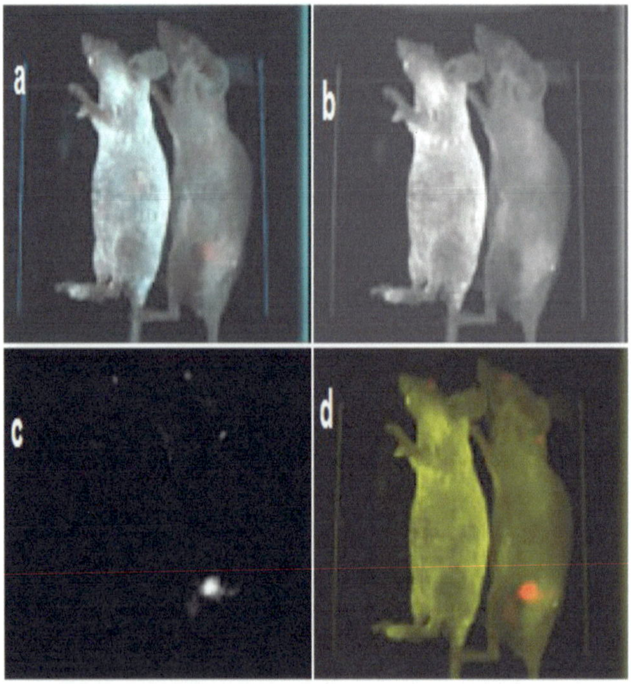


**Figure 16:** A light in dark places: Spectral imaging of quantum dots within the live mouse. Orange fluorescence signals are indication of the tumor growth [19].

Multifunctional $Fe_5C_2$ nanoparticles: A targeted Theranostic Platform for Magnetic Resonance Imaging and Photoacoustic Tomography-Guided Photo thermal Therapy [24], as shown in Fig. (**17**).

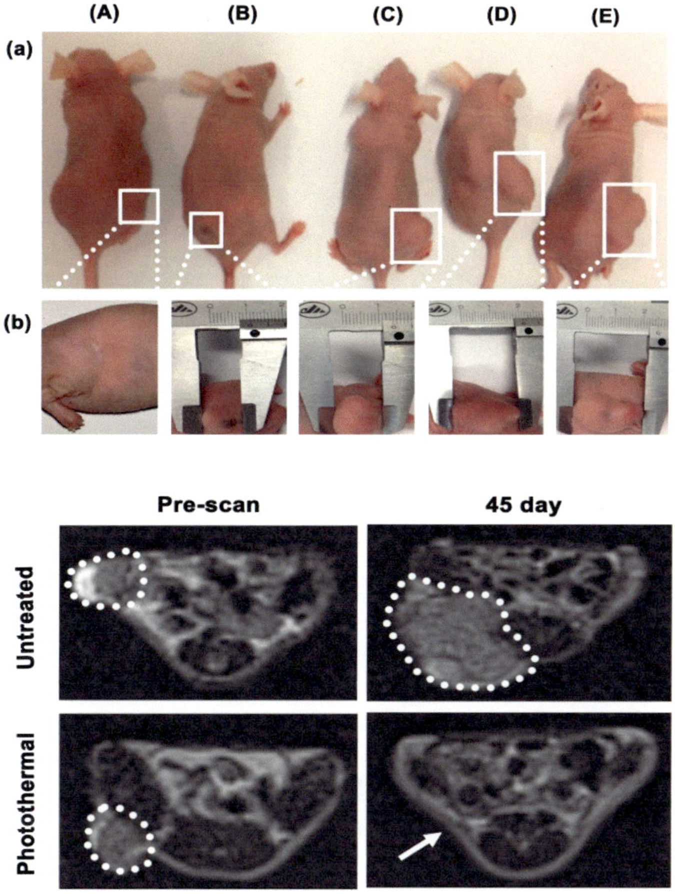


**Figure 17:** Magnetic Resonance Imaging and Photoacoustic Tomography-Guided Photo thermal Therapy [24].

### *3.2.3. Preventing Diseases*

#### 3.2.3.1. Prevention of Heart-Attack

Nanobots are useful in preventing heart-attacks. Fat-Deposits in the Blood vessels are the cause of Heart-Attacks. Nano-Bots can be crafted for removal of these fat deposits [20, 21]. Fig. (**18**), shows nanobots preventing Heart-attacks (Heart View). The figure shows nanobots eliminating fat sums on the inner side of blood vessels.

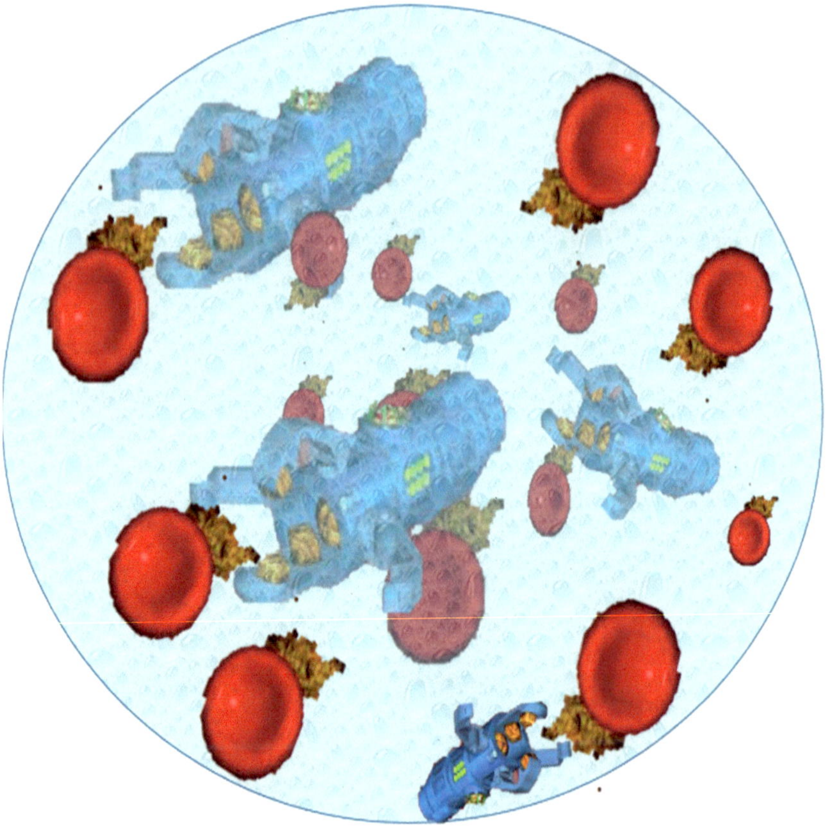

**Figure 18:** Nanobots preventing heart-attacks (Heart view).

### 3.2.3.2. Applications of Nano-Technology in Frying Tumors

Another important application of the nanomaterials is cancer treatment. The treatment is named as "Frying tumors" rule. Nanoparticles such as iron are swallowed as oral pills and they get attached to the tumor. The application of magnetic field causes the heating of the nanoparticles to finally burn the whole tumors [20, 22].

### 3.3. Tissue Reconstruction

Nanoparticles also have great significance in the field tissue engineering. It can design nanomaterials with a structure same to the human body part for example bone structure, as shown in Fig. (**19**). Ultrasound of the required bone shape is taken and same nanoparticles are shaped [25]. The shape nanoparticles in the form of thick paste are entrenched inside the body [22]. Where they gathered in well order form at the site of the fracture and become the part of the bone [20, 22].

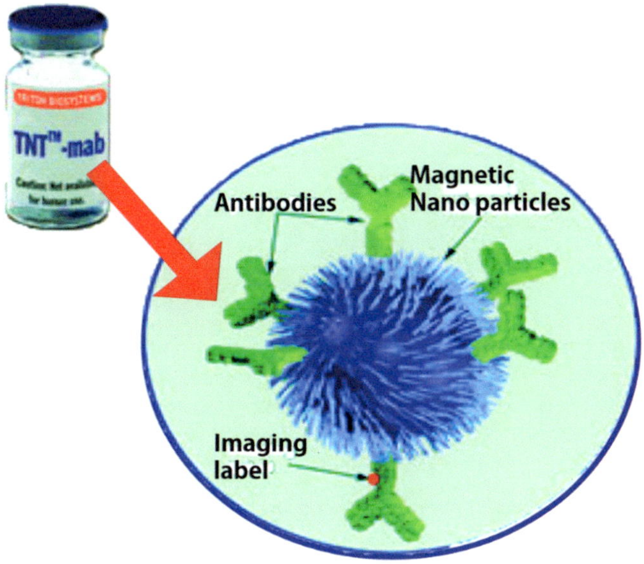


**Figure 19:** Nanoparticle bone structure.

Other key applications of nanoparticles include nerves treatment. Samuel Stupp and John Kessler at Northwestern University in Chicago introduced amphiphiles (tiny nanofibers). They are amino acids doped nanofiber that boost the neutron growth and resist mutilated tissue formation. Scientific research have proved that various animal like rat and mice having spinal injuries has been recovered with these nanofiber [23].

### 3.4. Medical Tools

Nano devices are stated to be as Nanoparticles that are made with the purpose to react with the cells and tissues and achieve very sophisticated tasks [25]. Imaging tool is the most familiar example of nano devices. Oral pills take a small cameras along with itself inside the body. These cameras have high resolution and can take the picture of the small body part even up to 01μ in width (for instance red blood cell whose size is 7 μ) [19].

Fig. (**20**), showed Nano cameras along with other nanoparticles inside the body at the site of disease [20].

Measuring device 'accelerometer' is kind of very small device that can be fit in the bone joint areas (like hip and knees ect) to measure the movement and level of strain at these joints. To avoid infection silver nanoparticles can be coated. It kills bacteria and decrease chances of infection [22].

**Figure 20:** Miniature cameras inside blood vessels (Blender battles).

### 3.5. Nanomedicines

The use of materials at nanoscale greatly enhances the production, quality and the performance of the industrial products. The health sector is no exception and nanomedicine has been one of the most influenced sectors of nanotechnology applications in recent years. It is continuing to expand for the treatment of serious diseases like cancer and aids at the cell level, ensuring a complete cure of these diseases as well as greatly improving the efficacy of existing diagnostic systems and the efficacy of existing drugs. Nanotechnology based applications include preparation of drugs against number of viral diseases like H5 NI bird Flu, seasonal influenza, tuberculosis (efficient and delayed release of medicines)., HIV, Hepatitis B & C, rabies and dengue fever which are most common in South East Asia. Nanomaterials are made for viral therapy. Nanoparticles and nanocrystalline materials are already commercialized as antibacterial & antifungal agents. Nanovaccines are prepared to treat hepatitis B & C without requiring the need of refrigerating conditions in the developing countries. Some application of nanotechnology in medicine are discussed by Kim *et al.* [26] reported the effect of silver nanoparticles against yeast, *Escherichia coli*, and *Staphylococcus aureus*. Sadiq *et al.* [27] reported the antimicrobial activity of alumina nanoparticles with a mean diameter of 179 nm, against *E. coli*. They reported that alumina nanoparticles showed a miner growth-constraining consequence, which was at very high attentions, that may be due to surface charge interactions between bacterial cells and the nanoparticles [27]. Gold nanoparticles are of different shapes like rods, shells, cages and spheres which are used to treat cancer nano-medicines for drug delivery and cancer treatment [28, 29]. Dengue is stated to be caused by one of the type of dengue bugs. There is quite similarity of symptoms of Dengue with those of influenza: For example, abscessed patents mostly cause severe headaches, inflammation and pain in muscles and joints.

Zhang and coworkers at the Institute of Microelectronics invented biosensor consist of a silicon nanowire which can find and sense the 'polymerase chain reaction involving reverse transcription' in short time of just 30 min [30]. This silicon nanowires biosensor attached with peptide nucleic acid (PNA) to detect paired DNA fragments of dengue type 2 [31, 32].

## 4. NANOTECHNOLOGY IN WATER PURIFICATION

Fig. (**21**), represent that without water life in the world is impossible. Water is extremely essential for the existence of life. Uncountable uses of water include field of agricultural, industry, daily common use, recreational, leisure and worldwide activities. It is authoritative to keep in mind that only 2.5% of water on earth is fresh water, and almost 2/3 is ice-covered form found in glaciers and polar ice caps. The big difference is raised in many parts of the globe between demand and supply of water, and lot of area on the earth is expected to have imbalance in the future. A temperature variation is significantly impacting water resources throughout the world due to close connection between the climate and hydrologic cycle. Expansion of human population has resulted in competition of water, and this competition has led to depletion of the World's chief geological formation. The massive need of neat and clear water, with increasing population, environment cruelty and lower water system, is periodically leading dangerous source of problems among the various countries in the globe [33].

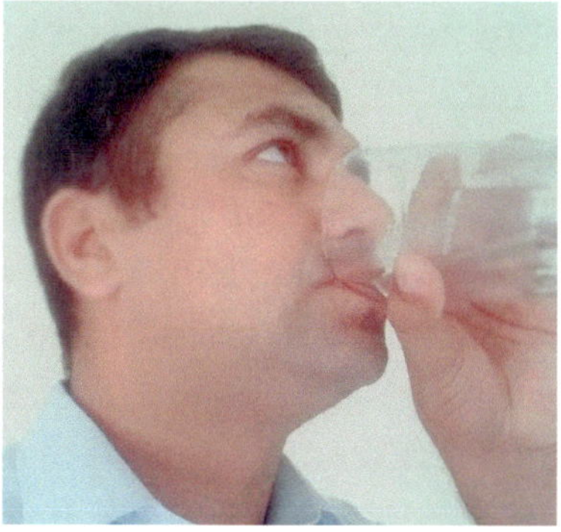

**Figure 21:** Pure drinking water.

Furthermore, the melting glaciers due to global increase in temperature have changed the shape of the sphere, and, in this case, we have a surely going towards disaster. The chase to ensure that everyone have right of entry to clean drinking water is unspoiled in the UN's millennium development goals, purpose to divide the big part of people without sustainable availability to safe drinking water by 2015 [18, 34]. One approach in the direction of achieving this goal is through the application of nanotechnology for providing clean and cheap drinking water. We list some of the nano techniques that are either under development or are already in use for providing clean and affordable drinking water to some of the very poor inhabitants of the third world.

### 4.1. Nano-Filtration Membranes

Nano-filtration membranes are using extensively to take away dissolved salts and micro-pollutants, soften water and treat waste water. The membranes with nano size pores used in the capacity of a physical barrier and prevent particles and microorganisms which are larger than the pores to go over and done with & by denying such substances selectively [18]. Research works are being carried out to remove salt from water and bring down the huge and prohibitively high cost of desalination [35]. North-West University of South Africa has been worked on treatment plant which integrates ultra- filtration membranes to neat saline groundwater in a rural area. This plant has the capacity to take away impurities such as chloride, nitrate, phosphate and sulphate to obtain safe and pure drinking water for domestic and community use. In another development, the water filter is source of clean and safe water supply for the residents of Sri Lanka, in particular the survivors of tsunami of 2004. The filter contains pot reinforced with clay paddy case at the base and coated with nano silver to solution out almost all harmful bacteria and parasites [36]. Its capacity is about 08 liters having a tap at its base. Due to filter availability there will be no need to boil water for making it safe and so fuel cost will be reduced. Cases of suffering from diseases such as diarrhea have considerably been reduced. The USA Red Cross, inaugurated making of clay filter in Sri Lanka in start of 2007 and has circulated more than 12,000 units up till now, aiding more than 55,000 Sri Lankans [36, 37]. Nano-facilitated expertise for water treatment is already available in the market with nano filtration which, at present, appears to be the most mature [38]. It is believed that the future generation water treatment apparatus will take advantage of new properties of Nano-Scale materials and may attest to be of interest in equally developing and developed countries [38].

### 4.2. Nano-Porous Materials

Researchers are developing new modules of Nano-Porous Materials that would be more effective than conventional filters. For example, let's talk of Nano-Filtration Membranes which can produce safe drinking water from salty ground water. An Indian and US group of scientists have established carbon nanotube (CNTs) filters which can get free of viruses and bacteria more effectively than traditional filters [18, 39]. Zeolites and natural attapulgite Clays are also used in Nano-Filters. They are easily available in all local markets throughout the world and have exists in natural nanometer size pores [18, 40]. Zeolites have the ability of separating detrimental organics from water and eliminating heavy metal ions like arsenic that is dangerous and harmful for health. At cheap price condensed water purifier are available that has been inaugurated in Asia like India, Pakistan *etc.*, last year, was proved to save millions of lives from water-borne diseases. In India, its name is Tata Swach — swach which mean clean in Hindi language and it is only two feet tall. This project was promoted by Tata group of industry for making water pure and clean. No motional water is need for the filter, and no electricity or boiling whatsoever are needed. Rice husk ash, which is biochar derived from the rice husk is use as matrix for silver nanoparticles and it have the ability to kill the bacteria is mounted in such a filter system [36]. Developing countries of India and Pakistan who produce millions of tons of rice husk can make benefits from nanotechnology to meet the needs of drinking water to millions of people affordable costs. Tata Company a chemical was firstly given a yield of one million filters a year. And after that, the purpose is to incline the production up to 3 million filters per year within the next 5 years with additional exports of filters to other countries [41].

### 4.3. Nano-Catalysts, Magnets and Detectors

Magnetic nanoparticles retain large surface areas compare to their volume and can easily bind up with chemicals. Magnetic nanoparticles of iron oxide ($Fe_2O_3$) when suspended in water, bind arsenic, which is afterwards obtained from the magnet [18, 42]. It is projected that 200-500 milligrams of Nano-Rust perhaps will give a liter of water. In Pakistan, India, Bangladesh and many other developing countries, arsenic poisoning linked to poisoned wells, is a major health problem.

### 4.4. Nano-Sensors

Nanotechnology can also detect marine contents. New sensor technologies are being developed through mixture of micro and nanofabrication to obtain small, moveable and highly perfect sensors. These sensors have the potential to identify and sense single cells of chemical and biochemical substances in water [18, 43, 44]. Preserved Nano-Wires could aid as very sensitive toxin or pathogen gauges. Scientists at Penn State University have discover a companion using electric field in which single nanowires is place on a silicon chip. They deposited an array of nanowires on the top of the electrode. This is a big breakthrough towards sensing devices that could sense hundreds of pathogens and toxic chemicals is seconds. Manufacturing nanowire detectors covers their coating with molecules which slog to specific mark molecules, like viruses. When a target molecule come across to the coating, the nanowire conductivity gets changed. The detection of this electrical indication clues to sensing device that are small in size, cost effective, and very sensitive compare to the currently using diagnostic chips that use large microscope to detect fluorescent molecules attached to the target molecule [45, 46].

## 5. BIOMEDICAL NANOTECHNOLOGY

Biomedical nanotechnology is principally suitable to biomedicine in drugs, prostheses, diagnostic methods and grafts. Interest is bring upsets in biomedical applications for use in body externally, like as diagnostic sensors and "labon-a-chip" skills, that is applicable for detecting blood samples, used as analytical tools in drugs for the purpose of research and development. Manufactured to be used within the body, some companies are promoting anticancer drugs, entrenched insulin pumps, and gene therapy. Nanotechnology also focus towards the prostheses and artificial implants [30, 47]. The nanotechnology field is currently one of the most energetic and promising fields of technology. While there's been exchange of nanotech applied to everything from wearing clothing to sports golf balls, the main area of R & D funding and investment is obviously centered in materials research pragmatic to semiconductor technology. One of the prime challenges for nanotechnology today is fabrication technique. The price to build a new semiconductor fabrication reaches easily into billions of dollars ($). This has directed many within the nanotechnology industry to find what is possible with already existing semiconductor technology. On the other hand, cutting-edge semiconductor handling is already going on well below 100nm, the dawn for nanotechnology. Additionally, consumer products are becoming richer in respect of their topographies, with customer reluctant to pay noticeably more for these properties. The hurdle that the designers of device is facing, is to put more attractive features into the same or smaller devices without meaningfully increasing costs and weight. With the trying of getting more functionality, it is needed that the devices must shrink even more. Nanotechnology developments guaranteed to help our present industry break out of the basic physics barriers we are facing with diminishment, along with addressing the heat glitches associated with device shrinkage [48, 49]. International nano biotechnology market is round about to $4.94 Billion by 2015, listed in accordance with recent survey of researcher [50]. Nano biotechnology is gathering pharmaceutical and biotechnology industries. This technology can sense the drugs growth levels, right from -inventions for optimum drug delivery to diagnostic diseases in various clinical trials. The emergence upset towards the nanomedicine, advanced analysis and nano medicines drug delivery gadgets will be done to realism having more advertised and idealized expansion small tools like nanobots that are accomplished of functioning minimally violent process. This area of nanotechnology optimally composed to enhance the prompt elderly globe and the paralleling growing requirement for better health care problems. The light of nano biotechnology blows into drug invasions and its delivery, tissue regeneration, analysis, grafting tools, magnetic resonance imaging, and environmental problems, medicals products will majorly be seen to the top in the incoming years. The nano biotechnology advances laboratory level research to more practical application ambitious pharmaceutical and biotechnological industries that have previously shown uncertain unwillingness toward this highly expensive, unknown and uncharted direction, that are now viewing more interest in appointing into this space. Nano biotechnology is now applied as a developing tool for profitable answers for all over the globe ranging from decorative medicine, health care to variable market, and environmental remedies. Furthermore, the present economic state of the globe has directed the view of big investors, who invest in industries with the biggest potential, and nano biotechnology is hope to arise into valuable profit of this trend. Worldwide nano biotechnology has made research directions to invest in the time casing of 3 to 6 years, will be the minimum influence with the most corrected wedged being middle- term investments within the time duration of 1-2 years. In umbrella of government ventures are hope to go on undeterred, while the private project are still has no chance to continue. The nanomaterials that is most wide market of advance nanotechnology group in recent area, is important factor causal towards the income influx for the nano biotechnology. The USA is the biggest center for nano biotechnology globally, while other markets are consider to be growing fast. Growth in other nanotechnology market mostly arises from the Asian region, especially people Republic of China, South Korea and Taiwan. Aduro BioTech, Arrowhead Research Corporation, Calando Pharmaceuticals, Inc., Unidym, Inc., Agilent Technologies Inc, Asklepios Bio Pharmaceutical Inc, Biosante Phosphate Pharmaceuticals, Inc., Elan Pharmaceuticals, Flamel Technologies, Inc., Life

Technologies Corporation, Bio Trove, Inc., Invitrogen Corporation, Nano Bio Corporation, Nanogen, Inc., Nanophase Technologies Corporation, Sivida Ltd, Sigma Aldrich Company, Starpharma Holdings Ltd., Air D-Fence™ & Nanolok™ by InMat (tennis), nano- CFC® by Holmenkol (skiing), N-FUSED™ by Easton (archery), NSi™ from St. Croix (fly-fishing), Nanopreme™ by Yonex (golf), Double Core™ & BLX™ by Wilson (tennis),Dendritic Nanotechnologies, Inc., and Zyvex Instruments LLC are the key players in the worldwide markets [51, 52].

## 6. NANOTECHNOLOGY IN ELECTRONICS

Nanotechnology is also playing crucial role in nano electronic applications. For example, CNTA field-effect transistors (FETs), Si nanowire FETs, and planar III–V compound semiconductor (*e.g.* InSb, InAs) FETs, are emerging nano electronic devices that are promising device to be integrated to enhance the circuit potential and also for extending Moore's Law [53], as shown in Fig. (**22**).

CNT field-effect transistors (FETs), Si nanowire FETs, and planar III–V compound semiconductor (*e.g.*, InSb, InAs) FETs nano electronic devices, is the state- of-the-art planar and non-planar Si devices (both Tri-gate and double-gate Fin FET transistors are changed the previous electronic industry totally [54].

**Figure 22:** Integrating nano circuit.

# 7. NANOTECHNOLOGY IN SPORTS GOODS

Nanotechnology brings a warren of intricacies in sports field, but elevates the efficiency of an athlete and reduces the happening of damage, making sport the source of leisure for the audiences and the athletes. In contemporary times, the sports market has emerged as easily provided hi-tech industry where more advance sports materials are available. Furthermore, global championships for instance Olympics and World Cups have introduced the sports tools with high standard, making it an attractive business for all the people [55, 56].

## 7.1. Impact of Nanotechnology on Sporting Equipment's

Like other filed of life nanotechnology has also a great influence on the sports as well. Inside the position of sport equipment's, this technology deals a number of impacts & great potential in improving sports equipment's to provide safer zone to athletes, contented and facility nimble than ever [55, 56], as shown in Fig. (**23**).

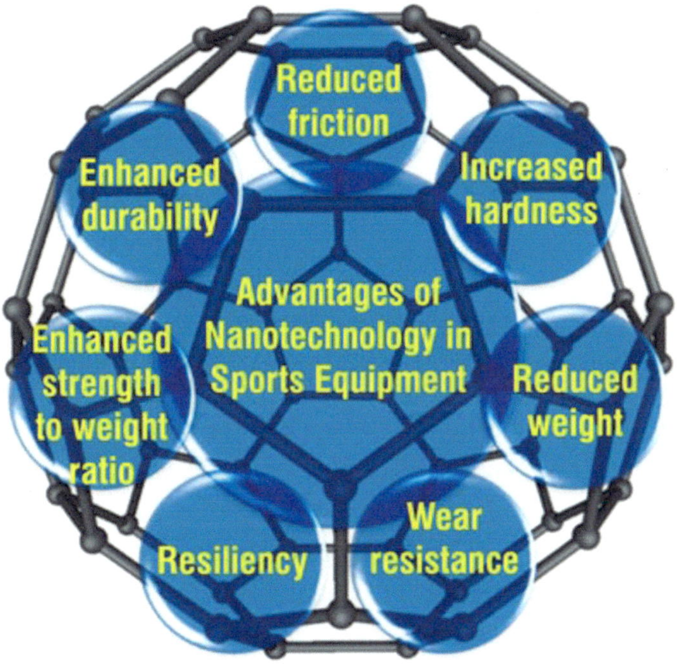


**Figure 23:** Importance of nanotechnology in sports equipment. (Image: Wikimedia Commons)

Bats, racquets of tennis and badminton, hockey sticks, archery arrows, bicycles, golf and clubs balls, skis, fishing rods, *etc.* are various sport equipment's, whose mechanical and elastic properties has been improved with advance nanotechnology [55, 56]. The famous tennis player, Roger Federer, as shown in Fig. (**24**) (a) has won many tournaments including Wimbledon using racquets (Wilson nsix-One Tour 90, ncode & K Factor). These racquets are reinforced with SNPs using nanotechnology techniques. These racquets possess more strength and have 50% more hitting power than commonly used racquets. These nanoparticles are able to impart more elasticity and strength for skiing [55, 56].

A number of racing champion such as Cadel Lee Evans and Floyd Landis, as shown in Fig. (**24**) (b) won various tournaments including Tour de France have bicycles using this advance technology. The main framework of these racing cycle are made from CNT. The BMC SLC01 Pro Machine bicycles used by these cyclists were manufactured by the Swiss company, bicycle manufacturing Co. (BMC), the weight of these cycles was just 1000gm (size 51), and have mechanical strength 500 times more than steel [55, 56].

**Figure 24:** (a) Roger Federer have silica nanoparticle containing racquet and (b) Oscar Pereiro Sio have bicycle containing CNT at Tour de Roman die 2007 riding a Pinarello bicycle. (Images: Wikimedia Commons).

Since last few decades due to improvement in this new technology there is reduction in size with increase in the mechanical strength. For instance CNTs are the most frequently used nanomaterial now a days in the sports equipment [55, 56]. The best property of the CNT is that they have 100 times more strength then steel but 5 times light in weight, which make them an ideal component for making sporting equipment's where low weight and high strength are of prime importance. Handlebars for mountain-bikes are composed of CNTs because of their super strength durability and light weight [55, 56].

Various nanomaterials such CNTs, silica nanoparticles, nano clays fullerenes, *etc.* are used in number of sports equipment's to improve the strength and reliability, light weight, stiffness ect. as listed in Table **1** [55, 56].

**Table 1. Added advantages of incorporated nanomaterials in sporting equipment's.**

| Nanomaterial | Sport | Benefits |
|---|---|---|
| Carbon nanotubes | Tennis/ Badminton | Increase stiffness, consistency, durability, resiliency, impact, repulsion power and vibration control of racquets |
| | Golf | Reduce weight, lower torque/spin of clubs |
| | Kayaking | Enhance abrasion/crack resistance and easy paddling in kayaks |
| | Archery | Better vibration control in arrows |
| Silica nanoparticles | Tennis/ Badminton | Increase stability, power & durability of racquets |
| | Skiing | Decrease torsion index & facilitate transition in skis |
| | Fly-fishing | Enhance hoop & flex strength of rods |
| Fullerenes | Tennis/ Badminton | Reduce weight and twisting of racquet frames |
| | Golf | Facilitate flexible club whipping |
| | Bowling | Reduce chipping & cracking of balls |
| Carbon nanofibers | Cycling | Reduce weight and increase stiffness of bicycles |
| Nanoclay | Tennis/Golf | Increase resiliency and bounce of balls |
| | Watercraft | Reduce weight and enhance speed of water-boats |
| Nano-titanium | Tennis/ Badminton | Resist deformation and increase strength & durability of racquets, transmit more power to shuttlecock/ball, more accurate shots |
| Carbon nanoparticles | Road Racing | Decrease rolling resistance, increase grip & mileage in tires |
| Nano-nickel | Golf | Increase moment of inertia & stability of clubs |

Various nanoparticles like $ZrO_2$, ZnO, CuO, *etc.* are incorporate ion the oil and lubricants for the purpose to reduce wear and friction, in order to decrease the resistance and aerodynamic drag nanomaterials are added to the paints. Nanophosphate lithium ions batteries which are lightweight, have high charge/discharge capacity better cycling ability and efficiency are mostly used in kinetic energy recovery systems (KERS), [55, 56].

Number of sporting equipment's such as golf shafts, badminton/tennis rackets, skis rods, snowboards used the high strength CNTs the most prominent nanomaterial. In 2010, the global market of CNTs was around US$ 200 million and is expected to boost at a CAGR of 25% to reach US$ 500 million by 2020. The total annual consumption of CNTs almost 15% is used in sports equipment's. Due to advancement of nanotechnology the quality of sorts equipment's has been increased by and due to this reason the selling rate has been increased by 10-20% in the market. Despite of the various advantages provided by the nanotechnology there still exist various demerits of this technology for instance it is very expensive and new filed to be explore. CNTs are too much expensive nowadays in the global market approximately 100-150 US$ per 7kg. Another drawback is that sports companies need new machine to install for the production of new nano sport goods. The market is still in the process to determine the need of these advance sports goods and want to sort out this problem in the right way. Most of these advance nano enhanced sports goods are very expensive and rare so they are not easily available for the sports man. For example, the price of the advance Nano golf clubs in the market nowadays is round about around US$ 600-800 compare to a regular golf club that may cost somewhere in between US$ 100-150. However, with further advancement in this technology it is hope so that these nano enhanced products will be soon available to a common sports man at quite cheap price. In the sport industry the wrinkle and stain resistant nanotexile is another big future prospect of the nanotechnology [55, 56].

## 8. NANOTECHNOLOGY IN ENVIRONMENTAL ISSUES

Nanotechnology researchers and creators are succeeding towards making pollution free environment *via* synthesizing special nanomaterials. For instance using silver nano catalysts lead to decline the byproducts generated in production of propylene oxide. Propylene oxide is the commonly used component for plastics, paint, detergents and brake fluid *etc.* [57].

Most of iron and iron nanoparticle nanocomposites are used as catalyst for the photodegradation of various organic dyes to clean the ground water. These nanoparticles scatter in overall water and decay the organic dye. This is cost effective method for cleaning the water then to be pumped out of the ground. The nano catalyst can breaks down volatile organic compounds (VOCs) from air at ambient temperature. These catalyst are mostly composed manganese oxide of doped with gold nanoparticles [57].

## 9. NANOTECHNOLOGY IN DEFENSE SECTOR

Nanotechnologies have also a great impact in the field of defense sector. Technologies are to be merged inside the same area which are directly related to the defense sector, including, aerodynamics, mobility, stealth, sensing, power generation and management, smart structures and materials, resilience and robustness *etc.* Furthermore, nanotechnologies will effect battle space systems related with information and signal processing, autonomy and intelligence. Nanotechnology is more important regarding to information technology, especially great advantages are expected to be achieved regarding the threat detection, new electronic displays systems, as well as a vital role for the development of small unmanned combat vehicles and robotics. Finally, nanotechnology are expected to develop novel nano-enhanced materials with new structure, better properties and greater strength (for example through novel nano-energetic catalysts), reducing cost (for example through wear reduction, self- healing and self-repair), enhanced functionality (for example through adaptive advance nanomaterials), and new types of materials with improved electronic / optical /magnetic properties [58]. Nanotechnology can help to control the rate of crimes for example the small tags–only that are invisible to naked eye just 25-30μ in diameter are coated onto gun containers so that they attached firmly to the hands of the person who handle it. Analytically, some of these 'NANOTAGS' also linger on the cartridge even after it has been ablaze. This should make it possible to establish a robust forensic relation between a cartridge fired during a crime and whoever handled it. At present it is problem to detect the fingerprints or collect significant amounts of DNA from cartridge surfaces, which are bright and flat [59]. Nano tags, are therefore consider a successful war against the gun crimes. The upset in the crime control has been gained by a collaborative efforts of chemists, engineers, management scientists, sociologists and nanotechnologists from Surrey, Brighton, Brunel, Cranfield, and York Universities, with funding from the (EPSRC) Engineering and Physical Sciences Research Council [59]. USA has entered in a big way in this field particularly defense related nanotechnology. Annual Department of Defense (DoD) funding on nanotechnology has steadily increased from $ 70 million in the year 2000 to $ 436 million in 2006 and still increasing.

Fig. (**25**), shows, the total annual investment in nanotechnology; 2006 estimated. DoD “Research and Development Programs of Nanotechnology Defense sector”, May 8, 2006.

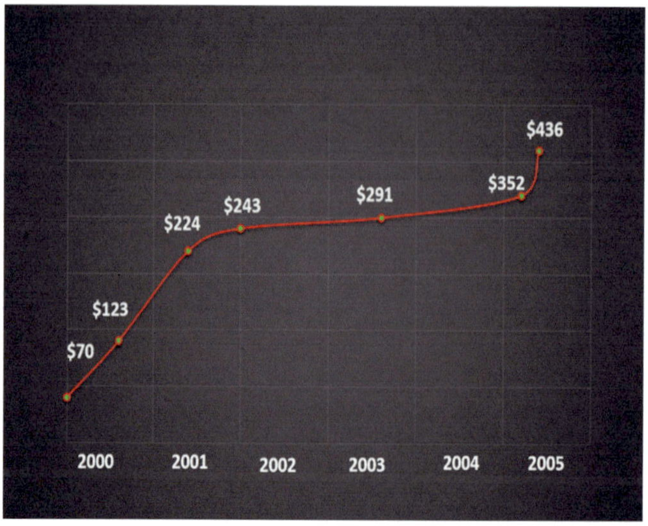


**Figure 25:** Total annual investment in nanotechnology; 2006 estimated. DoD "Research and Development Programs of Nanotechnology Defense sector", May 8, 2006.

Several universities, Armed Forces Laboratories and Nuclear Research Laboratories are participating in this program.

DoD is also collaborating in this programme with other agencies such as NASA, NSF, DHS, DoE. It is not well known what the other advanced countries are spending on defense related nanotechnology R&D but it is not significant comparable with USA & Japan, as shown in Fig. (**26**).

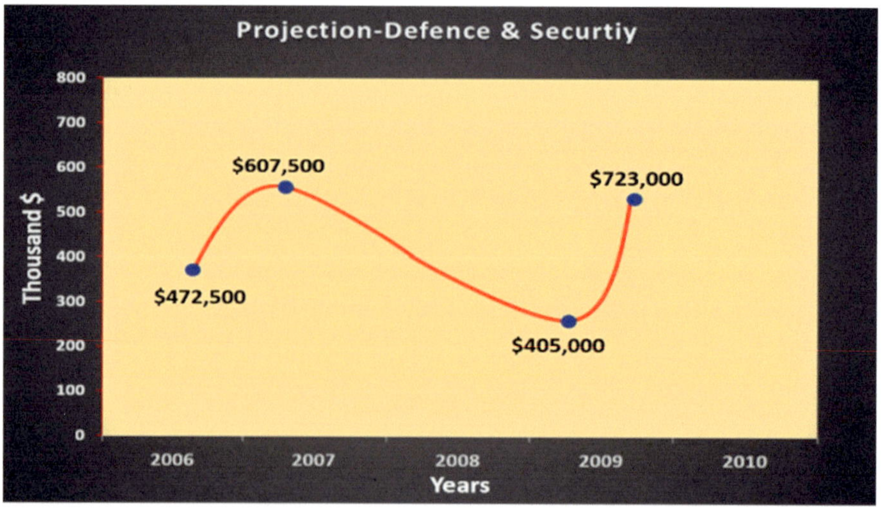


**Figure 26:** Projection-defense & security.

Several universities, Armed Forces Laboratories and Nuclear Research Laboratories are participating in this programme. DoD is also collaborating in this programme with other agencies such as NASA, NSF, DHS, DoE. It is not well known what the other advanced countries are spending on defense related nanotechnology R&D but it is not significant comparable with USA & Japan, as shown in Fig. (**26**). Fig. (**26**) On the nanotechnology field as a whole European Union, Japan and Russia are spending amounts comparable to USA. R&D on nanomaterials has led to the development of lightweight and strong materials – stronger than steel but lighter than aluminum. Nanosensors has been developed which are more sensitive and much smaller in size. As a result of the developments in the field of nanotechnology many defense related applications may emerge, such as,

1. High performance weapons sector (forb better missile): Replacement of depleted uranium missile launching.
2. Kinetic Energy (KE) penetrators with enhanced lethality (replacement of depleted uranium penetrators.
3. Chemical and biological warfare defense.
4. Antiterrorism, rescue efforts can be improved with the use of nanopaints.
5. Warfare's dirty work can be control by killing machines.
6. Electronic surveillance systems could boost by introducing special nanoparticles.
7. Nano sensors: More sensitive and lightweight.
8. High performance information technology.

It will allow miner unmanned and autonomous weapons systems to boost their mark with the aptitude to adjust to unexpected deviations in climate and sense and counter intimidations absorbed at them. A number of solder tools, for instance nano textile, armor weapons and communication and information devices with better and enhanced qualities, to meet changing conditions by a mechanism [60].

## 10. NANOTECHNOLOGY IN TEXTILE/TEXTILE

Fig. **27 (a and b)**, showing manufacturing composite textile composed of nano particles or fiber that perfection of fabric quality without enhancing the weight or stiffness like other commonly used techniques. For *e.g.*, adding whiskers like nanomaterial into the pants fabrics make it ultralight water and stain resistance [61].

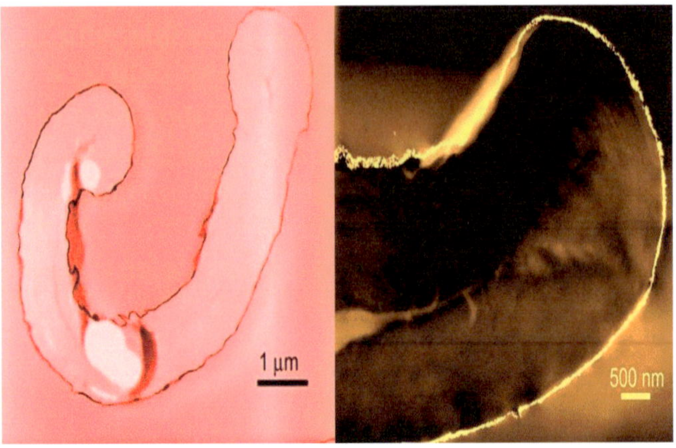


**Figrue 27(a):** Transmission electron microscopy (TEM) images of gold (L) and palladium (R) doped cotton fibers [63].

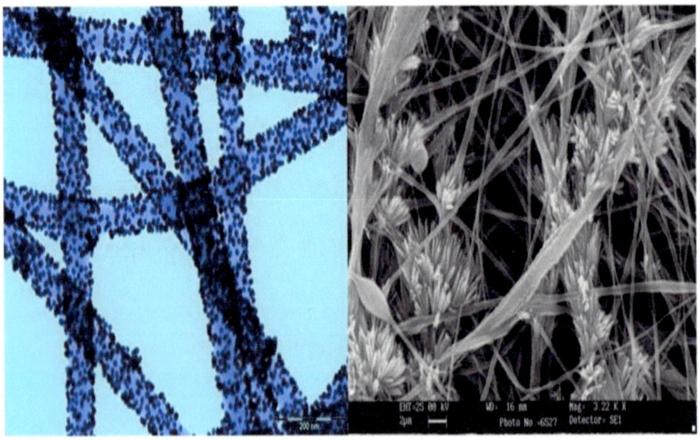

**Figure 27(b):** TEM and FESEM images of gold (L) nanoparticles doped nylon nanofibers.

A lot of industry areas that nanotechnology impressions upon, a possibly astonishing leader in acceptance the latest ideas the nanotechnology world has to offer, is the textile manufacturing. Not only does this industry look to nanotechnology to give it an edge in fashion, but, as we read NANO magazine, there are some surprising and exciting innovations in the form of nano-based textiles for a range of personal protection and healthcare applications, emerging top universities of the world. In line with our focus on textiles, we hear about smart fabrics that can sense blood, and how a sensing capability is also being incorporated into Kevlar. In Nicholas Kotov's lab, in the University of Michigan, work is taking place on the design and engineering of new conductive materials with blood and other biosensing capabilities. This has vital applications for individuals working in extreme conditions where it may be difficult to detect bleeding. Other research on embedding nano clays into Kevlar fabrics is leading to better ballistic-impact resistant materials, with sensing potential which is now being commercialized by Nico Technologies Inc. In the field of textiles for personal protection, nanotechnology offers protection and comfort, with some additional surprising applications in improving air quality. In a wide- ranging article, international expert, Professor Kay Obendorf, from the College of Human Ecology, Cornell University, explains how the application of nanotechnology is providing chemically and biologically protective materials allied with improved comfort and performance. Some of these new materials are also being utilized to improve air quality in the built environment, benefiting hospitalized patients with lowered immune systems.

## 10.1. Fabric: Current Nanotechnology

Nano whiskers (synthetic fiber) that lead to water beading, and cause the fabric water and stain unaffected. In order to make the fabrics odor- unaffected silver nanoparticles are added to it that cause the killing of the bacteria. Nano pores providing larger lining for shoe inserts in cold weather. 'Lotus plant' effect for fabric used awnings and other material left out in the weather provided by nanoparticles, cause the cleaning of the fabric in the rain [62]. Textile nanotechnology study work can be divided into three (03) main shoves: Latent applications consist of catalytic mantles, coloring of the fabrics without dyes) and antibacterial elastic substance. Changes & improvement of present fabrics using electrostatic self-assembly & atomic layer deposition (ALD) method to make new, multifunctional & optimized surfaces on common fabrics. Researcher has been put polyelectrolytes, inorganic & metallic nanolayers along with a number of nanoparticles, catalytic and variable morphology mixture over fabrics with nanoscale accuracy leading to brand new progress avenue for the smart textiles [63]. Potential applications include active and catalytic infiltration of risky gases and industrial toxic chemicals additionally anti-counterfeiting devices.

By using various techniques like directed assembly as well as external fields to create fibers with unique catalytic, optical, electronic and magnetic and properties of nano manufacturing of polymeric nanofibers. To creating novel nanocomposites materials magnetic and electric fields are used to gain accurate direction control of multifunctional nanoparticles in the polymeric fibers.

Therefore, anti-counterfeiting, antibacterial, anti- odor, anti- stain and wear resistance and self-decontaminating applications are all nanofibers applications. Additional coloration and control of near infrared signatures for active camouflage applications of properties of these nanofibers [63].

Scanning probe microscopy (SPM) personalized to detect nanoscale phenomena on surfaces of low energy having high radius of curvature for instance fabrics fibers is development of metrology tools. This is a ground breaking use of (AFM) as an investigative microscopy to find the influence of electrical charge degradation on electret filter media. Nanotechnology uses a novel technique Atomic Force Acoustic Microscopy (AFAM) to investigate the mechanical strength of nanofibers [63], as shown in Fig. (**28**).

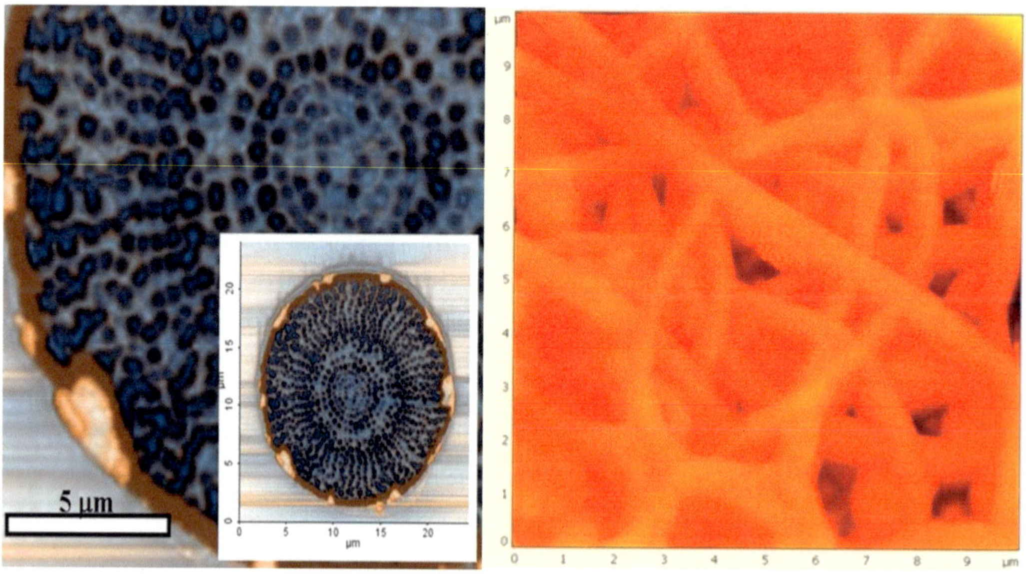


**Figure 28:** AFAM results of a conjugated fiber show that it contain 1220 polyester nanofibers in a sea of polyethylene (L) [63]. Electrostatic Force Microscopy (EFM) is used in electret filtration media (R) for a nonwoven nanofiber web [63].

## 11. DRAWBACKS OF NANOTECHNOLOGY

In real this technology is boundless gathering of various other technologies. 'But the self-same properties that to novel applications in various fields are also the very properties that can cause their destructiveness' [64]. Advance nanotechnology, directs a little attention towards the environmental and health aspects. For example in the year 2004, the US government invest roughly $0.5-1 billion on this technology, approximately $8 million (less than 1%) was invested on the environmental and health destructiveness [65].

### 11.1. Environmental Problems

The great danger to the global warming is the continuous large scale invention, production and advancement of nanoparticles [44]. Rice University published a research report that various nanoparticles that are discard to water and cause the destruction of bactericidal that are the basis component of the ecosystem [65]. Researchers also concluded that nanoparticles can cause the destruction of the ozone layer [19]. Many people investigated that nanoparticles are the main cause of the 'grey goo' that are produced *via* self-replication. Additionally, some scientists have proposed it as a scientific fantasy [64-66], as shown in Fig. (**29**).

**Figure 29:** Grey Goo covering the earth, although just a scientific fantasy (Earth buried in Nanobots).

### 11.2. Health Problems

Nanoparticles are quite hazardous for the health of human beings., Director of the University of Michigan (James Baker) says 'any time we put a material into something as complex as a human being, it has various special influences' [19]. Nanoparticles come in connection to human body *via* the lungs, intestines and skin [66].

#### *11.2.1. Risk to Lungs*

Nanoparticles are ultra-light in weight and effortlessly move in air here and there. So these nanoparticle can easily be inhaled *via* breathing. Inside the body when they come in contact with different body cells they cause different diseases like asthma and bronchitis and can be fatal to body [64]. The followings Fig. **30**, shows the inhalation process of the nanoparticles that how it travels throughout the body. Fig. (**30**): Presenting the traveling process of nanoparticles into the brain, lungs and the blood stream [64].

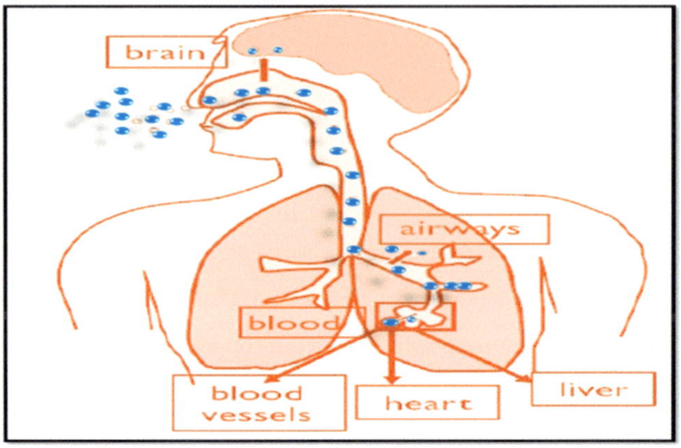


**Figure 30:** Presenting the traveling process of nanoparticles into the brain, lungs and the blood stream [64].

### *11.2.2. Effects of Nanoparticles on Brain*

Inhaled nanoparticles move up *via* nose and breathe canal into the base of the brain part. Where it causes the damage of the brain and the nervous system that is very dangerous and fatal. (Donaldson, Stone, 2004)[66].

### *11.2.3. Problems Cause by Nanoparticles to Blood*

Nanoparticles cause the clotting of the blood system that can cause a heart-attack. These small and light in weight nanoparticles can cause the damage to organs like the heart or the liver and can affect their functionality [64, 66].

## 11.3. Feasibility Problems

### *11.3.1. Investment on Nanotechnology*

Research in the field of nanotechnology is quite luxurious. The Nanotech Report of 2004 have claimed that investment on nanotechnology has reached: $10 billion throughout the global: Total investment. $4.6 billion from government. $3.8 billion from corporate research and development, $200 million from venture capitalists [19]. Recently the devices and techniques used for improvement and enhancement of nanotechnology are very basic and there is big gap still need for boosting this nanotechnology.

### *11.3.2. Lack of Knowledge and Research*

For the development of nanotechnology money is not a big problem. There is also lack of qualified and educated individuals who can improve this technology. Many of the advance techniques and methods are require to characterize nanomaterials because these are still in beginning stage of development [65, 66]. A review from North Caroline University in Raleigh found that more than 80% of Americans know the skills of this new technology [66]. But more professionals are still need for the development of the nanotechnology.

## 11.4. Moral Dilemma

Nanotechnology helps us to control the separate molecules. Most of the pathological type process has origin of molecules, and due to this elementary information the development of nanomedicine has been arises [25]. Researchers hope that this technology could change the living style of the people. Giving them a better quality of life, with full control over the disease, more strong devices can save their money and many more [25]. Inappropriately these potentials are linked with moral consequences that should be measured, before nanotechnology can cover the whole globe. Now the question is, who is in regulate? Nanotechnology promise things that are not natural or foreseen, for instance as to modify the genes of the organisms. At present there is no well- known system to control this up growing technology and there is no precise entity to control it. With the aptitude to recognize & operate precise genetic arrangements, people will search the belongings of virtuous genes. With the help of this technology Individuals are trying to have the eyes and hair color according to their choice. In doing this people risk losing their individuality. Although there are numerous advantages of this new technology are but before boosting this technology to top we should think about the consequences and we should try our best for the development of a safe nanotechnology industry [66].

## 11.5. Potential Safety Concern in Nanotechnology

While nanotechnology is clearly offering enormous benefits to mitigate the harmful effects of climate change on health, it is important to keep in mind that any new technology is always associated with harmful side effects. So may be the case with nanotechnology, especially when it is used for the treatment of diseases. However the overriding decision to use a technology is dictated by the logical fact that if the technology offers excessively large benefits as compared to small damaging effects it is decisively preferable to use. Such happens to be the case with the use of Nanotechnology which is the reason of very fast financial investment in the development of Nanotechnology on global level. The benefits far excessively outweigh the drawbacks. However it does not absolve us of the responsibility of doing serious research in minimizing the harmful the side effects, whatever, to the life on earth. Because of the smallness of the nanoscale sizes of toxic chemicals it is important to do research on toxic effects of nano pharmaceuticals and other chemical industries as not enough is available on this new technology from this point of view. The other important matter is the observance of safety rules and regulations as well as the safety precautions which should be strictly observed by the industrial workers and the industrial managers should provide adequate facilities for such safety requirements. In doing so the harmful effects would be minimized and correspondingly the beneficial effects of Nanotechnology to the society will be enhanced. We have enough experience that Nuclear technology being dangerous too, by observing the safety precautions and having Nuclear Regulatory Bodies as watchdogs, we have been able derive lot of benefits from Nuclear Power Reactors, use of nuclear radiation in diagnostic of cancer and its treatment, in agriculture and in several industrial applications. Further the research and regulations are the two critical tools which can help society to better understand and manage the potential risks. Research should be better coordinated, both at national and international level, while the many diverse applications of nanotechnology need a coherent regulatory framework [67, 68], particularly in the regime of Intellectual property rights for Nanotechnology as this technology deals with control and use of atomic and molecular sizes of materials and the knowledge of atomic science is required for the persons who manage the IPR organizations like the Patent Offices.

## CONCLUSION

This new technology is still in its initial phases. The advantages and applications conferred here have already been established & are already benefiting to the patients all over the world. With the passage of time as more research will carry on in this field, new things will be naked. Various things that have no solution today may be successfully cured by nanotechnology in the future [66]. In Pakistan we need to be very supportive and active for the rapid development and use of nanotechnology for the welfare of the people, to solve the problem of poverty and welfare. Conscious of this need the government of Pakistan has already established a Commission of Nano Science and Technology (NCNST). Several projects of nanotechnology have been launched at some universities and centers of research. Much more is needed to be done as far as possible well before "Western Embargoes" and their commercial exploitation.

Some of the worries were also deliberated but with main use of this technology these problem can be solved. Some researchers are in contradiction with the use of this latest technology they decide that development in new technology should remain due to its enormous benefit, but research in this area should guarantee the safety of the people. If it leads to a sound research well, in future this can be found in our daily life [66]. Frankly, nanotechnology has the latent to re-create the globe we live in, from future arenas to tomorrow morning's shave. Slight wonder truly! [69].

## ACKNOWLEDGEMENTS

Declard none.

## CONFLICT OF INTEREST

The author confirms that he has no conflict of interest to declare for this publication.

## REFERENCES

[1] Department of Trade and Industry-U.K, Foresight Conference: *Opportunities for Industry in the Application of Nanotechnology*; Institute of Nanotechnology: Stirling, U.K., 2000; **44** pp; available at www.nano.org.uk/contends.htm.
[2] www.NanoPartz.com.
[3] www.newagepublishers.com.
[4] www.egyptsearch.com.
[5] Ineke Malsch Industrial physica, 17 (www.malsch.demon.nl).
[6] www.egyptsearch.com.
[7] Wolfgang Luther, "Application of nanotechnologies in the energy sector", Volume of the series Aktionslinie Hessen-Nanotech of the Hessian Ministry of Economy, Transport Urban and Regional Development, August 2008.
[8] www.nanoconsulting.com.sg.
[9] H. Vidam, W. Kong, Singapore Nano Energy report, NanoConsulting Pte. Ltd, 2010.
[10] hypertextbook.com/facts/2000/EvanAbel.shtml.
[11] Curbing global energy demand growth: The energy productivity opportunity, McKinsey Global Energy Institute, May 2007.
[12] Nanotechnology for sustainability: Energy Conservation, Storage, and Conservation by C. Jeffrey Brinker and David Genger, CH; **6**.
[13] http://www.eia.doe.gov/oiaf/ forecasting.html.
[14] www.corridor.anl.gov.
[15] Potential Impacts of Nanotechnology on Energy Transmission Applications and Needs, ANL/EVS/TM/08-3 by Deborah Elcock, Environmental Science Division-Argonne National Laboratory, November 2007.
[16] D. Mori, K. Hirose, "Recent challenges of hydrogen storage technologies for fuel cell vehicles," *Int. J. Hydrogen Energy.*, Vol. 34, pp. 4569-4574, 2009.
[17] C. Liu, F. Li, Lai-Peng Ma, and Hui-Ming Cheng, "Advanced materials for energy storage," *Adv. Mat.*, Vol. 22, pp. E28-E62, 2010.
[18] ijaeee.uacee.org.
[19] Perkel, M Jeffrey. "The Ups and Downs of Nanobiotech." The Scientist **30** Aug. 2004: 1-8.
[20] student.kfupm.edu.sa.

[21] Roman, Harry. "Micro and Nanotechnology, The Next Big Tiny Thing?" *Mercer. Business*. **1** Jan. 2005: 1-4.
[22] Adhikari, Raju. "Nanobiotechnology: Will It Deliver?" Healthcare Purchasing News. Jan. 2005: 1-3.
[23] Weiss, Rick. "Nanomedicine's Promise Is Anything but Tiny." Washington Post 31 Jan. 2005: 1-3.
[24] Yanglong Hou Group, College of Engineering, Peking University, P. R. China.
[25] Silva, Gabriel A. "Introduction to Nanotechnology and Its Applications to Medicine." *Surg. Neurol.*, 61.3 216-220, Mar. 2004.
[26] Kim *et al.*, 2007. Antimicrobial effects of silver nanoparticles. *Nanomedicines*. **3(1)**: 95-101.
[27] I. M. Sadiq, B. Chowdhury, N. Chandrasekaran, A. Mukherjee, Nanomedicine., Vol. 5: 282-286, 2009.
[28] Y. Liu, H. Miyoshi, and M. Nakamura, "Nanomedicine for drug delivery and imaging: A promising avenue for cancer therapy and diagnosis using targeted functional nanoparticles, " *Int. J. Can.*, vol. 120, 2527–2537, 2007. doi: 10.1002/ijc.22709.
[29] Z. Liu, *et al.*, "Drug delivery with carbon nanotubes for *in vivo* cancer treatment," Cancer Res., vol. 68, 6652, 2008.
[30] www.understandingnano.com.
[31] www.nanotec.org.uk.
[32] http://www.nanowerk.com/news/newsidphp.
[33] N. M. Butt, T. Ahmed. "The 3rd International Symposium-Cum Training Course on Molecular Medicines and Drug Research (MMDR-**3**)", 3-6 Jan 2011.
[34] The Millenium Development Goals and water. World Water Assessment Programme (2005).
[35] T. Hillie, and M. Hlophe, "Nanotechnology and the Challenge of Clean Water", *Nat. Nanotech,* vol. 2, no. 11, pp. 663-664, 2007.
[36] www.nssa.com.au.
[37] www.irinnews.org/Report.aspx?ReportId=79042
[38] www.nature.com.
[39] Efficient filters produced from carbon nanotubes through Rensselaer Polytechnic Institute-Banaras Hindu University collaborative research. Rensselaer (2004).
[40] K. Khider, D.E. Akretche and Larbot A., "Purification of water effluent from a milk factory by ultrafiltration using Algerian clay support". Desalination, vol. 167, 141, 2004.
[41] G.P. Gillman, "A simple technology for arsenic removal from drinking water using hydrotalcite" Science of the Total Environment., vol. 366, 926, 2006.
[42] C. T. Yavuz, J. T. Mayo, W. W. Yu *et al.* « Low-field magnetic separation of monodisperse Fe3O4 nanocrystals, » *Science*, vol. 314, 964, 2006.
[43] T. Hille, M. Munasinghe, M. Hlope *et al.*, "Nanotechnology, Water and Development". Report of Meridian Institute (2006).
[44] Patel P, Technology Review, **20** Jan. (2009).
[45] www.technologyreview.com/computing/page2/.
[46] www.nano.org.uk.
[47] NIH Bioengineering Consortium (BECON). *Nanoscience and Nanotechnology: Shaping the Future of Biomedicine*; report of a symposium on June 25–26, 2000; http://www.becon.nih.gov/nanotechsympreport.pdg.
[48] The Role of Nanotechnology in the Semiconductor Industry Linda Rae, Keithley Instruments, Inc.
[49] www.nanomagazine.co.uk.
[50] cleanroom.net.
[51] www.strategyr.com/Nanobiotechnology.asp.
[52] C. Jeffrey Brinker. "Nanotechnology for Sustainability: Energy Conversion, Storage, and Conservation", Nanotechnology Research Directions for Societal Needs in 2020, 2011.
[53] www.pentium.co.il.
[54] Robert Chau *et al*, "Benchmarking nanotechnology for high-performance and low-power logic transistor applications," *IEEE. TRANS. ON. NANOTECHNOLOGY,* VOL. **4**, NO. 2, MARCH 2005.
[55] http://www.nanotechbuzz.com/50226711/carbon_nanotube_bike_now_available.php2
[56] http://www.nanowerk.com/spotlight/spotid= 30661.php (Nanotechnology in sports equipment: The game changer).
[57] www.understandingnano.com.

[58] A. El-Fatatry., “Educational Notes RTO-EN-AVT-**129”**, Paper 5. Neuilly-sur-Seine, France: RTO, (2005) (pp. 5-1 – 5-6).
[59] www.nanotechwire.com.
[60] www.nanotec.org.uk.
[61] Ben Anderson. "Hope for nanotechnology: anticipatory knowledge and the governance of affect", Area, 6/2007.
[62] www.understandingnano.com.
[63] www.semiconductor.net.
[64] Donaldson, Ken, and Vicki Stone. “Nanoscience Fact Versus Fiction.” Association for Computing Machinery (Nov. 2004): **113**. Full Text (ABI/INFORM. ProQuest).
[65] Balbus, John, *et al*, Issues in Science and Technology 2005: 1-4. http://www.findarticles.com/p/articles/mi_qa3622/is200507/ai_n14716314 > (7 Nov. 2006).
[66] www.pharmainfo.net.
[67] Robichaud C, Tanzil D, Weilenmann U. and Wiesner M, Environmental Science and Technology, 4 Oct. 2005.
[68] Roco M. and Bainbridge W, Boston, Massachusetts: Kluwer Academic Publishers 2001.
[69] www.metrocorpcounsel.com

# Development and Prospective Applications of Nanoscience and Nanotechnology

## *Volume 1: Nanomaterials and their Fascinating Attributes*

## CHAPTER 2

# Multifunctional Polymer Nanocomposites in Next Generation Smart Structures

**Zaffar Mohammad Khan[1,*] and Muhammad Rizwan Saleem[2,3]**

[1]*Department of Aeronautics and Astronautics, Institute of Space Technology (IST), Islamabad Highway, Pakistan;* [2]*Institute of Photonics, University of Eastern Finland (UEF), P.O. Box 111, FI-80101, Joensuu, Finland and* [3]*Center for Advance Studies in Energy (CAS-EN), National University of Sciences and Technology (NUST), Sector H-12, Islamabad, Pakistan*

**Abstract**: The development of nano-structured materials have opened a new paradigm whereby the nano-reinforcement and matrix resin can be tailored to optimize the physical and mechanical properties of resulting reinforced nano-composite materials manifold. The potential applications of this revolutionary class of nano-composite promise a significant opportunity for improving structural dynamic properties. This has led to emergence of new frontiers in mechanical, electronic and aerospace structures focusing on multi functionalized, self assembling and self healing materials. Such kind of materials can sense and autonomously respond to changes in environmental conditions, flight envelope and structural health, ultimately enhancing the durable performance characteristics and mission survivability for next generation aerospace vehicles. This chapter reviews the latest developments in the area of nano-composites with reference to nano-clays, nano-fibers and nano-tubes and their applications in smart aerospace structures. The classification, characterization and applications of various types of nano-composites have been discussed in aerospace industry in detail. It will highlight the applications of nano-composites in relaunch vehicles, space ladders, ablatives, internal motor casing and nano-modified carbon/carbon composites. Successes in this field have sprawled the subsequent attempts to emulate biological, bio-mimetic and smart materials. This will result in paradigm shift in space doctrine and philosophy for establishing space stations in low earth orbit in times to come.

**Keywords:** Composite materials, nanoclay, nanocarbon tubes, nano fibers, smart structures, fiber optic sensors, morphing wings, nanophotonics, photonic structures.

## 1. INTRODUCTION

The quest for improved materials in engineering industry is never ending. Mankind has been looking for improved materials from Paleolithic period (Stone

***Corresponding author Zaffar Mohammad Khan:** Department of Aeronautics and Astronautics, Institute of Space Technology, Islamabad Highway, Islamabad, Pakistan; E-mail: zaffark7@gmail.com

**Sher Bahadar Khan, Abdullah M. Asiri & Kalsoom Akhtar (Eds)**

Age), through the emergence of metallurgy at the dusk of Neolithic period (bronze/iron age) to novel materials in space age synthesized as nano-composites (smart materials age). Hominids selected natural ceramics for tools and weapons in the Stone Age. It is cited in the book of Exodus in the Biblical Old Testament that Israelites were required to manufacture bricks in clay and straw for the Pharaoh in Egypt [1]. According to Roman mythology, Icarus and Dedaulus tried to mimic the bird's flying to escape from Greek's captivity by affixing feathers to reinforced wax to their bodies. They however crashed into sea due to poor thermo-mechanical response of wax.

Subsequently the monolithic materials exploited the traditional mechanical alloying procedures and smelting technologies for the next 5000 years. In 1812, Sir George Cayley postulated that the streamlined geometry of a trout served the basis for the cross section shape of his low drag wing airfoil. The aileron employed to control aircraft was based upon ornithological studies performed by Wilbur Weight in 1900. The consequence of the research focused upon the development of synthetic engineering materials that mimic the attributes of naturally occurring biological materials over last hundred years. An independent approach to use these materials with self adapting and self correcting characteristics is to continuously achieve a performance under all operating conditions while functioning in an unstructured environment. This has led to aggressive development of new class of engineering materials such as multi functional nano-composites. In this connection carbon nanotubes were discovered by Iijima in 1991 by NEC Corporation, Japan. However, after the discovery of a new form of carbon, so called buckyballs $C_{60}$ in 1985, much research interest was focused on carbon nano-tubes immediately after Iijima observations. Moreover, the research works on the synthesis, characterization and theoretical investigations for carbon nano-particles, nano-tubes and nano-fibers have grown exponentially. An independent research discovered single walled carbon nano-tubes jointly worked by Iijima group and an IBM team in 1993. The tremendous mechanical and electronic properties have found a significant impact on the future technological applications. This prompted many Governments on worldwide basis ranging from USA to Saudi Arabia to initiate nano-technology programs. The US has initiated National Nanotechnology Initiative (NNI) with yearly budgetary outlay of close to $1 billion. The Pakistan government has allocated close to $ 70 million for promotion of nano-technology and jump start their economy. The Saudi Arabia has initiated major programs to promote nano-composite materials science and technology in academia, industry and R&D organizations.

In the field of material science, a paradigm shift has occurred from conventional materials to the revolutionary nano-structured material. The methods to integrate structural, sensory and actuator characteristics of components over macroscopic and mesoscopic scale have been replaced by methodologies that will allow their precise integration at microscopic & nano-scopic levels in engineering design as shown in Fig. (**1**) [1].

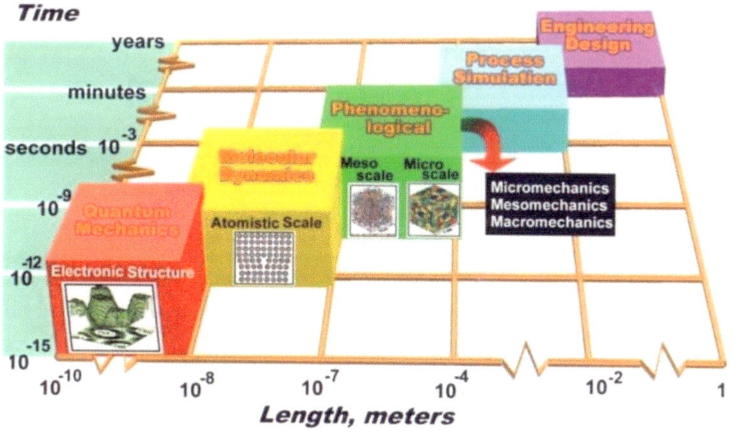


**Figure 1:** Courtesy of picture: the analogy between macro-, micro- and nano-scopic methodologies with time [1].

The science of nano-materials deals with the control and manipulation of materials properties, particularly of nano-particles with sizes ranging 1–100 nm ($10^{-9}$ m) to design and synthesize novel materials and structures possess superior physical and mechanical properties due to their extremely small size (of the order of nm). This reduction in size carries out a transition from micro to nano-particles and thus increases surface area to volume ratio, thus enhancing the surface energy of the materials constituents (nano-particles). In addition, an increase in surface energy results in exponential increase in surface atoms that generates more reactive surface sites available for reaction with neighboring atoms, catalysis, bonding *etc*. This results in improved properties such as enhanced wear resistance, thermal & electric conductivity, flame retardance, stiffness and mechanical strength. Owing to small dimensions of nano-particles below critical wavelength enables to show transparent behavior as well as possess aesthetic implications. One can control the fundamental properties of nano-structural materials such as band gap, interfacial energy, charge density, magnetic properties, melting temperature, and color without changing chemical composition. The nano-composites are composed of nano-clay, nanotubes, nano-fibers, nano-fibrils and nano-particles. One common class of commercial nano-clay is ceramics that can be classified into two main groups: silicates or silicon oxide, found in nano-scale flakes of clay, and metallic oxides of titanium, zinc, aluminum and iron [2].

## 2. NANO-CLAYS AND METAL OXIDES

The most widely used layer silicate is the montmorillonite (MMT) as shown in Fig. (**2**) [3]. MMT is composed of triple layer where silica tetrahedral sheets are bonded together with an edge-shared octahedral sheet of alumina or magnesium as shown in Fig. (**2**).

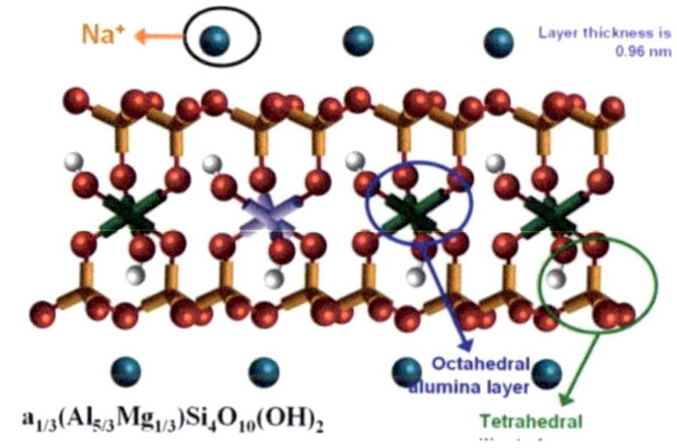


**Figure 2:** Courtesy of picture: the micro-structure of montmorillonite clay [3].

The stacking layer sequence of individual silicates sheets exhibit hydrophilic nature that is different to that of hydrophobic nature of polymer matrix. Dispersion of layered silicates sheets and their compatibility to the host matrix polymer is an important part in making nano-composites. The surface of the individual sheets of the layered silicate is modified with additive materials called surfactants such as alkyl ammonium salt by ion exchange process to become compatible with the matrix polymer. The addition of organic onium ion to that of long hydrocarbon chains on layered silicate surface expands the structure gallery. The interaction between individual sheets decreases which enable the penetration of the polymer or polymer precursor into gallery easier. When the interfacial interaction between the polymer and layered organo-silicate is strong enough, the intercalated or exfoliated nano-composite is made. The MMT can be delaminated into individual sheets with a thickness of one nanometer across a length of 70–150 nm which results in a radical enhancement of surface to volume ratio. The originally modified nano sized MMT platelets are called Cloisite. The MMT particles without separation are often called tactoids. Thus, the classification of nano-composite is based on separation of tactoids. The partially separated tactoids in polymers matrix are referred as being intercalated as shown in Fig. (**3**) [3].

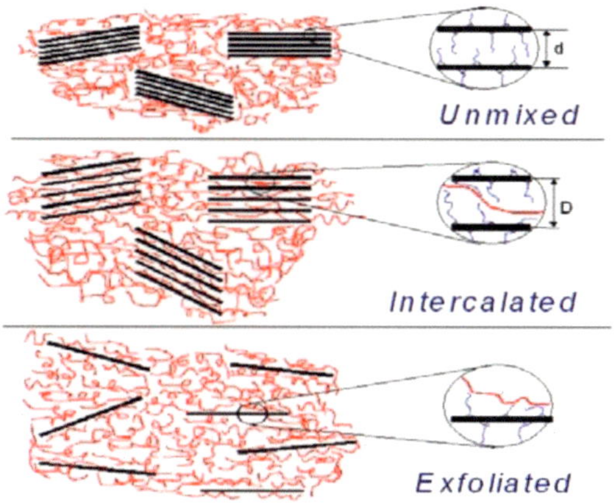


**Figure 3:** Courtesy of picture: the nano-composite classification based on separation between tactoids [3].

However, after a complete separation, individual platelets are known as exfoliated. Thus, the first step for the synthesis of a nano-composite is to change the structure of clay by treating surface with organic moieties for complete dispersion and miscibility in the polymer matrix. Subsequent to surface organic treatment the tactoids are exfoliated into platelets. This can be achieved either through shear stresses which are applied during compound melting or using chemical methods during matrix polymerization as shown in Fig. (**4**).

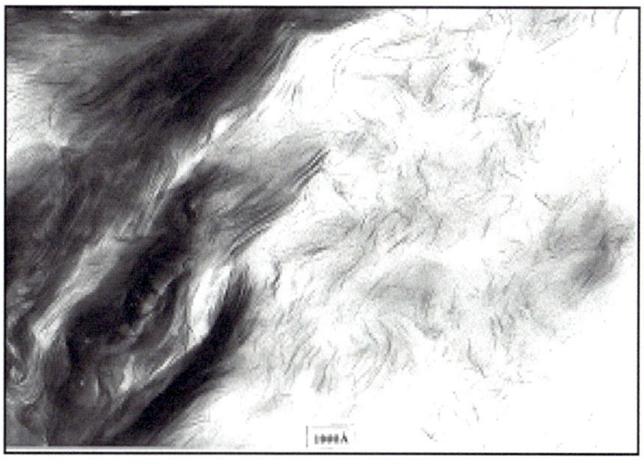

**Figure 4:** Courtesy of picture: shear/ peel mechanism of platelets in nano-composite during dispersion [3].

The cloisite leads to increase mechanical properties, such as scratch resistance, melt strength, modulus, heat resistance, abrasion resistance, flame retardation and dimensional stability. The degree of improvement depends on the extent of exfoliation. About 40 % MMT platelets have been exfoliated completely while rest form tactoids of 2–10 platelets. Polyhedral Oligomeric Sil Sequioxane (POSS) represents a multifunctional polymer additive acting simultaneously as molecular level reinforcement, processing aid and flame retardants. Nylon 6, Cyanate Ester and Epoxy are also extensively used as polymer nano-composite.

Aluminum oxide nano-particles are produced through physical vapor synthesis. For the production of nano-particles, arc energy is supplied to a metallic precursor compound to create vapors at high temperature. The generated vapors are combined with a reaction gas to form a mixture at a well controlled cooling rate to cool down. The cooling mixture condenses and produces nano-particles which are then deposited with a thin polymeric shell to ensure compatibility with other fluids and polymers. During Nano-phase tests, 1 % loading of particles lead to 5 to 10 fold increase in scratch resistance with less than 0.5 % haze. Superior aircraft automotive finishes require high degree of scratch resistance. The use of sufficient quantities of silica nano-particles or nano-clay fillers in polymer matrix can improve the scratch resistance but make the coating hazy with a decrease in viscosity. While $Al_2O_3$ nano-particles are harder than that of $SiO_2$, resulting in coating haziness. The fumed titanium oxide ($TiO_2$) nano-particles are produced through flame hydrolysis of titanium tetrachloride in oxygen-hydrogen gas flame. It offers excellent thermal resistance.

Many different techniques are used for characterizing polymer nano-composites: most commonly used techniques for characterizing nano-composites are Wide Angle X-ray Diffraction (WAXD), Small Angle X-ray Scattering (SAXS), Thermal Gravimetric Analysis (TGA), Transmission Electron Microscopy (TEM), Mechanical Testing and Nuclear Magnetic Resonance (NMR), Atomic Force Microscopy (AFM) and Scanning Probe Microscope (SPM) *etc*. WAXD measures the spacing between ordered crystalline layers of the clay. Spacing changes (increase or decrease) can help to determine the type of nano-composite, such as immiscible (no d-spacing change), decomposed/de-intercalated (d-spacing decrease), intercalated (d-spacing increase), and exfoliated (d-spacing outside of WAXD). TEM measures overall clay dispersion in the sample. Clay dispersion and microstructure observed under the microscope can determine the nature of clay nano-composite such as immiscible, intercalated or exfoliated [2, 3].

## 3. CARBON NANO-TUBES AND NANO-FIBERS

Nano-carbons are discrete, ordered structures grown from specific catalytic sites on a substrate and include carbon nano-tubes and nano-fibers grown axially along their respective lengths away from the substrate surface. Carbon nano-tubes may take a number of forms, including single-wall carbon nano-tubes (SWCNTs), multi-wall carbon nano-tubes (MWCNTs), carbon nano-fibers or graphite nano-fibres (GNFs) as shown in Fig. (**5**).

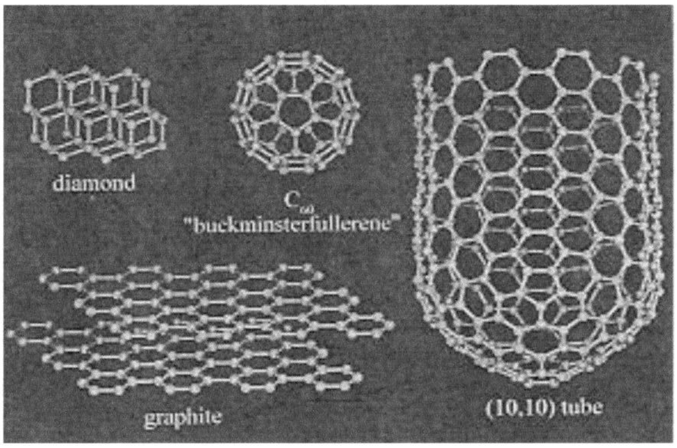


**Figure 5:** Courtesy of picture: the structure of a carbon nano-tube, bucky ball and graphite planes.

SWCNTs are essentially cylindrical forms of graphite, whose walls are one atom thick; MWCNTs are concentric collections of SWCNTs as shown in Fig. (**6**).

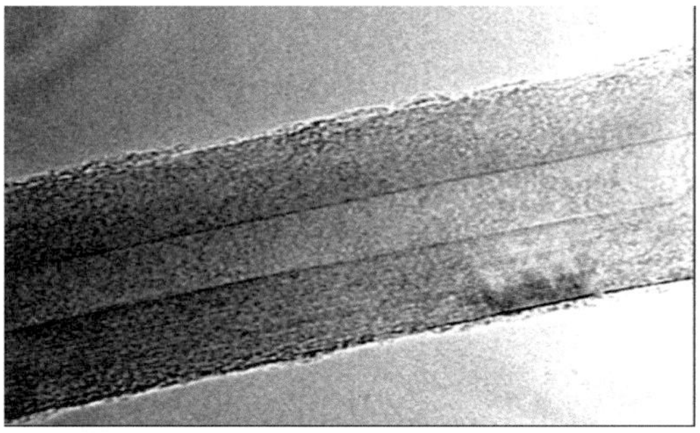

**Figure 6:** Courtesy of picture: concentric walls of MWCNT [3].

These tubular forms of carbon nano-tubes may be end capped, uncapped, side walled and may be filled with other compounds. In GNFs, the graphene planes (one atom thick layers of graphitic carbon) are stacked on top of each other in either a planar or herring-bone configuration, the latter being thought to represent a cross-section through the actual structure comprising a stack of conical graphene sheets. In GNFs, hydrogen is thought to adsorb interstitially between the graphene planes. The carbon nano-fibers are unique form of vapor grown fiber that bridges the gap in physical properties between large usual PAN or pitch based carbon fibers, five to ten micron and smaller than single wall and multi-walled carbon nano-tubes that are one to ten nm as shown in Fig. (**7**).

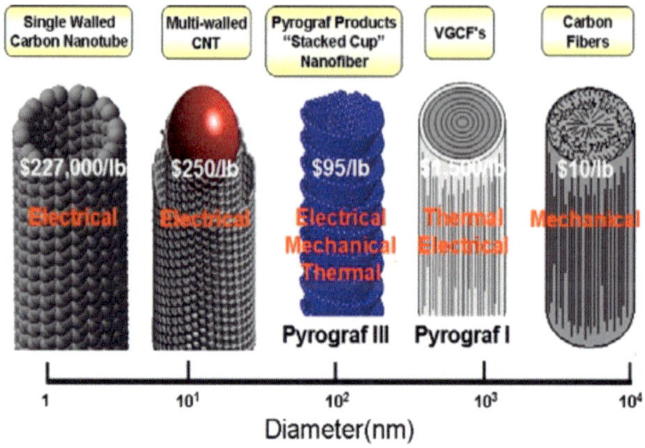


**Figure 7:** Courtesy of picture: sizes from carbon nano-fiber to carbon nano-tubes [3].

The multifarious properties of carbon nano-fibers, tubes, clays and fabrils are shown in Table **1**. Pyrograf-III® is a very highly fine carbon nano-fiber (graphite) which has an average diameter between 70–200 nm and a length estimated to 50-100 micron [4, 5].

## 4. FABRICATION TECHNIQUES

There are three main methods to fabricate carbon nano-tubes such as vapor grown, electric arc and laser ablation or deposition. They usually rely on the presence of metal powder catalyst, such as iron, cobalt or nickel, on which the carbon atoms combine to form nano-structures. Growth occurs in the vicinity of the substrates such as graphite foil, nickel grids, carbon cloth, tungsten coil *etc*., in an ordered manner, with the growth plane disposed within the substrate surface. It is usually desirable for the diameter, length, and sometimes the alignment of the tubes and stacks to be controlled. The resulting nano-tubes may contain significant amounts of the metal particles, which may require removal by acid washing. Chemical vapor deposition involves exposure of catalytic substrate to a gas that undergoes a chemical reaction to form a coating on the substrate; the coating may be formed where the gases impinge on the catalytic substrate or elsewhere, assisted by diffusion of atoms through the catalyst. In this case, the gas is a carbon bearing gas such as ammonia, with the result that a non-carbon substrate may be used. The process need not be plasma-assisted, and hence has the advantage that it may be conducted at atmospheric pressure. Plasma-enhanced chemical vapor deposition (PECVD) process for growing carbon nano-tubes on catalytic metal substrates where the catalytic metal film preferably has a thickness of 0.5 to 200 nm, and is described as forming "iclands" to promote nano-tube growth. It is possible to achieve higher growth rate by using PECVD as opposed to thermal CVD, and by treating multiple batches of samples.

**Table 1. Comparison of mechanical, electrical and thermal properties of carbon nano-fibers, nano-tubes and nano-clays [3].**

| | **Carb-on Black** | **Graphite Flake** | **Nano-clay** | **Single Walled Carbon Nanotubes** | **Multi Walled Carb-on Nano-tubes** | **Carbon Nano-Fibers** |
|---|---|---|---|---|---|---|
| U.S. 2004 Production | 3 Billion | 30 Billion | 10 Billion | 100 | 20,000 | 70,000 |
| Price, $/lb | $0.25 to $10 | $5 to $10 | $5 to $10 | $50,000 to $180,000 depending on purity | $250 to $3,000 depending on purity | $100 now; <$10 with WCl project |
| Unpurified Metal Content, WT % | 0.2%0.5% | 2%-10% | Unknown | 40%-90% | 10%-40% | 0.8%-1.4% |
| Purified Metal Content, WT% | 0.2%-0.5% | 0.5% to 2% | Unknown | 5% to 10% | 2% to 5% | <100 ppm |
| Lead Time, Months | 1 | 1 | 1 | 6 | 3 | 1 |
| Mechanical Properties | Toughens rubber, degrades plastic | Degrades strength in plastics | Toughens thermo-plastics | Degrades strength in plastics | Degrades strength in plastics | Improves both strength and modulus in rubber and plastics |
| Electrical Conductivity | At high loadings | At moderate loadings | None | At low loadings | At low loadings | At low loadings |
| Thermal Conductivity | No improvement | Some fire retardancy | Some fire retardancy | None | Some fire retardancy | Conductive, some fire retardancy |

Electric arc deposition involves passing a direct current between axial extension of carbonaceous anode rod and successively repositioned cathode surface. An arc discharge occurs with the simultaneous formation of carbon nano-tubes on the exposed portions of the cathode surface, and the carbonaceous deposits are then scraped and collected. This method is susceptible to continuous operation, although this would be dependent upon the longevity of the anode and the re-usability of the cathode surface. Electrical heating has the advantage that it can be carefully controlled, depending on the current and wire thickness selected, and provides a uniformly heated substrate. Longer deposition lengths and higher reproducibility is therefore attainable. Thus, DC heating of a wire substrate can provide an effective reactor length of several meters, which would not be realizable by passing the wire through a furnace or using RF/plasma CVD, and results in improved production rates and uniformity of the product. Advantageously, an electrically conductive wire is used, and the use of a metallic wire having a diameter of 100–600 microns is preferred. It would normally be provided on a reel in a length of up to, or exceeding 1 km, and may be provided in lengths up to, or exceeding, 5 km or even 10 km, enabling many hours or days of continuous production. The wire is preferably formed of steel, iron, nickel or alloys thereof, but it may comprise other commercially available alloy wires such as, for example, ones containing any one or more of iron, nickel, cobalt and copper [6-7].

The nano-particles/fibers are dispersed in a solvated thermoset by solution intercalation, melt intercalation, ball/roll milling or polymerization process. The resin is sold in solution as a ready mixing pre-preg kit. The resin can be cast, applied or sprayed onto the carbon, glass or UHMPE fabric and cured by using VARTM, RTM or LRI or autoclave based process. The physical characterization of nano-composites may be carried out by using X-ray diffraction, TEM, WAXD, SAXS and mechanical characterization may be carried out by using tensile, flexure, compressive, ILSS, IPSS techniques to meet the electromagnetic, mechanical and aerospace requirement as shown in Fig. (**8**).

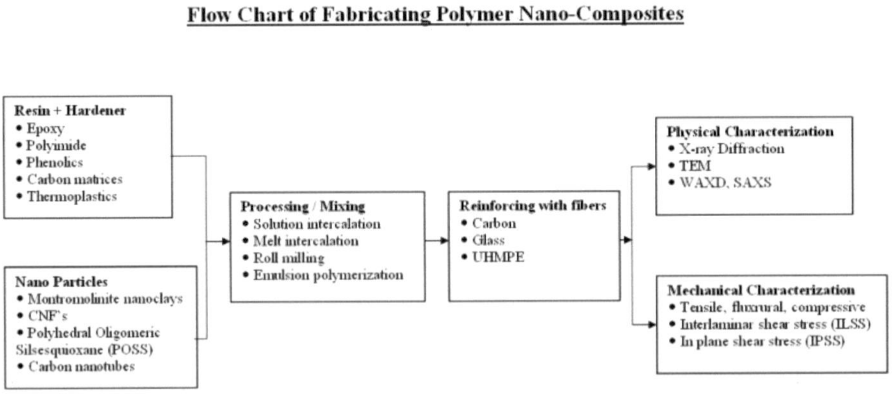


**Figure 8:** Courtesy of picture: the fabrication process of polymer nano-composite materials.

## 5. CONDUCTIVE NANO-COMPOSITES

The design and synthesis of affordable nano-scale materials is desirable to facilitate and solve the electrical conductivity problem of composite materials for smart structures. The electrical properties of such nano-composite materials are tailored which are demanding in a wide range of applications such as sensors, actuators, organic LED, micro- and nano-fluidic devices, biomolecular sensors, automotive fuel systems for electrostatic discharge, micro-electronic circuit boards, grounding resistive heating, electromagnetic interference (EMI), and lightening strike effects in aircrafts. In particular, newly synthesized nano-particles have superior electrical conductivities than previously available one to improve conductivities of polymer nano-composites. Newly synthesized nanomaterials show different geometries to those of previously demonstrated conductive particles. Particle geometry is one of the parameter affecting the behavior and nature of nano-composite after adding conductive nano-particles to non-conductive polymer matrix. The already reported conductive particles (additives) in various forms such as platelets, coated spheres, milled powders *etc*., with aspect ratios of the order of 1–10 while newly developed nano-particles (carbon nano-fibers) have more than 2 orders of magnitude to previous ones. This nano-filling shows and facilitate to get same conductivity levels of nano-composite with less material consumption both in terms of weight and volume percent.

The mechanical and electrical properties of single-walled nanotubes (SWNTs) such as electrical conductivity of 1 billion amp/cm$^3$, modulus more than 1 tera Pa, and strain of about 10 % are superior to that of multi-walled nanotubes (MWNTs) due to complex structure and variety of forms. MWNTs are more cost-effective (an order of magnitude less cost than SWNTs) and are used in potential commercial applications since last 10 years as a replacement of carbon black in thermoplastics to get suitable interesting electrical properties. An extensive research work is carried out to use MWNTs in thermosets for electrical properties and improve also mechanical properties. Carbon black and nano-tubes almost have the same electrical properties (graphitic nature), however, the former has spherical geometry while MWNTs have rod-like structure with high aspect ratio which is sufficient to get a conductive bridge in a non-conductive polymer matrix with suitably small amount of reinforcement of nano-particles/-tubes. Typically, the reinforcement percentage of carbon black ranges 7–18 in polymer matrix to acquire desire electrical conductivities while the same can be achieved by 1–3 loading percentage of CNTs that makes nano-composite more flexible (high toughness). This enables conductivity-loading curve of nano-composite with CNTs to be more gradually smooth which can easily target a particular conductivity level compared to carbon black. At such percent of reinforcement for CNTs with promising electrical conductivities are use in a wide range of electrical applications such as resistive heating, electromagnetic interference (EMI), protection from electrostatic discharge (ESD), electrostatic discharge nano-devices, resistive heating, and lightning strikes effects *etc*., which conventional composites lacks to have. The electrically conductive composite enclosures address these applications with approximately a weight of one-third to that of their metallic counterparts. Analogously, the use of new class of materials: conductive coatings, conductive sealants, and conductive adhesives *etc*., supports their use in several applications ranging from commercial to defense market.

Advanced composite materials have replaced the metallic flight structures of Boeing and Airbus aircrafts at reduced production costs. Cylindrical carbon molecules are combined to form carbon nano-tubes (CNTs) of a few nm in diameter, showing properties of about 50 times stronger and 10 times lighter than steel. CNTs not only possess superior mechanical properties but also have electrical and thermal conductivities about 100 and 3 times to that of copper, respectively. Such multifunctional properties (mechanical, electrical and thermal) of composite materials (carbon fiber reinforced polymers) are being used for the development of structural components. For the manufacturing of these materials, nano-particles are inserted into substrates for the onset growth of nano-tubes and/or spinning of micro-fiber from CNTs to synthesize nano-tubes with hybrid fibers [8].

The index of refraction of a material is a measure of light confinement and affecting the speed of light passes through. The forest of CNTs has a low refractive index near to that of air and effect the propagation of light in a similar fashion. As a beam of light from a light source impinges on a layer CNTs from air, little reflection and scattering is observed due to low refractive index.

This optical property of CNTs is exploited by Michigan research group to hide the structure of an object visually. The description is presented in Journal of Applied Physics Letters; a 3D image was constructed by scientists of silicon. On illumination of this 3D image by white light, the reflection of light revealed the image features as tank's contours. However, on covering the structure with a forest of CNTs, the illuminate light is absorbed by surface coating, which revealed only a black sheet. In fact absorption instead of scattering could cloak the structure of an object against a black background sheet due to coating by CNTs such as a deep space. In this experiment the presence of CNTs forest behave as a perfect magical black cloth to hide the complete features of a 3D structure.

The research group supervised by author at National University of Science and Technology/ Institute of Space Technology, Pakistan has investigated the effect of addition of carbon nano-particles in polymer matrix on microwave absorption in the frequency range of 11–18 GHz. A band-stop of Electromagnetic Interference (EMI) and gaining Electromagnetic Counter-measure (EMC) are the main goals of this study which are applicable in industry dealing with high speed electronic devices such as wireless local area networks, wireless data communication systems, mobile & satellite communication system *etc*; operated in GHz range due to standard practice being employed. Such composite materials are used as Microwave Absorbers in civil and military applications. The electromagnetic wave attenuations, created intentionally to disable the access of Wi-Fi signals of mobile phone in certain areas are a result of microwave absorption. In this regard, it is important to treat the problem as a light-matter interaction, keeping in view different materials for various applications. Earlier only the visible region was used to detect the objects but with the evolution of wide range electromagnetic spectrum with broad frequency range is used for the detection of objects [9].

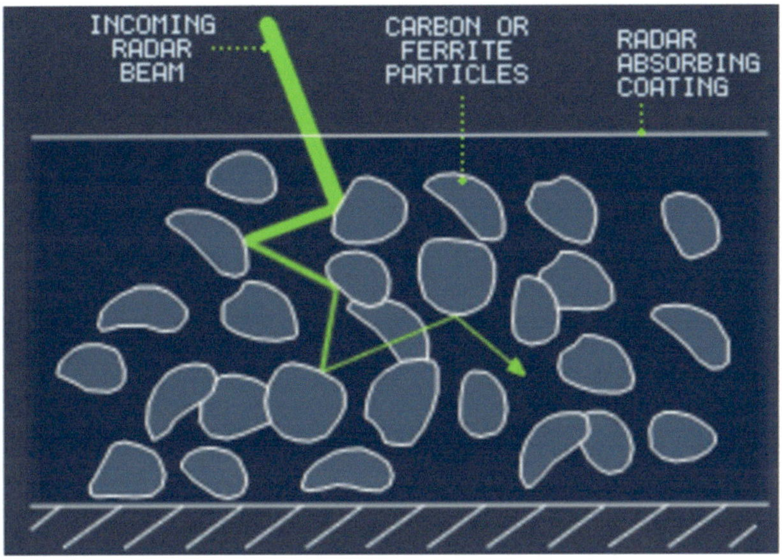


**Figure 9:** Courtesy of picture: schematic presentation of a RAM [12].

The use of adequate microwave absorbing materials in an electronic device can greatly help to control the smooth functioning, unnecessary interference due to external EM waves, attenuation, and suppression of unnecessary emissions of EM waves [10]. Two concise approaches were considered to study the stealthy properties, namely: (1) evaluation and synthesis of Radar Absorbent Materials (RAM) as thin films and (2) Radar Absorbing Structures (RAS). Fig. (**9**) shows the schematic functioning of a RAM.

The fundamental working principle to predict radiation absorbing properties of materials at a particular energy depends on the atomic and molecular band gap structures of the material, that is, the allowable available electronic energy states. A perfectly applied material for this purpose is the one whose atomic/molecules vibrate in the microwave frequency regime (11–18 GHz) and/or the electrons of the materials can absorb energy that is emitted by Radar in the form of photons (fundamental particle in light streams) [11]. Typically, RAMs are synthesized in the form of sheets containing insulating polymer resin matrix loaded (reinforced) by loss inducing agents (carbon nano-tubes, perm alloys, ferrites, carbon black).

In general, the EM absorbing tendency of a material is accessed by magnetic, ohmic, and dielectric losses. We consider the dielectric absorbing property irrespective to the heavy magnetic materials [13-15]. Furthermore, the chief properties that enable materials to absorb microwaves are the dielectric constant (complex permittivity of the material, $\varepsilon^{'} - i\varepsilon^{''}$) and the energy dissipation factor [11, 16, 17].

*In this research, RAM samples of epoxy resin containing nano sized carbon black in different weight ratios were investigated for scattering parameters and complex permittivity measurements. Also, for the RAS part, sandwich composite structures with the combination of same face sheet and different core materials were compared with respect to same parameters as discussed for RAM.*

*For the synthesis of RAM, the polymeric composite and accompanying epoxy matrix were considered primarily due to simple processing and attractive chemical and physical properties. As we are focusing on dielectric RAM, therefore, carbon nano-particles are used as reinforcement or filler/loader (dispersant phase of the polymeric composite). Fig. (**10**) show the scanning electron microscopy (SEM) and transmission electron microscopy (TEM) images of the reinforced carbon black (CB) nano-particles of size 50–70 nm in polymer epoxy matrix which reveals porous aggregated morphology with suitable electrical conductivity properties.*

## 6. DISPERSION OF CARBON BLACK PARTICLES IN EPOXY

In general, the dispersion of CB particles in epoxy resin was analyzed at a range of weight percentages (0–5 %) as depicted in Table **2** with name-codes.

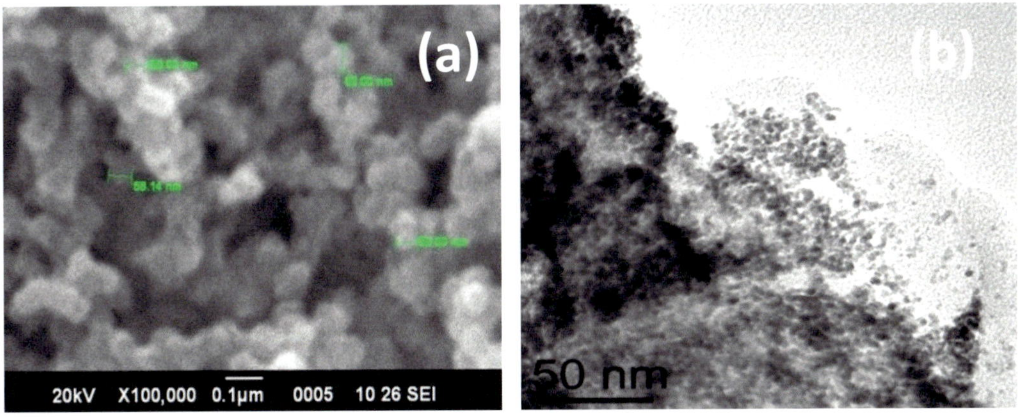


**Figure 10:** Courtesy of picture: SEM and TEM photographs of CB particles.

**Table 2. The filler concentration (per hundred resins) in the composite material.**

| *Sr. No.* | *Composite Code* | *Concentration (% Resin)* |
|---|---|---|
| *1.* | CB0 | 0 % |
| *2.* | CB1 | 1 % |
| *3.* | CB2 | 2 % |
| *4.* | CB3 | 3 % |
| *5.* | CB4 | 4 % |
| *6.* | CB5 | 5 % |

*In this work relating to synthesis of RAM samples, the epoxy resin of Araldite DBF and Hardener Aradur HY-956 EN of Huntsman were purchased from Nihaal Polymers, Karachi while carbon black nano-particles were taken from China. A suitable weight fraction of CB nano-particles dispersed in ethanol in ultrasonic bath for one hour at room temperature. Epoxy resin was mixed with ethanol in ultrasonic bath where a process of sonication was performed for next 45 min at elevated temperatures. The mixture was placed on a hot plate with a magnetic stirrer. Ethonol was evaporated within 30 min under controlled agitation at a temperature of 80 °C. Following the ethanol evaporation, hardener was added to the remaining mixture which was sonicated further for 10 min. After cooling down, the final mixture was coated on Al plate with size of 6 × 6 square inches and thickness ranging 0.3–0.5 mm using Doctor's blade. The coated Al plate was placed at room temperature for 4 hours and then in a vacuum oven at*

*80 °C for further 4 hours to dry properly [18-20]. Here, Al is used as a reference material to compare observed reflections during these measurements with EM waves.*

## 7. CHARACTERIZATION OF SAMPLES

*SEM was carried out to investigate the dispersion of CB nano-particles in the polymer matrix. Fig. (**11**) shows the SEM images of surface topography (surface morphology) of various samples of dispersing with varying (increasing) weight ratio content of CB nano-particles (0 %, 1 %, 2 %, 3 %, 4 %, and 5 %) in polymer matrix.*

*The dielectric properties causing absorption of EM waves were characterized using HP8514B Vector Network Analyzer to calculate the reflection/transmission characteristics of the samples under test. The Network Analyzer System treats reflection and transmission as scalar, vector, or phase only quantities e.g., the return loss retrieved from reflection is a scalar while impedance is a vector. The determined ratios permit to perform for reflection and transmission measurements independent on source power variations with frequency and/or absolute power. The reflection coefficient (complex quantity) possesses both the magnitude (real part) and phase (imaginary part) which is a ratio of reflected to incident EM waves on the sample surface under fixed experimental conditions. Mathematically, reflection coefficient is expressed by Eq. (1).*

$$\Gamma = \frac{V_{\text{reflected}}}{V_{\text{incident}}} = \frac{Z_L - Z_O}{Z_L + Z_O} \tag{1}$$

The magnitude of reflection coefficient is expressed in logarithmic terminology (decibels) in terms of reflection loss/return loss (the number of decibels at which the measured reflected signal is below the incident signal).

$$\text{Re}_{\text{Return Loss}} = -20\log_{10}|\Gamma|\,(dB) \tag{2}$$

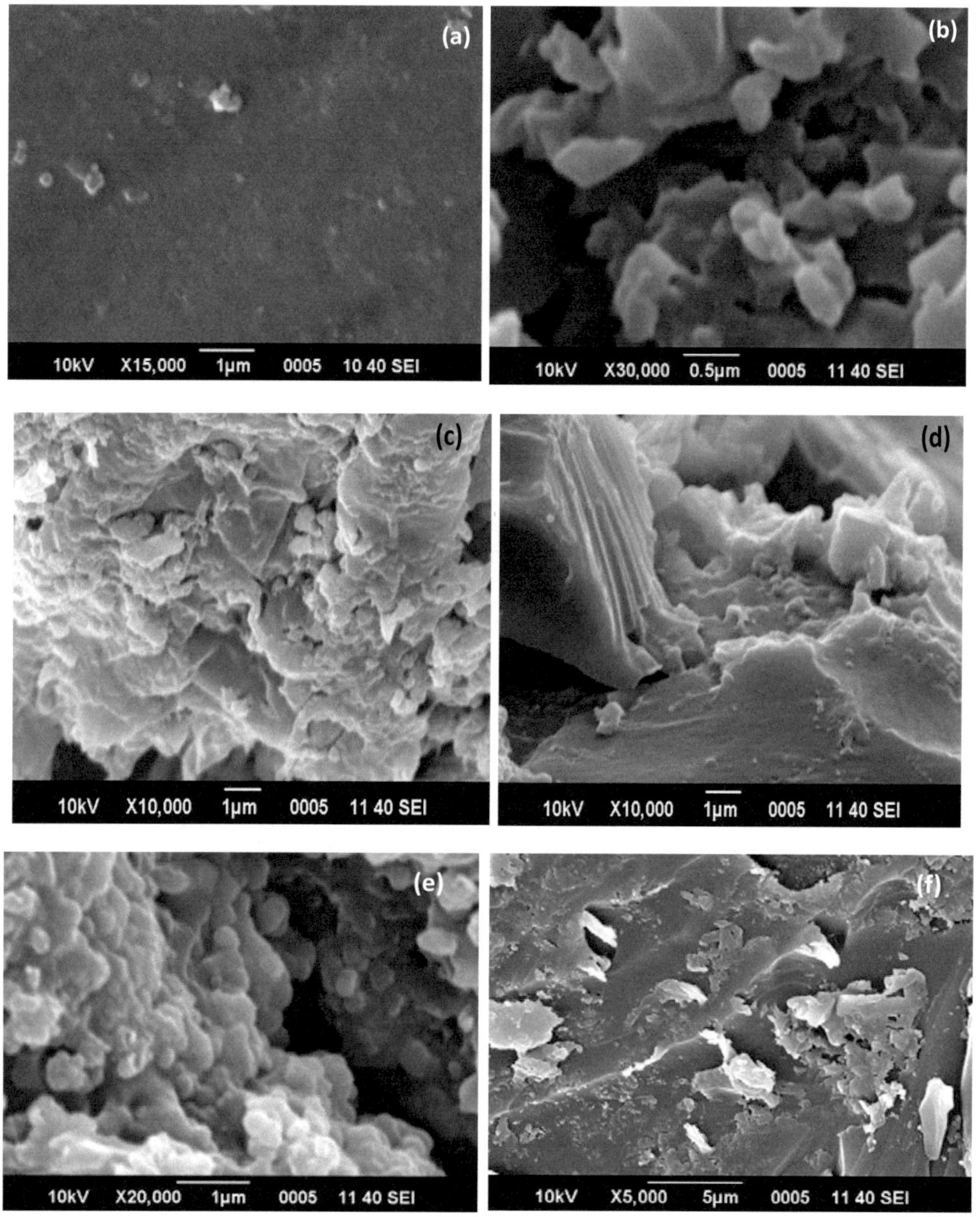


**Figure 11:** Courtesy of picture: SEM photographs of the samples. (a) CB0 (b) CB1(c) CB2 (d) CB3 (e) CB4 (f) CB5.

Scattering parameters S11 and S21 are described as input complex reflection and forward complex transmission coefficients, respectively while S12 and S22 are reverse components of a two port S parameters.

The free space measurement system employed in this testing has a pair of spot focusing horn lens antennas, a sample holder, network analyzer, and a PC to retrieve data computations. The S-parameter tester of network analyzer was integrated with spot focusing horn lens and circular to rectangular waveguide adapters were connected to coaxial cable. The schematic diagram of precise measurement of S-parameter in presence of network analyzer setup is shown in Fig. (**12**).

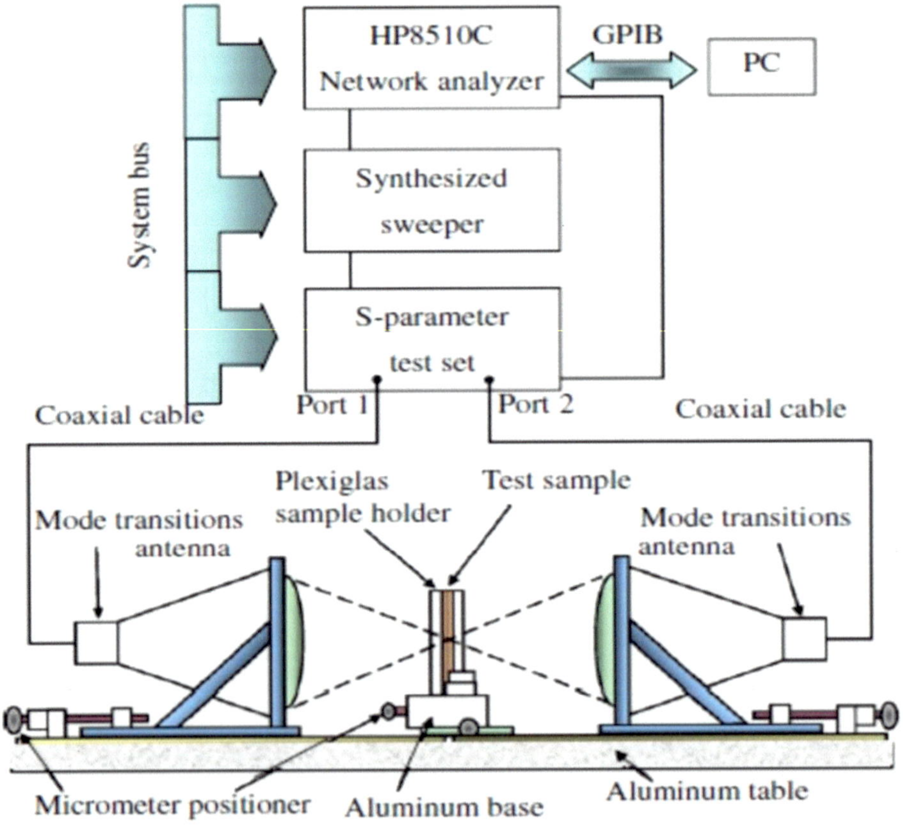


**Figure 12:** Courtesy of picture: schematic diagram of testing set up.

During measurements, couple of spot-focusing horn lens antenna joint back to back which transmits and receives electromagnetic waves in the frequency range of 11–18 GHz. The sample was placed in the middle of the two antennas mounted on a micrometer carriage of resolution 1µm. The antennas are separated by a distance equal to twice the focal length. The sample holder is placed at focal plane to hold planar samples of dimension about 150 mm.

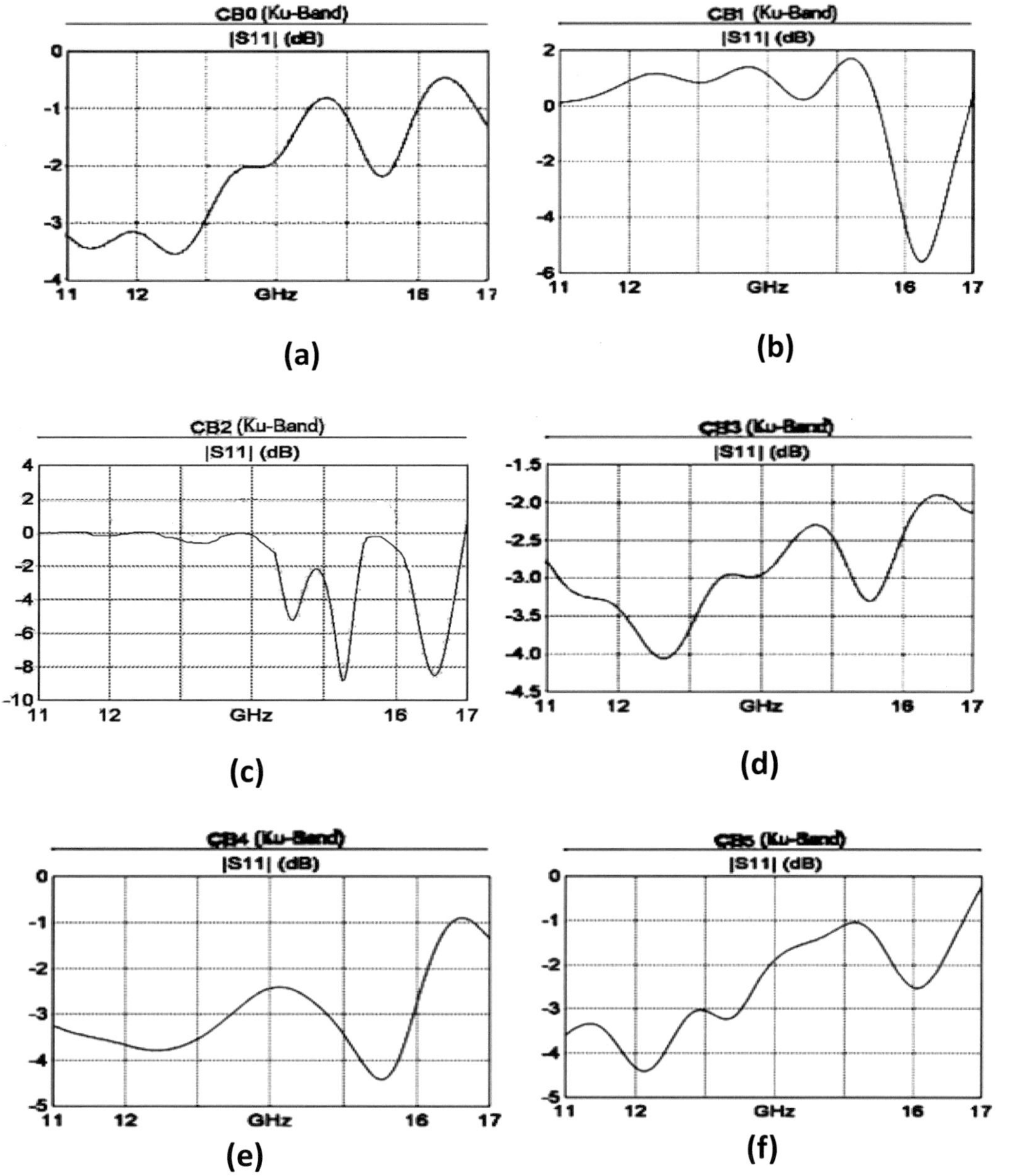


**Figure 13:** Courtesy of picture: the measured reflection losses of the samples (a) CB0 (b) CB1 (c) CB2 (d) CB3 (e) CB4 (f) CB5.

The radiation-matter interaction measurements can take two possible options: (1) through-mode and (2) reflection-mode measurements. The through-mode measures the dielectric and magnetic properties while the reflection mode measures only the dielectric properties of the metallic backed samples [21]. The reflection loss measurements are shown in Fig. (**13**) where a reasonable estimate of the reflection loss with the increase in weight percentage of reinforced CB particles is accessed.

A maximum reflection loss of 3.5 dB is found for pure epoxy matrix (sample CB0) in the frequency range of 12–13 GHz. The maximum reflection loss increases by 5.5 dB (for sample CB1) in frequency range of 16–17 GHz and 9.0 dB (sample CB2) in frequency ranges of 12–13 GHz and 16–17 GHz. These results shows that with the increasing content of CB particles in epoxy resin the absorption of EM waves increases which ultimately reduces the reflected energy spectrum. A further increase in weight percent of CB particles (sample CB3), a maximum reflection loss of 4 dB was observed in the frequency range of 12 – 13 GHz (see Fig. **13** d). A further increase in filler weight percent (CB particles) for samples CB4 and CB5 shows almost constant reflection loss because a percolation threshold was achieved already during reinforcement of sample CB2. Thereby, a maximum absorbing power of 87 % was observed at a loading content of 2 % CB particles in epoxy resin.

The complex permittivity and permeability of the composite materials are measured by network analyzer which enables to calculate the relaxation time and tangent loss of the composite material under test. The measurement of complex permittivity of dielectric absorber material leads one to predict absorption properties of under investigated composite materials.

The measured complex permittivity of the composite material with varying percentage of CB particles is shown in Fig. (**14**). For samples CB1 and CB2 the real part of the complex permittivity holds a unity which shows the presence of impedance matching and enunciates significant absorbing power that has already been measured as 87 %.

This research reports on EM waves (microwaves) absorption in Ku-band through the synthesis of increasing percentage of reinforced carbon black particles in epoxy polymer matrix. A maximum reflection loss of 9 dB was found at 2 % loading content of CB particles. However, after increasing weight percentage of CB particles in the epoxy matrix, the return loss reverses and could not show any further increase in loss. This was attributed to the fact that the excess CB particles (more than threshold limit) could not dispersed appropriately in matrix and results in agglomeration (shown in SEM images). The reported research was carried out in narrow bandwidth which is proposed to conduct over a wide band region in future.

Furthermore, matrices with more polar components are reinforced with properly surface treated CB particles so that they could be dispersed homogeneously to form an electrical network bridge which is essential to effective attenuation of electromagnetic waves. This can be achieved using filler-particles of both dielectric and magnetic nature to show return loss over a wide frequency range. The thickness of absorbing material coating also has a significant impact on the absorption of EM waves.

In our case the thickness for different samples ranged from 0.3 mm to 0.5 mm as evident in Fig. **14**.

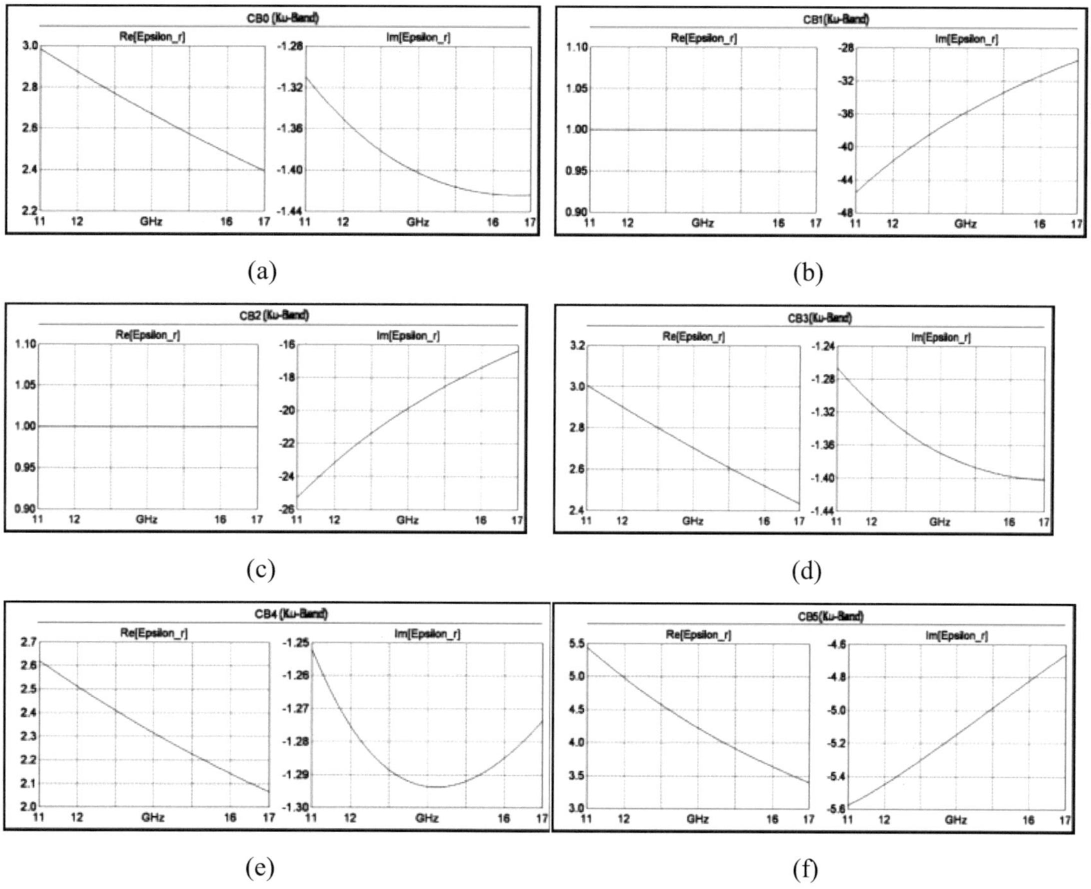


**Figure 14:** Courtesy of picture: Various samples with measured complex permittivity of real and imaginary components of CB reinforced particles in epoxy resin: (a) CB0 (b) CB1 (c) CB2 (d) CB3 (e) CB4 (f) CB5.

## 8. NANO-COATINGS AND NANO-PHOTONICS

High transmittance optical filters in visible range while high reflectance and low absorption in near and far infrared regions, respectively, have found attractive applications in imaging systems related to TV and other video transmissions. These filters are used to protect human eye from the harmful rays of the solar radiations and can also be utilized in aircraft windows and openings. A photo detector detects radiations up to 1200 nm with bulk of the important information lying within the spectral range of 450 to 700 nm. Radiations emitted in the IR range of 700−1200 nm produce undesirable signals owing to degradation in contrast and image quality. Further they can saturate the sensor directly affecting its performance. These signals are generated from the photocurrent produced from the radiations in this range and carrying no valuable information about the image color. Therefore, the radiations in the range 700−1100 nm must be blocked for quality image production. Nano-composite science is also used for development of Complex optical filters by sputter deposition. A multilayer thin film optical filter (infrared heat reflector) consisting of alternate nano layers of $ZrO_2$ and $SiO_2$ is designed and fabricated on a BK7 glass substrate by RF-magnetron sputtering. The process parameters for the fabrication of such sensors are shown in Table **3** [22].

**Table 3. Courtesy of Table: deposition condition for $ZrO_2$ & $SiO_2$.**

| Deposition conditions | Value |
|---|---|
| Base pressure | $5 \times 10^{-4}$ Torr |
| Oxygen flow rate | 42 sccm |
| Argon flow rate | 130 sccm |
| Deposition rate | 1.5 Å/s |
| Forward power $P_f$ | 2.74 kW |
| Reflected power $P_r$ | 0.06 kW |
| Chamber pressure | $3.30 \times 10^{-3}$ Torr |
| Substrate temp. | Room temp. |
| Target | Zr |

| Deposition conditions | Value |
|---|---|
| Base pressure | $5 \times 10^{-4}$ Torr |
| Oxygen flow rate | 21 sccm |
| Argon flow rate | 190 sccm |
| Deposition rate | 1.1 Å/s |
| Forward power $P_f$ | 2.70 kW |
| Reflected power $P_r$ | 0.17 kW |
| Chamber pressure | $4.5 \times 10^{-3}$ Torr |
| Substrate temp. | Room temp. |
| Target | Si |

Prior to deposition, target and substrate surfaces were cleaned for 10 min inside the deposition chamber. The chamber pressure during the pre-cleaning process was $4 \times 10^{-4}$ Torr. Argon and oxygen flow rates were 130 and 70 sccm, respectively. Drum speed was adjusted to 60 rpm during the deposition process. Zirconium and silicon targets were used as starting materials for deposition of $ZrO_2$ and $SiO_2$ films. Pure argon was used as sputtering gas while oxygen was used to achieve stiochometric films of $ZrO_2$ and $SiO_2$. Structural, optical, mechanical and roughness studies for the deposited film were made using

Siemens/Bruker D 5000 X-ray diffractometer, HR2000 spectrophotometer and Atomic Force Microscope (AFM). Micro-structural and cross-sectional images of the deposited samples were analyzed by Hitachi S4100 Scanning Electron Microscope (SEM).

The structural and surface quality inspection of the substrate materials were carried out by XRD, SEM, AFM before deposition to fabricate a multilayer optical sensor. It is evident from spectral measurements with an average transmission more than 90 % and less than 2 % in wavelengths regions 450–700 nm and 700–1100 nm, respectively to optimize design parameters of filter. The XRD characterization of multilayer structure revealed few patterns of $ZrO_2$ and a solitary pattern for $SiO_2$ indicating the presence of some crystallinity structure of $ZrO_2$ in the layered structure. The initially predominant indentation phase as predicted by hardness analysis was ductile which further showed a ductile to brittle transition with increase in penetration. Rutherford Back Scattering RBS by employing Tandem 5 Mev Ion Accelerator was carried out to study interface analysis of multilayered structure. RBS studies revealed sharp interfaces in the multilayer structure without interface mixing or interlayer diffusion. The optimized deposition conditions for $ZrO_2$ and $SiO_2$ provides a very good control over the deposition of multilayer structure as shown in Fig. (**15**) [22].

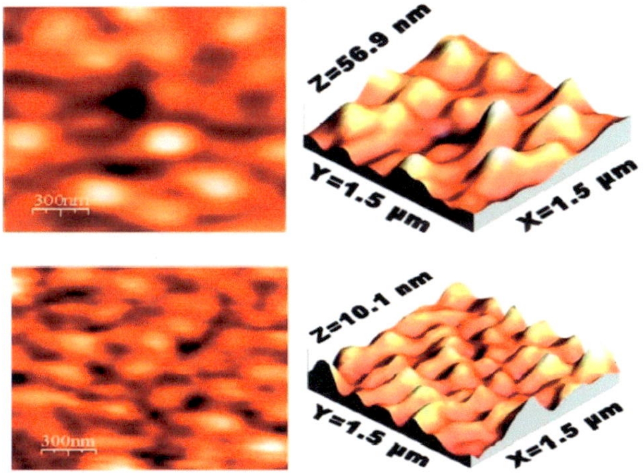


**Figure 15:** Courtesy of picture: AFM image of single layer $ZrO_2$ (top) & $SiO_2$ (Bottom).

The characterization of the filter shows that the structure is dense and both $ZrO_2$ and $SiO_2$ are repeatable with good accuracy. Studying interfacial behavior by RBS technique shows no sign of interface mixing or interlayer diffusion in the multilayered filter structure [22].

The thin film stacked layer optical sensor properties can also be achieved by resonant waveguide structures in the form of narrow spectral lines. Most of the research work of resonant gratings focuses the design and fabrication on the inorganic dielectric materials [23]. However, for some potential applications the use of organic transparent materials is desirable due to their flexible nature as for nano-encapsulation for efficient optical devices which are accompanied through Nano-imprinting Lithography (NIL) technique [24-30].

The replication of grating structures through NIL in organic polymer materials (transparent materials) is highly promising in advance nano-optical devices. For example, cost-effective guided mode resonance filters (GMRFs) are accompanied through replication technology with well-defined dimensional nano-features (linewidths and structure height) have potential use in nanophotonic devices. The GMRFs are in fact Diffractive Optical Elements (DOEs) which consist of a periodic structural profile with alternative refractive indices and features ranging from sub-wavelength (periodicity is less than wavelength) to tens of microns. Such grooved profiled structures are fabricated in polymer thin sheets using various replication techniques, for example, hot embossing and injection molding techniques.

In this research work we repot on a most economic, precise and bulk production method to manufacture GMRFs in polycarbonate materials (as substrates) using hot-embossing replication. The first step for the nano-replication is the fabrication of a master stamp that is fabricated on silicon substrate using a high resolution, binary, negative tone inorganic electron beam resist that is hydrogen silsesquioxane (HSQ) [31]. The specular reflectance characteristics of the replicated GMRFs are in agreement to that of theoretical calculations by Rigorous Coupled Wave Analysis. Fig. (**16**) shows an ideal binary replicated profile of a GMRF on polycarbonate substrate, using an HSQ resist stamp [32].

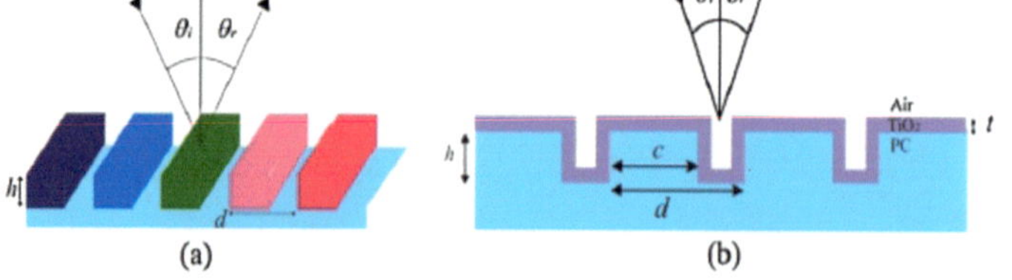


**Figure 16:** Courtesy of picture: A replicated, ideal binary profile of a GMRF by HSQ resist in polycarbonate substrate: (a) without $TiO_2$ layer and (b) with $TiO_2$ cover layer; *c* is the linewidth and *d* is the period of the grating profile [32].

The replication stamp is manufactured by spin coating the silicon substrates with HSQ resist for suitable spinning time and speed to get desired resist layer thickness. The coated resist films are illuminated by e-beam exposure followed by development to get grating structure. The so fabricated grating structures are heat treated at 300 °C for three hours to acquire sufficient hardness so as to achieve mechanical properties analogous to fused silica to facilitate imprinting process.

A grating fabrication study was carried out with and without etching process. However, the requirements for the imprinted structure fulfills, such as of high quality stamp by thermal NIL, low-cost fabrication process without etching, uniform imprinted density over wide grating surface area, high imprint fidelity, and smooth sidewalls with very low residual material contaminants. This means that the exclusion of reactive ion etching process results in a structural profile with low degree of inaccuracies in the structure depth and sidewalls. Thus, the simple stamp manufacturing and replication methods lead to high efficiency fast prototyping of nano-optical sensors with high pattern imprint fidelity and optical quality spectral appearance and applicability to large scale prototype devices [32]. The SEM images of master HSQ stamp and replicated nano-structure in polycarbonate substrate are shown in Fig. (**17**).

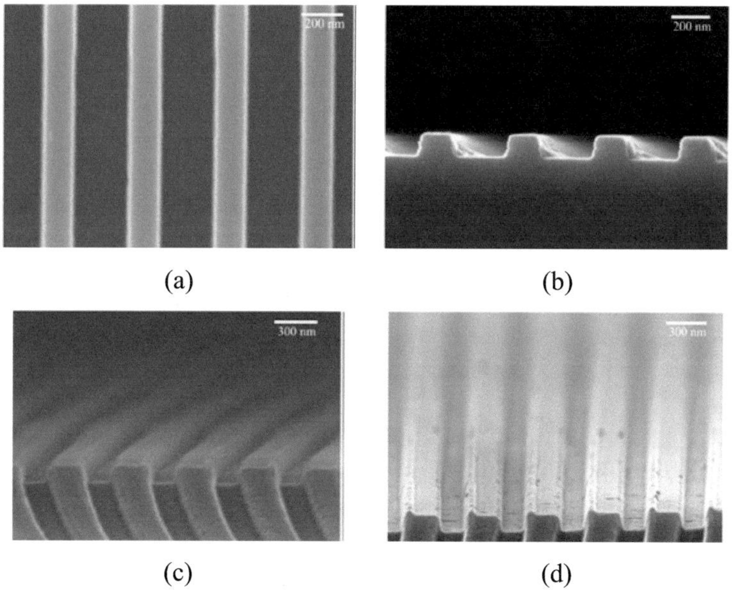


**Figure 17:** Courtesy of picture: SEM images of HSQ resist master stamp (a,b) and replicated structures in polycarbonate substrates (c,d).

The specular reflectance of replicated GMRFs are measured and found in agreement to that of theoretically calculated ones as shown in Fig. (**18**). The gratings with periodicities of 425 and 368 nm show spectral peaks at 826 and 709 nm, respectively [32].

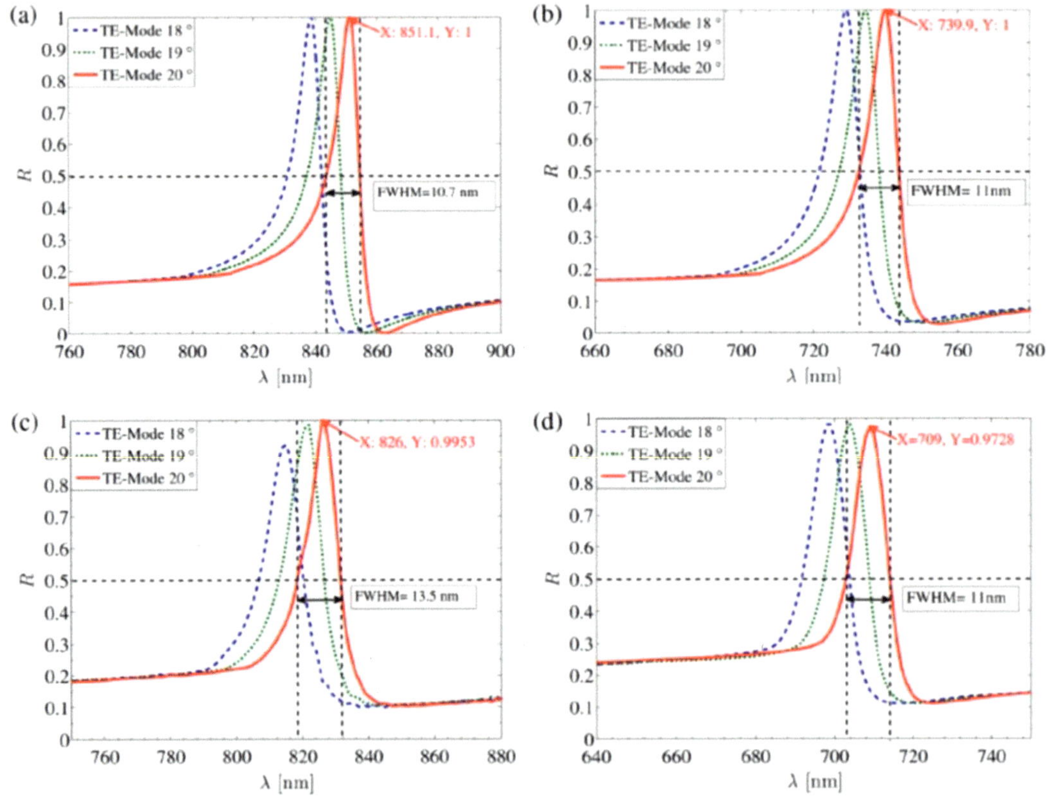


**Figure 18:** Courtesy of picture: Specular reflectance of replicated GMRFs, (a),(b): theoretical calculated for periodicities 425 & 368 nm, and (c),(d): experimentally calculated for periodicities 425 & 368 nm.

## 9. MECHANICAL PROPERTIES

A paradigm shift has occurred from conventional materials to nano-structured, functionalized, self-assembling and self-healing materials. The breakthrough in synthesis of nano-polymer has led to developments of highly tailored composite materials having significantly enhanced mechanical properties. The computationally designed materials have bridge the gap between quantum physics and continuum mechanics on nano- and micro-levels. Computer models are generated at the atomic and molecular levels that model the relationship between the atomic structure of materials at the nano-scale and the physical properties at the macro-scale as shown in Fig. (**19**).

These new computational tools have the potentials to optimize material chemistry for specific applications and justify trade off studies. If small-diameter single-walled tubes are reinforced to the polymer matrix with compete dispersion, the resulting material structures could have a significantly high strength to weight ratio compared to those manufactured from aluminum alloys and carbon fiber composite materials used in conventional aerospace applications. The mechanical properties of SWCNTs and MWCNTs have been determined from computer simulations using quantum/ continuum mechanics, atomistic simulation (molecular dynamics) and limited experimental measurements of properties are based on (AFM) and Raman spectroscopy. The computer simulated and experimental results show that small-diameter SWCNTs have elastic moduli more than 145 Msi (1 TPa), and strengths approaching 29 Msi (200 GPa) as indicated in Table **4**.

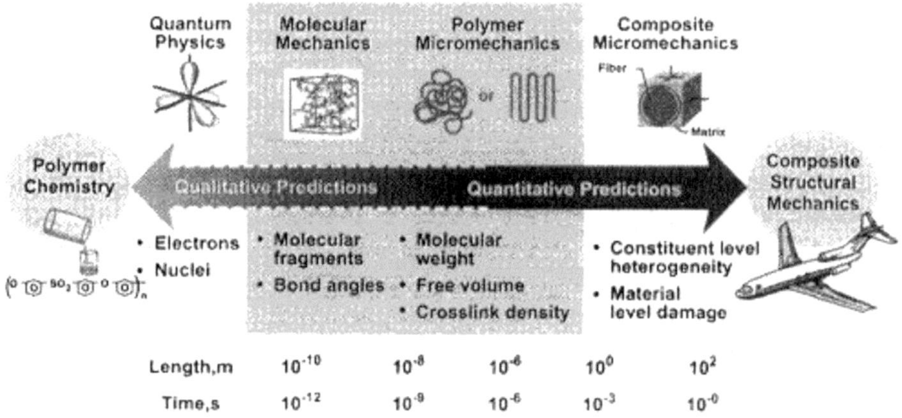


**Figure 19:** Courtesy of picture: Micro- and nano-scale computationally designed materials and structures [7].

**Table 4. The predicted mechanical properties of CNTs [7].**

| Parameter | Range |
|---|---|
| Tensile modulus | 44-260 Msi (300-1800 GPa) |
| Tensile ultimate strength | 0.9-26 Msi (6-180 GPa) |
| Bending strength | 0.9-3 Msi (6-22 GPa) |
| Elongation (strain to failure) | 6-15% |
| Thermal conductivity | 1750-5800 $W.m^{-1}.K^{-1}$ |

Comparison of specific modulus and specific strength of aluminum 2219, High Modulus Carbon Fiber Reinforced Plastic (CFRP) and Single Wall Nano-tube Fiber Reinforced Polymer Composite (SWNTFRP) reveal that composite reinforced with nano-tubes offer significant advantages as shown in Fig. (**20**).

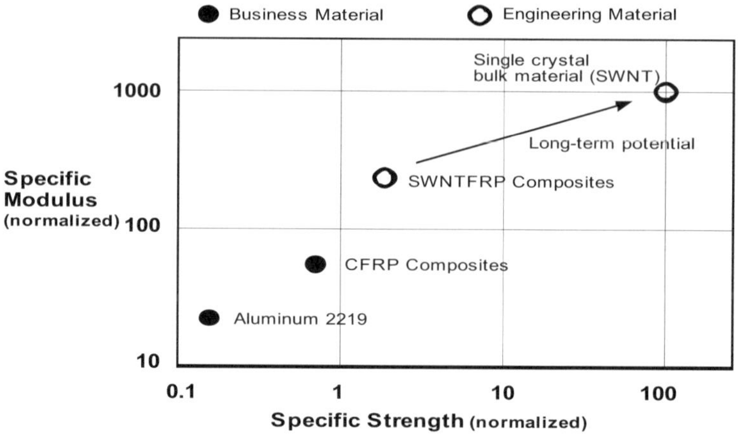


**Figure 20:** Courtesy of picture: properties of carbon nano-tube and composite materials [7].

The comparison of dry weights of single stage re-launch vehicle, RLV using aluminum, carbon Fiber Reinforced Plastics (CFRP) and Single Wall Nano-tube Fiber Reinforced Plastic (SWNTFRP) show that it is possible to achieve 82 % reduction in weight by replacing the aluminum components of the airframe structure with the SWNTFRP composite materials as shown in Fig. (**21**).

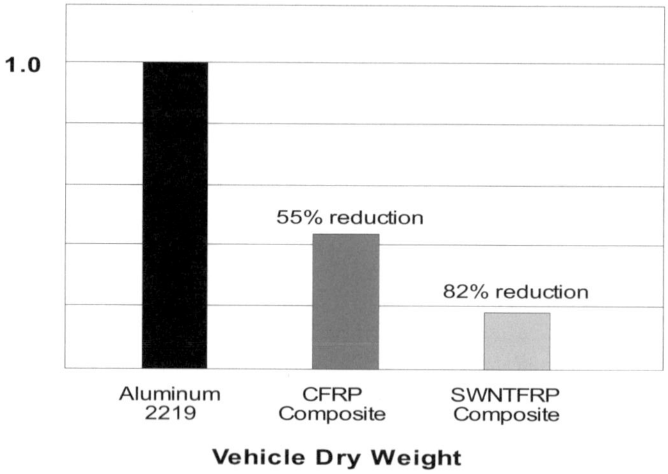


**Figure 21:** Courtesy of picture: System analysis of aerospace vehicle dray weight using aluminum, CFRP and SWNTFRP composites [7].

## 10. THE NEXT GENERATION AEROSPACE APPLICATIONS

The design philosophy of next generation aerospace vehicles is based on autonomous, resilient, evolvable, self-sufficient, highly distributed, ultra efficient flight configuration for deep space missions with high degree of precision. They should be ultra efficient in terms of having optimal specific strength and stiffness to travel around the earth in most cost effective manner. They must be able to think, act and respond independently when faced with uncertainties. They should be resilient, highly durable, damage tolerant in harshest hot-wet environments and be able to perform self-diagnoses and repairs. They must be adaptive to meet the different aerodynamic requirements during takeoff, cruise and descend and overcome unanticipated problems. They should be highly distributed in terms of broad continuous presence and coverage in interactive networks as shown in Fig. (**22**).

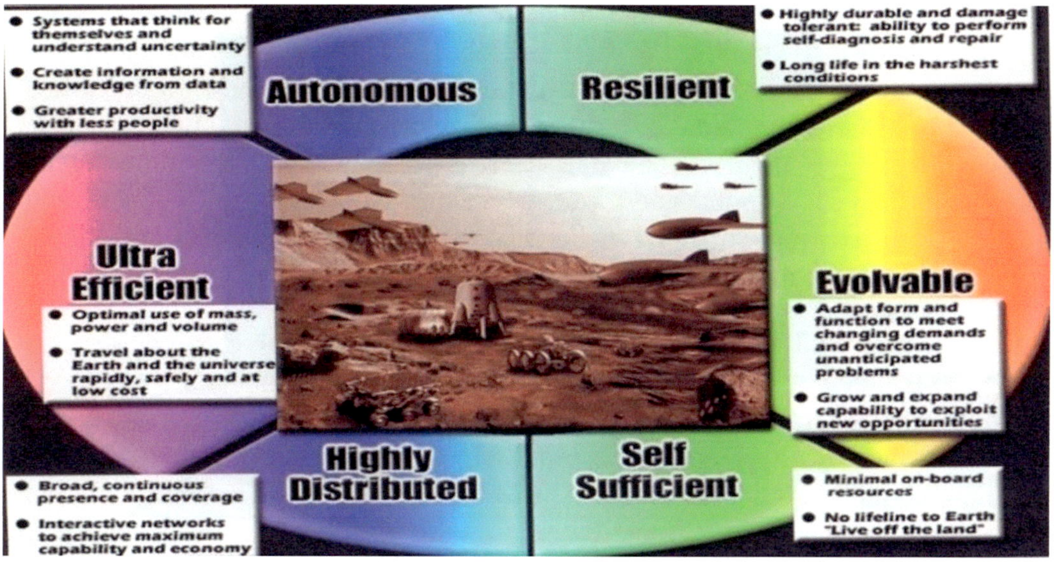


**Figure 22:** Courtesy of picture: the requirements of next generation aerospace vehicles [33].

## 11. SHAPE ADAPTIVE SMART STRUCTURES

Design concepts of interest in shape-adaptive structures include the wing warping, camber shaping/control surface deformation, variable stiffness structure, inlet blunting and lip wall shaping. Use of smoothly contoured surface delays flow separation, improves lift to drag ratio (L/D), stall angle, maneuverability, aero elastic effects, takeoff gross weight, range and stealth characteristics as shown in Fig. (**23**) [33].

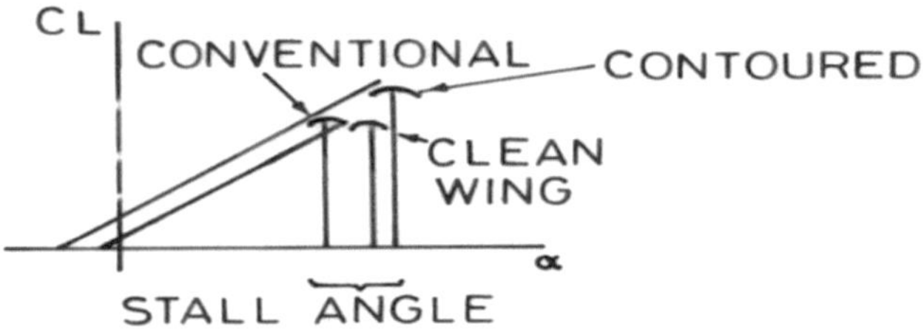


**Figure 23:** Courtesy of picture: The effect of shape adaptive wing on stall angel (α).

In general, smart materials and structures can be developed by two approaches for shape adaptive aircraft structure. The first approach is based on development of composites containing active sensors and actuators embedded in structure in discrete global sense and second approach accompanies the synthesis of advance materials with efficiently smart functionality at the atomic/molecular scale. Such materials functionality in terms of high strength structure, actuator or sensor response is being incorporated into smart structures in discrete global sense. Sensors presently include Braggs Gratings, Mach-Zehnder or Sagnac fiber optic, piezoelectric ceramics and polymers while the actuators take forms of piezoelectric, electrostrictive ceramics, shape memory alloys (SMAs), and magnetostrictive materials. The multiplexing (wavelength, frequency or intensity) with a sensor/signal processing network and control system give rise structural parameter to optimize and target operational performance as shown in Fig. (**23**) [34].

Air Force Research Laboratory in the Directorate of Adaptive Structures Air Vehicles conducted research on several novel areas. The thermally activated reconfiguration system was supported by fabrication of nano-particle doped shape memory polymer (SMP). Various nano-particle doped volume fractions in SMP were prepared with varying thermal conductivities to investigate passive actuation time control. The micro air-vehicle concept was in continuous investigation and testing in terms of its initial planned shape and size for a prototype development of a perching mechanism [35].

Joint NASA/ Wright Laboratory demonstrated that the smooth continuous airfoil of mission adaptive wing on an F-111 test aircraft greatly improved its stability, control and aerodynamic performance. McDonnell Douglas initiated active damage interrogation (ADI) program which used array of piezoelectric sensors and actuators embedded in composite flex beam of MD-900 explorer rotor. A prognostics and health management (PHM) system is being developed for the joint strike fighter (JSF). The JSF program is examining prognostic methods for artificial neural network, fuzzy logic and genetic algorithm. JSF is also supporting efforts to develop wireless PHM sensors for monitoring airframe health. Algorithms have been developed to extract relevant features from the data for classification. By using artificial neural network, the ability to classify maneuver in real time has been demonstrated [33].

Very often health of aerospace structures is inspected by using traditional non-destructive techniques. In recent years, implementation of a range of technologies and precise sensing techniques have advanced damage detection in composite and metallic material structures. Such new developments in technology allow one to continuously detect as well as integrated and on-line damage monitoring systems. These modern technologies have potential applications and are found attractive and efficient developments for manufacturers and civil and military aircraft operators to reduce life-cycle costs, reduce inspection/maintenance works, extended structure life and improved safety. A schematic of fiber optic sensor, used as online health monitoring detection system for smart aerospace structures is shown in Fig. **24** [34, 35].

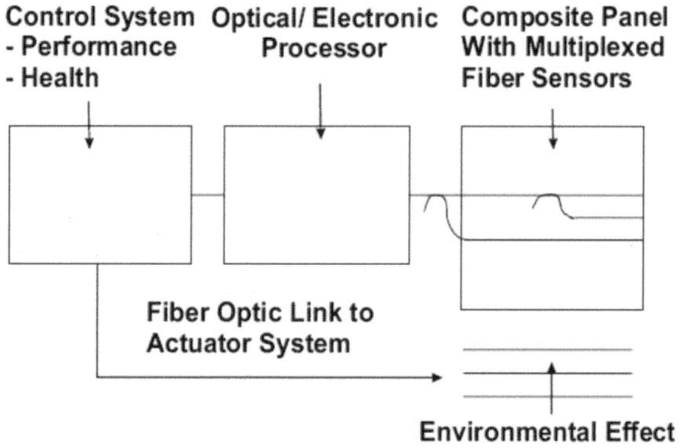


**Figure 24:** Courtesy of picture: the schematic of fiber optic sensors for online health monitoring of smart aerospace structure [34, 35].

A jointly funded project on Structural Health Monitoring (SHM) by Air Force, Clarkson University and Indian Institute of Science was initiated to work on propagation of waves through damaged composite structures by establishing wavelet-based spectral finite elements (WSFE). Such compactly supported wavelet allows application of finite boundaries by employing highly efficient computational models, for example, WSFE approach to solve inverse problem of SHM. Use of compactly supported wavelet basis functions allows treatment of finite boundaries and yields very efficient computational models. The WSFE approach is highly suitable for solving the inverse problem of SHM. The structural damage has been detected by using guided wave approach where residual useful life prediction is evaluated base on- online and off-line estimation. Use of such computational tools has applications in complex composite structures with braided architecture, satellite structure, and aircraft hotspot [36]. Measurement of axial/transverse strain, temperatures and pressures using multi-axis fiber grating strain sensor can be used for aeroelastic tailoring of smart wings from manufacturing to design process. This could be extended other applications which affect its spectral peaks such as measurement of corrosion, hydraulic leaks and ice formation *etc*., of aircraft structures through multiplexing in Braggs Grating Sensors as shown in Fig. (**25**). These sensors can be multiplexed using wavelength/frequency division multiplexing to not only support health management against foreign body object and curing temperatures during manufacturing processes but also to monitor the flight data and other controls during flight.

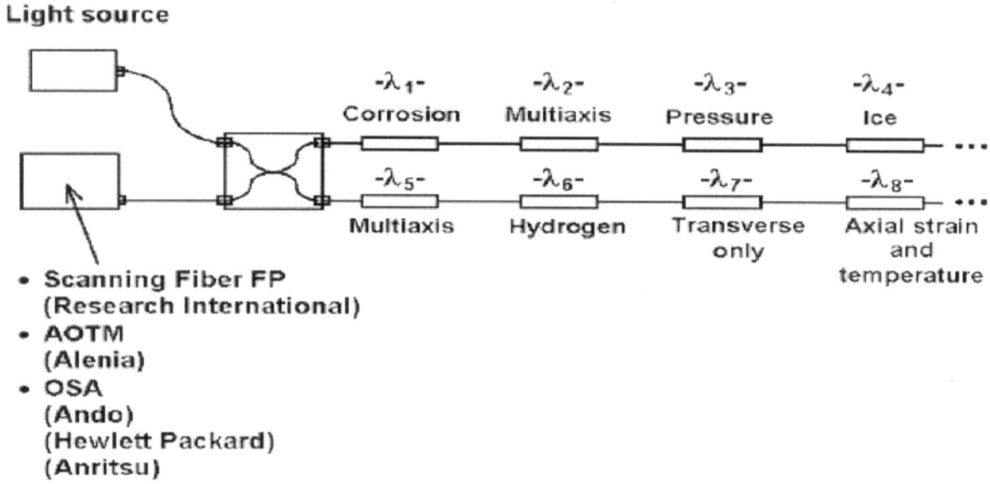


**Figure 25:** Courtesy of picture: multiplexing of Braggs Grating Fiber Optic Sensor [35].

## 12. ACTIVE AEROELASTIC WING TWIST (AAWT)

Wing deformation is used to enhance the roll performance at all dynamic pressures without using usual aileron system to predict the aerodynamic rolling moment. The aerodynamic wing profile is twisted and cambered in differential mode on the right and left wings to target the acquired rolling moment for a specific steady roll. AAWT and camber control are being developed under the DARPA Smart Wing Program. DARPA, AFRL, NASA and Boeing have completed flight testing of an F/A-18 research aircraft modified with aeroelastic wing technology at different Mach/altitude combination. The project aims to study the prospective of aerodynamically twisted flexible wings to improve aircraft roll maneuverability performance under transonic and supersonic speeds. The wing twist can be achieved by air stream energy with slightly surface control motion. Traditional control surfaces were used as active trim tabs to aerodynamically induce twist. The flight test and simulation data aircraft proved that wing aeroelastic twist can be exploited for vehicle roll while controlling wing structural loads. The test data will help in designing lighter and flexible high aspect ratio wings for next generation aircrafts.

A state of the art smart structure in next generation aerospace structure feature a load bearing carbon fiber composite structure with piezoelectric embedded discs and shape memory alloy wires for sensor and actuation purposes through micro processors in smart composite structure. These embedded material functionalities (sensors, actuators, structures) are accessed at atomic level. For example, the carbon fibers are coated with piezoelectric materials to synthesize a smart composite material on micro- or nano-scale. Similarly, electro-rheological fluids may be filled in hollow nano-carbon tubes for sensing and structural reinforcement, respectively. These attributes permit a structure to be synthesized with self adaptive smart abilities as shown by blended wing all composite next generation aircraft as shown in Fig. (**26**) [37-40].

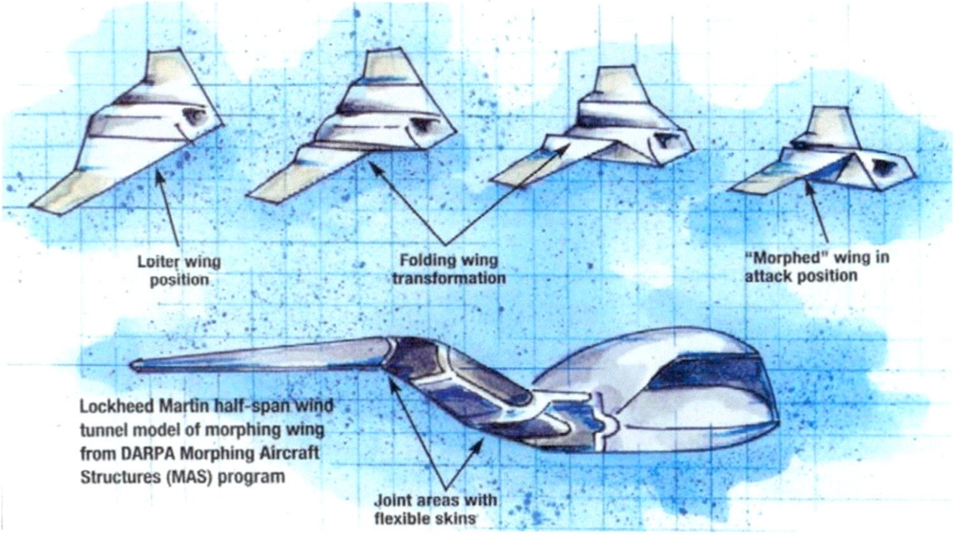


**Figure 26:** Courtesy of picture: the lockheed Martin's half span wind tunnel model of morphing wing aircraft [42].

## 13. NANO-COMPOSITES IN ABLATIVES

A funded research by the U.S Air Force Office of Scientific Research (AFOSR) was conducted at the University of Texas at Austin to investigate the solid rocket nozzle ablatives in terms of improvements in erosion resistance and heat transfer characteristics by reinforcement of nano-fillers (nano-particles and fillers) in polymer matrix. Three different types of nanoparticles and nanofibers were dispersed in phenolic: Cloisite Montmorillonite (MMT) nanoclay, polyhedral oligomeric silsesquioxane (POSS), hybrid silica/silicone nano-particles and vapor grown Pyrograf III carbon nano-fibers as shown in Table **5** [3]. Subsequently, different loading percentage of MWCNTs were blended in phenolic resin and exfoliation for ablative properties is ensured by using high shear mixing technique.

**Table 5. The nanoclay, POSS and carbon nanofiber composites used in the ablative test.**

| Material/ Sample ID | Resin (wt %) | Reinforcement (wt %) | Filler (wt %) |
|---|---|---|---|
| MX-4926: mil spec ablative (control) | SC-1008 resole phenolic | rayon fabric | carbon black |
| | 35 | 50 | 15 |
| (HE)2MT: MX-4926 ALT with nanoclay particles, no carbon black | SC-1008 resole phenolic | rayon fabric | Cloisite® 30B |
| | 47.5 | 50 | 2.5 |
| | 45 | 50 | 5 |
| | 42.5 | 50 | 7.5 |
| SO-1458: MX-4926 ALT with POSS® nanoparticles, no carbon black | SC-1008 resole phenolic | rayon fabric | SO-1458 POSS® |
| | 49 | 50 | 1 |
| | 47 | 50 | 3 |
| | 45 | 50 | 5 |
| PR-24-PS: SC-1008 with 65 nm diameter carbon nanofibers, no carbon black or rayon fabric | SC-1008 resole phenolic | Carbon nanofiber | none |
| | 80 | 20 | none |
| | 76 | 24 | none |
| | 72 | 28 | none |

The embedded thermocouple composite samples made of reinforced phenolic resin with carbon particles were bonded to a steel substrate to monitor backside heat soaked temperature. The prepared nano-composite samples together with industrial standard materials were placed in a simulated solid rocket testing facility. Such facility use abrasive aluminum oxide particles to produce exhaust plume with flame temperature and plume velocity of approximately 2200 °C and 2000 m/sec, respectively.

Test results suggested that nano-fillers greatly improved the erosion resistance of the nano-composite material and a lowest erosion rate of any sample was achieved at a loading level of 28 percent carbon nano-fibers as shown in Fig. (**27**).

Furthermore, the measured heat soaked temperatures of the synthesized nano-composite samples were less than baseline ablative materials. Thus, polymer nano-composites are capable to offer great promise for future high-temperature applications, heat shielding effects, and exhaust nozzles [3].

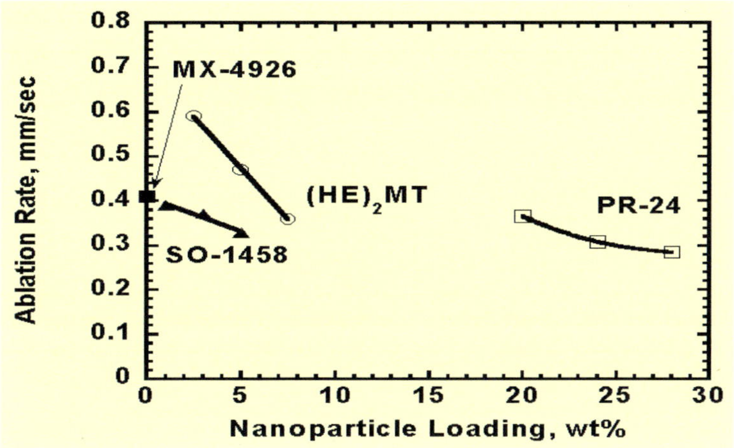


**Figure 27:** Courtesy of picture: Effect of nanoparticles and loading on ablation rate [3].

Ablation-resistant upper layers that can withstand to high entry atmospheric speed and high heat flux in combination with mass-efficient insulating bottom to protect structure and reduce its areal weight are made in real practice using advanced rigid ablators. Such advance ablators may facilitate transportation of commercial orbital and paved NASA missions in order to make it capable for high velocity returns (from asteroids of Mars). The advanced materials are evaluated thermally to 400–450 W/cm$^2$ in Laser Hardened Materials Evaluation Lab and Hypersonic Materials Evaluation Test System and possess much lower back-face temperatures and reduce recession in comparison to baseline PICA materials. The projects of NASA Exploration Program's Entry, descent, and landing were the main funding which ended in FY–11, however, in house development of ablative materials is still in progress to evaluate new resin systems and subsequent fiber enforcements. Furthermore, extensive research on polyamide 11 and 12 nano-composites for Selective Laser Sintering (SLS), additive manufacturing, nano-materials (nano-clay NC, carbon nano-fiber CNF), nano-silica, nano-alumina, MWCNT, nano-graphene platelet NGP *etc*., is in progress into PA11 and PA12 polymers.

Similarly, potential research results show that 2–3 % MWCNT in PA11 exhibit promising electrostatic dissipation levels with reduces reinforcement percentage of CNF, NC, NGP nano-composite, flame retardant additives are in development. In order to capture aerospace commercial market trends, conductive PA11 and FR PA11 nano-composites enable SLS additive manufacturing [3, 41].

## 14. CARBON NANO-TUBES IN SPACE ELEVATOR

The concept of reaching space without flying has become possible through space elevator. Payloads or satellites can be loaded onto climbers (unmanned cargo canisters) that ascend the paper-thin ribbon by compressing it in sets of electric rollers over 62,000 miles long ribbon as shown in Fig. (**28**).

**Figure 28:** Courtesy of picture: Unmanned cargo canisters, ascending from earth to gyrostatic station over the nano-carbon ribbon and powered by laser from the ground [42].

The ribbon was made of a nano-composite material having super strong carbon nano-tubes. In general large communication satellites or payloads may be deployed to geosynchronous orbit through climbers 22,500 miles away, above Earth. A long carbon nano-tube molecule was grown by scientists on silicon wafer using a dip pen. An iron-chloride catalyst was placed at one end of the silicon wafer (substrate) and exposed to hot ethanol vapor in a furnace. This method brings to start a chemical reaction for the onset growth and deposition of carbon atoms along the length of silicon wafer (substrate).

This paper-thin ribbon would be strongest enough to fit rolled up between the rockets deploying it. Climbers and other relevant components are developed by a straight forward approach using existing engineering tools. The laser light is beamed to light-sensitive gallium arsenide cells to convert light into voltage for driving the climbers from earth to gyrostatic orbital station [42].

## 15. FUTURE TRENDS

DARPA's current morphing aircraft system program is based on changing of wing shape to enhance performance. This program seeks to expand the wing morphing degrees of freedom to include large wing area, sweep and span changes. Such design reduces the weight of aircraft for undertaking complex missions and reduces the number of aircrafts required for a mission. Wings made from CNT with embedded tiny fiber optic sensors and actuators could morph and continuously deform during different phases of aircraft flight. A true morphing aircraft would require the development of nano-composites that would allow flexing in better way. In the long term, one seeks to design more flexible materials with embedded components that cover aircraft throughout and adjust aircraft itself continuously during a flight mission as shown in Fig. (**29**).

**Figure 29:** Courtesy of picture: morphing wings concept in next generation aerospace structures.

A state of the art smart structure in next generation aerospace structure feature a load bearing carbon fiber composite structure, embedded with piezoelectric discs and shape memory alloy wires for sensing and actuation purposes. They are electrically connected by a microprocessor.

Research is being carried out to explore material properties on embedding the material functions for sensor, actuator and structure responses at high local level. The smart composite materials for sensor applications can be synthesized by coating micro- or nano-carbon fibers with piezoelectric materials. Similarly, electro-rheological fluids may be filled in hollow nano-carbon tubes for sensing and structural reinforcements. These attributes permit a substance to be synthesized with self-adaptive abilities.

With the advancement in structural complexity, the correlation among design, manufacturing process, and analysis become inextricably intertwined and multidisciplinary. Such implications mandates the realization of sophisticated and precise process engineering driven design, methodologies for synthesis, and analysis for sensors, actuators, and micro/nano-structures, for example, Micro Electro Mechanical Systems (MEMS). The molecularly designed nano-composites such as Single/Multiple Carbon Fiber Nanotubes (S/MCNT) have the promise to be much lighter yet stronger. The advancements in design and manufacturing techniques at the atomic level will diminish the boundaries between sensors and actuators. These attributes permit a substance to be synthesized with self-adaptive abilities at the nano- and micro-levels. The application of fiber optic neural sensors has led to emergence of biomimetics, neuromimetics and neuromorphic. The integration of bio, nano- and info technologies with photonics and composites will enable the mankind to mimic the bird's flight.

## ACKNOWLEDGEMENTS

Thanks are due to Dr. J.H. Koo, The University of Texas at Austin for fruitful technical discussions during SAMPE Conference, Long Beach, California and Dr. Eric Udd for technical communications at Portland University, Oregonn which inspired me to initiate research in nanocomposites and write this book chapter. The assistance by Dr. Hammad Asghar, University of West Scotland and Mr. Rahim Jan, National University of Sciences and Technology, Islamabad to carry out collaborative research about nanostructured materials is greatly appreciated.

## CONFLICT OF INTEREST

The author confirms that he has no conflict of interest to declare for this publication.

## REFERENCES

[1] B.S. Thomson, Biomimetic Materials: Was Leonardo Mistaken, Part-I, SAMPE Journal, Jan/Feb, 39-43, 1996.

[2] Nanoparticles, Nanotubes and Nanocomposites, Composites Technology, 42-47 April 2005.

[3] J. M. Koo, Polymers Nanocomposites, Tutorial Notes, SAMPE 2006, Long Beach CA, May 1, 2006.

[4] J. C. Leslie, ACPT Properties of Advanced Composites, March 15, 2002.

[5] M. K.Zaffar, "Design and Development of Carbon Fiber Composite Materials in Aerospace Industry," Eng. Hor., Vol. IV, No. 5, pp. 5-7, 1993.

[6] Z. F. Ren, *et al*, "An Insight into Carbon Nanotubes Production," *SAMPE J.*, 64-67, Sep/Oct 2001.

[7] C. E. Harris, J. H. Starnes, Jr. and M. J. Shuart, "Design and Manufacturing of Aerospace Composite Structures, State-of-the-Art Assessment," *AIAA J Aircraft.*, Vol. 39, No. 4, pp. 545-559, Jul-Aug 2002.

[8] N. Hansen and G. Hansen, "From Inception to Insertion: Successful Products and Application using Nickel Nanostrands," *SAMPE J.,* Vol. 47, No. 3, May/June 2011.

[9] E. Lackey, E. Hutchcraft, J. Vaughan, R. Averill, "Composites Manufacturing-Official Magazine of the American Composites Manufacturing Association", May 2006.

[10] W. Jun Lee, J. W. Lee, C. G. Kim, "Characteristics of an electromagnetic wave absorbing Composite structure with a conducting polymer electromagnetic band gap (EBG) in the X- band", *Compos. Sci. Technol.,* vol. 68, pp. 2485–2489, 2008.

[11] J. B. Kim, S. K. Lee, C. G. Kim, "Comparison study on the effect of carbon Nano Materials for single-layer microwave absorbers in X-band," *Compos. Sci. Technol.,* vol. 68, pp. 2909–2916, 2008.

[12] www.f-22raptor.com/st_getstealthy.php

[13] K. Y. Park, S. E Lee, C. G. Kim, J. H. Han, "Fabrication and Electromagnetic Characteristics of Electromagnetic Wave Absorbing Sandwich Structures", *Compos. Sci. Technol.,* vol. 66, pp. 576–584, 2006.

[14] M. E. Achour, A. Mdarhri, F. Carmona, F. Lahjomri, A. Oueriagli, "Dielectric Properties of Carbon Black-Epoxy Resin Composites Studied with Impedance Spectroscopy", *Spectroscopy. Letters.,* Number 4-5.

[15] A. Ghasemi, A. Hossienpour, A. Morisako, A. Saatchi, M. Salehi, "Electromagnetic Properties and Microwave Absorbing Characteristics of doped barium hexaferrite", *J. Mag. Mag. Mater.,* vol. 302, pp. 429–435, 2006.

[16] A. M. Elbakly, "Design of Double-Layer Rubber Radar Absorbing Sheet", *The Online J. Elec. Elec. Eng (OJEEE).,* vol. 1, no. 1.

[17] S.M. Abbas, M. Chandra, A. Verma, R. Chatterjee, T.C. Goel, "Complex Permittivity and Microwave Absorption Properties of a Composite Dielectric Absorber", *Composites: Part A.,* vol. 37, pp. 2148–2154, 2006.

[18] C. Jason Jan, M. D. Walton, E. P. McConnell, W. Sik Jang, Y.S. Kim, J. C. Grunlan, "Carbon Black Thin Films with Tunable Resistance & Optical Transparency", *Carbon.,* vol. 44, pp. 1974–1981, 2006.

[19] J. Sandler, M.S.P. Shaffer, T. Prasse, W. Bauhofer, K. Schulte, A.H. Windle, "Development of a Dispersion Process for Carbon Nanotubes in an Epoxy Matrix and the Resulting Electrical Properties", Polymer Communication.

[20] A. Markov, B. Fiedler, K. Schulte, "Electrical Conductivity of Carbon Black/ Fibers Filled Glass-Fiber Reinforced Thermoplastic Composites", *Composites: Part A.,* vol. 37, pp. 1390–1395, 2006.

[21] I. S. Seo, W. S. Chin, D. G. Lee, "Characterization of Electromagnetic Properties of Polymeric Composite Materials with Free Space Method", *Compos. Struct.,* vol. 66, pp. 533–542, 2004.

[22] M.H. Asghar, M. Shoaib, Z.M. Khan, *et al.,* "Complex optical filter prepared by sputter deposition", *Eur. Phys. J. Appl. Phys.,* (2010).

[23] M. R. Saleem, S. Honkanen, and J. Turunen, "Effect of substrate overetching and heat treatment of titanium oxide waveguide gratings and thin films on their optical properties," *Appl. Opt.***,** vol. 52, 422-432, 2013.

[24] M. R. Saleem, S. Honkanen, and J. Turunen, "Thermo-optic coefficient of ormocomp and comparison of polymer materials in athermal replicated subwavelength resonant waveguide gratings," *Opt. Commun.,* vol. 288, pp. 56-65, 2013.

[25] M. R. Saleem, D. Zheng, B. Bai, P. Stenberg, M. Kuittinen, S. Honkanen and J. Turunen, "Replicable one-dimensional non-polarizing guided mode resonance gratings under normal incidence," *Opt. Express***.,** vol. 20, pp. 16974-16980, 2012.

[26] M. R. Saleem, P. Stenberg, T. Alasaarela, P. Silfsten, M. B. Khan, S. Honkanen, and J. Turunen, "Towards athermal organic-inorganic guided mode resonance filters," *Opt. Express.,* vol. 19, pp. 24241-24251, 2011.

[27] M. R. Saleem, S. Honkanen, and J. Turunen, "Mode-splitting of a non-polarizing guided mode resonance filter on substrate overetching effect," *Proc. of SPIE.,* Vol. 8974, pp. 897417, 2014.

[28] M. R. Saleem, S. Honkanen, and J. Turunen, "Non-polarizing single layer inorganic and double layer organic-inorganic one–dimensional guided mode resonance filters," *Proc. of the SPIE.,* Vol. 8613, 86130C-1, 2013.

[29] M. R. Saleem, S. Honkanen, and J. Turunen, "Partially athermalized waveguide gratings," *Proc. of SPIE.,* Vol., 8428 842817-1, 2012.

[30] M. R. Saleem, M. B. Khan, P. A. Stenberg, T. Alasaarela, S. Honkanen, B. Bai, J. Turunen, and P. Vahimaa, "Thermal behavior of waveguide gratings," *Proc. of SPIE.,* 8069 80690A-1, 2011.

[31] M. R. Saleem, P. A. Stenberg, M. B. Khan, Z. M. Khan, S. Honkanen, and J. Turunen, "HSQ resist for replication stamp in polymers," *Proc. of SPIE.,* Vol. 8249 82490G-1, 2012.

[32] M.R. Saleem, P.A. Stenberg, M.B. Khan, Z. M. Khan, *et al.*, "Hydrogen silsesquioxane resist stamp for replication of nanophotonic components in polymers", *J. Micro/Nanolith. MEMS MOEMS.,* vol. 11(1), 013007, Jan–Mar 2012.

[33] A. K. Noor, Structure Technology for Future Aerospace Systems, American Institute for Aeronautic and Astronautics, 2000.

[34] E. Udd, Fiber Optic Sensors, John Wiley and Sons, Inc, 1991

[35] E. Udd, Fiber Optic Smart Structures, John Wiley and Sons, 1994.

[36] H. H. Hilton, "Structural Dynamics", Aerospace America, December, 2011.

[37] Z. M. Khan, Fiber Optic Smart Structures for Assessing Layer Waviness in Smart Structures, Canadian Composite Conference Proceedings, CANCOM-2001, pp 794-801.

[38] J.R. Wilson, Active Aerolastic Wing: A New/Old Twist on Flight, Aerospace America, Sep. 2002.

[39] J.R. Wilson, A Twisted Approach to Wing Warping, Aerospace America, Oct. 2002.

[40] N.S Knot, J.V Zweber and D.E Valey, "Flexible Composite wing with Internal Activation for Roll Maneuver," *AIAA J. Aircraft,* Vol. 39, Number 4, Jul-Aug. 2002.

[41] S. Black, "Are High Temperature Thermosets Ready to Go Commercial", *High Performance Composites,* vol. 24-29, Nov 2004.

[42] Ben Iannotta, Nanocomposite; Lift hopes for Space Elevator, Aerospace America, Mar, 2006.

CHAPTER 3

# Applications of Nano Materials, Organic Semiconductor Complexes and Composites Based Devices

**Kh. S. Karimov[1,2,*] and Muhammad Abid[3]**

*[1]GIK Institute of Engineering Sciences and Technology, Topi, KPK, Pakistan; [2]Physical Technical Institute of Academy of Sciences, Rudaki Ave.33, Dushanbe, Tajikistan and [3]Interdisciplinary Research Center, COMSATS Institute of Information Technology, Wah Cantt, Pakistan*

**Abstract**: Nano materials, organic semiconductor complexes and composites have wide range of applications in various devices. In this chapter, use and potential applications of nano materials, organic semiconductor complexes and composites sensors, organic field effect transistors, thin-film field effect transistors, temperature sensitive field-effect transistor, solar cells, and light-emitting diodes have been summarized.

**Keywords:** Complex, composite, device, energy, field effect transistor, light-emitting diodes, nano material, organic, rectenna, semiconductor, sensor, solar cell, temperature, thin-film.

## 1. INTRODUCTION

During the last years, many efforts were devoted to the fabrication of new nanostructures and their composites with organic materials and semiconductors. Based on these materials, a number of electronic devices having sufficient sensitivity and efficiency for practical application were designed and investigated. In this chapter, results of authors investigations and obtained from the literature regarding effect of humidity to electric properties of $Cu_2O$-PEPC (poly-N-epoxypropylcarbazole) nanocomposite films, ITO/CuPc/NiPc/Al and ITO/NiPc/CuPc/Al junction cells with double junction, surface-type Al/CNT/Al pressure sensors, resistive-type displacement sensor based on composite of vanadium complex ($VO_2$ (3-fl)) and CNT, the CNT and $Cu_2O$ strain sensors based on composites are presented and discussed. Data on properties of thin film organic field effect transistors (OFETs), can be used not only for amplification of the

***Corresponding author Kh. S. Karimov:** GIK Institute of Engineering Sciences and Technology, Topi, KPK, Pakistan; Physical Technical Institute of Academy of Sciences, Rudaki Ave.33, Dushanbe, Tajikistan; E-mail: khasansangink@gmail.com voltage but, OFETs having Schottky junction, as MESFET, a number of sensors can be fabricated. In addition, properties of OFET as a temperature sensor are discussed in detail.

The structure and properties of organic solar cells with Schottky junction and p-n heterojunction, bulk heterojunction solar cells, and based on $T_iO_2$ nanostructure dye sensitized solar cells (DSSC) are discussed. Special attention was given to the properties of rectifying antennas, rectennas and based solar cells unlike to the conventional solar cells; if it will be fabricated using nanotechnology can convert the energy of light into electric power with very high efficiency.

In addition, structure and properties of a single layer organic light-emitting diode (OLED) fabricated on the base of Poly(p-phenylene vinylene (PPV), two-layer organic light-emitting diode which uses $Alq_3$ (tris(8-hydroxyquinolinato) aluminum, and NPB (N,N'-di(naphthalene-1-yl)-N,N'diphenyl-benzidine (e- electrons, h- holes), OLEDs with three CuPc/NPB/$Alq_3$ layers and five layers OLET with structure CuPc/NPB/CBP+$Ir(ppy)_3$/BCP/$Alq_3$ are discussed.

## 2. SENSORS

### 2.1. Humidity Sensor

Humidity sensors are used for the evaluation of the environmental parameters in different industrial applications [1-3]. There are different kinds of humidity sensors such as capacitive, resistive, hydrometric, gravimetric, optical and integrated types operating on different principles [4-6]. Capacitive humidity sensors are one of the simplest, less power consuming and reliable. Properties of these sensors depend on hygroscopic material used in the capacitance as insulator, and on the construction of plates. Unusually surface type electrodes are more popular [7].

For the fabrication of humidity sensors, a number of techniques and materials are used [8-15], *e.g.* for capacitive humidity sensors, surface and bulk micromachining techniques are popular for the production of porous silicon-based sensors [8, 9]. For the manufacturing of humidity and temperature sensors a thin-film surface micromachining technique was proposed by Park *et al.* [10]. Cellulose acetate butyrate [11, 12] and polyimide [13, 14] materials are commonly used as sensing materials for capacitive-type humidity sensors. Karimov *et al.* [15] fabricated a surface-type capacitive humidity sensor based on copper phthalocyanine (CuPc). As relative humidity is increased from 35% to 92%, the capacitance of the detector is also increased continuously by 200 times.

The complexes of poly-N-epoxypropylcarbazole (PEPC) as one of the photosensitive materials are known. The solar cells and photo capacitors are PEPC based on their good adhesive properties [16]. Another photosensitive semiconductor copper oxide ($Cu_2O$), p-type, with a band gap of 2 eV, is nontoxic and is available in nature in large scale is used for the fabrication of nanodots and nanostructure thin films [17-20]. Humidity sensors with high sensitivity, stability and linearity of characteristics, potentially can be fabricated with composites of different materials and can be realized by investigating the effects of humidity on copper oxide composites. The results of the investigations of the effect of humidity on the electric properties of the copper oxide nanoparticles and poly-N-epoxypropylcarbazole ($Cu_2O$-PEPC) nanocomposite films are presented in this chapter.

For experiments $Cu_2O$ nanopowder was commercially purchased from WINLAB UK. PEPC was synthesized in laboratory [16]. Fig. (**1**) shows molecular structure of PEPC. The glass substrates were prepared by cleaning ultrasonically by acetone for 10 minutes and then by plasma cleaning for 5 minutes. The blend of copper oxide nanopowder (3.0 wt.%) and poly-N-epoxypropylcarbazole (2.0 wt.%) in benzol was prepared. By drop-casting, a thin film on glass substrates was deposited with pre-deposited surface-type silver electrodes. The gap between silver electrodes was 40 μm for the fabrication of $Cu_2O$-PEPC nanocomposite thin films. It was found that deposited films were stable and had good adhesion to the substrate.

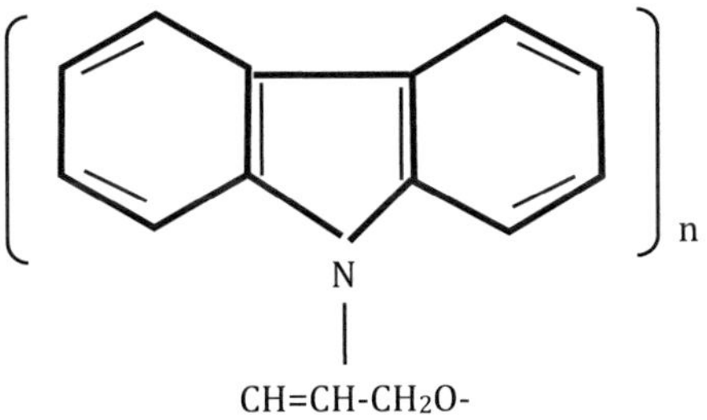


**Figure 1:** Molecular structure of poly-N-epoxypropylcarbazole (PEPC), n=4-6.

Scanning electron microscope (SEM) micrographs of $Cu_2O$-PEPC nanocomposite thin film shows porosity that make this material a potential candidate for humidity sensors fabrications and is shown in Fig. **2(a-b)** at different magnifications. The $Cu_2O$-PEPC nanocomposite thin films were investigated by X-ray differaction (XRD) technique. Fig. (**3**) shows the pattern of XRD of the nanocomposite thin film obtained from "Panalytical Xpert pro". It was found that with the ICDD reference card "00-001-1142" the indexed peaks of the XRD pattern are in agreement.

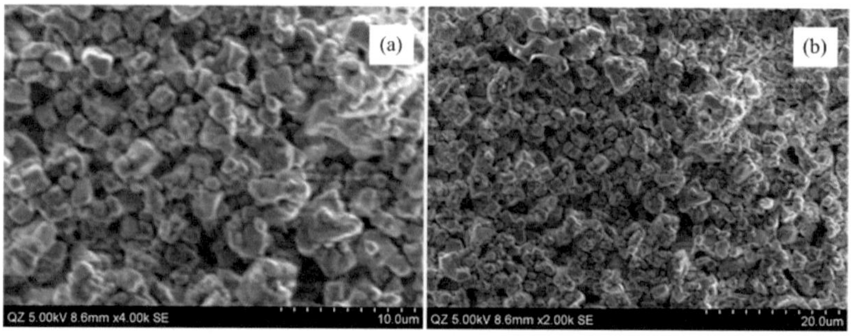


**Figure 2:** SEM images of the $Cu_2O$-PEPC film at different magnifications.

Fig. **4(a-b)** shows two and three dimensional atomic force microscope (AFM) images obtained using Agilent's Pico Plus under ambient condition, with a scan size of 5 μm.

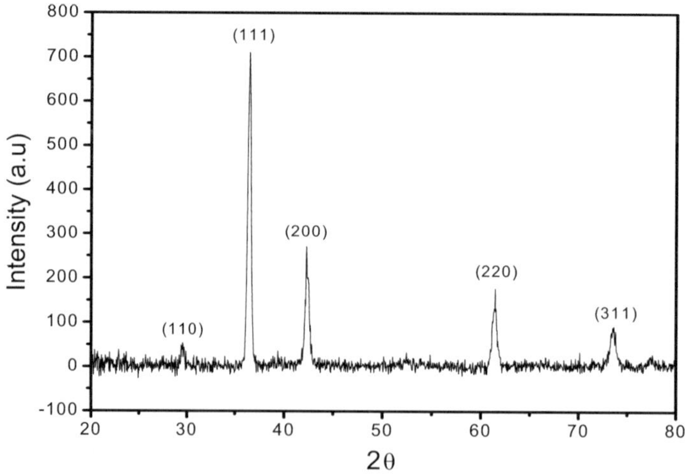


**Figure 3:** XRD pattern of the $Cu_2O$-PEPC nanocomposite thin film.

By the measurement of capacitance and dissipation the effect of humidity on the electrical properties of the nanocomposite films was investigated. The frequencies of the measuring applied voltage were 120 Hz and 1 kHz. Measurements were taken at room temperature. From the values of dissipation, the AC resistance of the samples was determined. DC resistance was measured directly. Fig. (**5**) shows cross-sectional view of the fabricated Ag/$Cu_2O$-PEPC/Ag sensor.

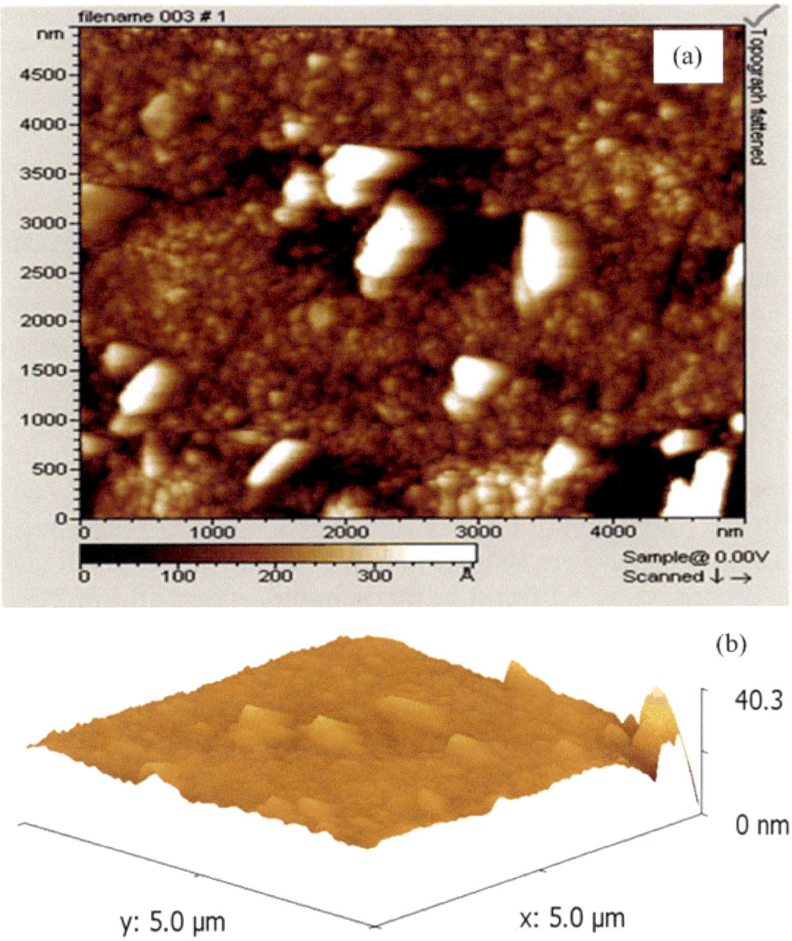


**Figure 4:** AFM micrographs of the $Cu_2O$-PEPC nanocomposite thin film: (a) 2D, (b) 3D.

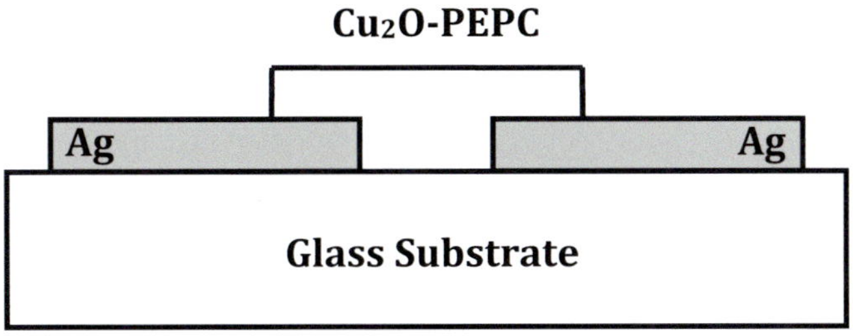


**Figure 5:** Schematic diagram of the Ag/$Cu_2O$-PEPC/Ag humidity sensing device.

Following expression was used for the calculation of resistance ($R$) from measurement of dissipation ($D$) [21]:

$$R = \frac{1}{2\pi fCD} \tag{1}$$

where $f$ is frequency and $C$ is capacitance.

During a number of measurements of capacitance and resistance, in the ageing process it was found that in average the effect of humidity to capacitance and resistance decreased approximately of 50-60%.

Fig. (**6**) shows the results of capacitance-relative humidity (RH) measurements for the Ag/$Cu_2O$-PEPC/Ag sample measured at 120 Hz and 1 kHz. Humidity was measured up to 86%. From Fig. (**6**) it is seen that capacitance of the sample is increased at 120 Hz and 1 kHz respectively. Fig. (**7**) shows that at 120 Hz and 1 kHz, resistance of the sample is decreased by 345 and 157 times respectively, with respect to the initial condition. Fig. (**7**) shows that DC resistance-humidity relationships of the sample were sharper than the AC resistance relationships. The DC resistance decreased by 63 times in the humidity range of 46-52%., whereas AC resistance decreased by 3 and 2.2 times respectively at 120 Hz and at 1 kHz. Absorption of water vapor in semiconductor thin film, formation of charge transfer complexes (CTC) and doping of the nanocomposite by $H_2O$ are observed responsible for the observed phenomena and in particular for the increase of capacitance and decrease of resistance of the sensor.

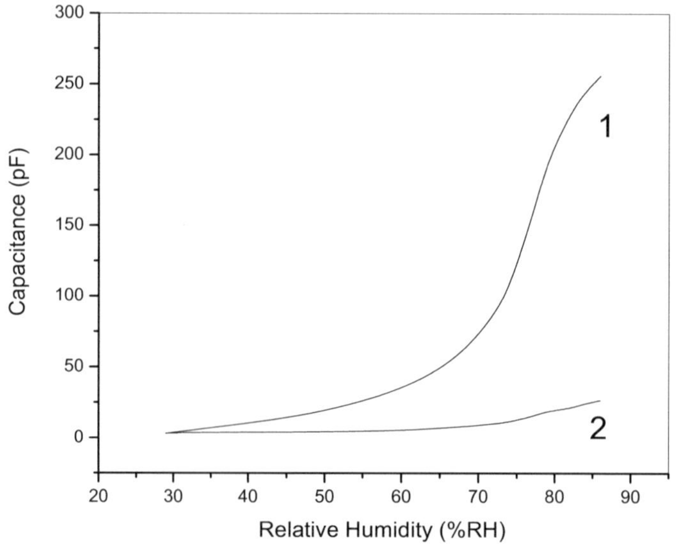


**Figure 6:** Capacitance-relative humidity (RH) relationships of the Ag/$Cu_2O$-PEPC/Ag device measured at; (1) 120 Hz, (2) 1 kHz of the applied voltage.

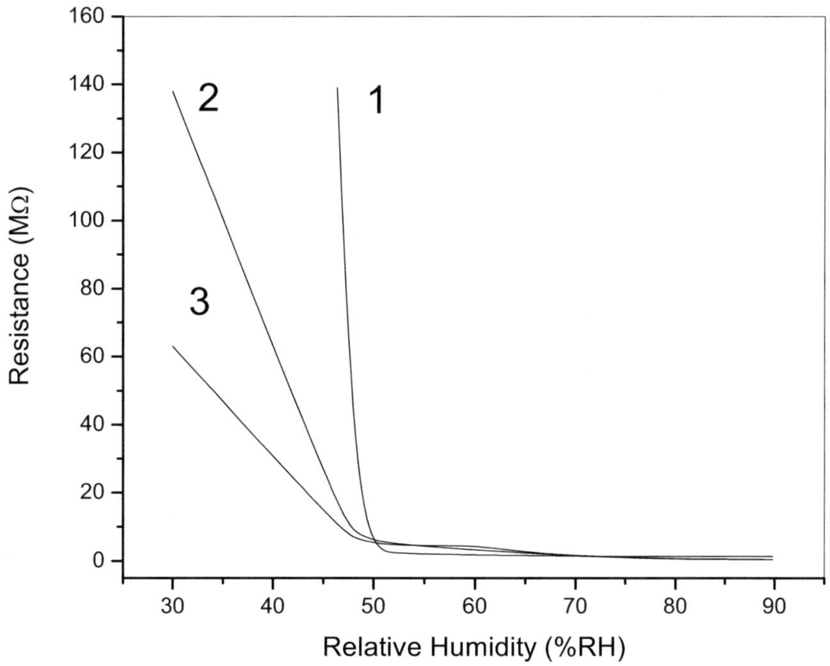


**Figure 7:** Resistance-relative humidity (RH) relationships of the Ag/$Cu_2O$-PEPC/Ag device measured at; (1) DC, (2) 120 Hz, (3) 1 kHz of the applied voltage.

For practical applications of humidity sensors, response (in case of absorption) and recovery (in case of desorption) times are observed important. At room temperature conditions, response time of the sensor was measured from 30~90%. For the capacitive and resistive measurements, response and recovery times were almost 10 s, 30 s, 12 s and 34 s respectively. Response times are observed shorter than the recovery times. For capacitive and resistance humidity measurements, sensitivity (*S*) were determined using equation 2:

$$S_C = \frac{\Delta C}{\Delta (RH)} \qquad (2)$$

and

$$S_R = \frac{\Delta R}{\Delta (RH)} \qquad (3)$$

For capacitive measurements (Fig. **6**) sensitivity of the sensor was 0.83 pF/% and 0.17 pF/% at 120 Hz and 1 kHz respectively. For resistive measurements (Fig. **7**), sensitivity of the sensor for AC resistance at 120 Hz and 1 kHz was 8 MΩ/%, and 3 MΩ/% respectively and 65 MΩ/% for DC resistance.

At relative humidity of 60% and 46%, respectively, capacitance-frequency and resistance-frequency relationships for the Ag/$Cu_2O$-PEPC/Ag sample were measured (Figs. **8** and **9**). It is seen that with increase in frequency the capacitance and resistance decrease 14 and 7 times respectively. One of the reason for this is observed relatively longer relaxation time [22], resulting in decrease of displacement current that is due to the orientation and deformation of $H_2O$ molecules under the effect of an electric field [23].

It is known that the parameters including capacitor's plate area, distance between plates and relative dielectric constant of the materials between plates influence the value of capacitance. By taking into account the different types of polarizabilities, the increase in relative capacitance with increase of humidity can be explained. Basically there are several sources of polarizability *i.e.* dipolar ($\alpha_{dip}$), ionic ($\alpha_i$) and electronic ($\alpha_e$) [22]. The relative displacement of the orbit of electrons provides electronic polarizability. As effect of humidity on the capacitance and resistance of the sensor is strong. It is assumed that the absorption of $H_2O$ increases dipolar polarizability ($\alpha_{dipH}$) in the $Cu_2O$-PEPC nanocomposite with respect to the dipolar polarizabilty ($\alpha_{dip}$) due to $Cu_2O$, PEPC and some uncontrolled impurities. Probably at low frequencies (120 Hz, 1 kHz) the dipolar, ionic and electronic polarizabilities affects the value of capacitance. At higher frequencies the electronic polarizability influences. It had been reported in [24-26] that the polarizability due to the transfer ($\alpha_{tn}$) of charge carriers as electrons and holes was observed at normal conditions. Therefore, the following equation can be written for the total polarizability ($\alpha_n$) at normal conditions (RH is 30%):

$$\alpha_n = \alpha_{dip} + \alpha_i + \alpha_e + \alpha_{tn} \tag{4}$$

If the samples are under effect of humidity (RH>40%), the following equation can be used for the description of total polarizability ($\alpha_h$):

$$\alpha_h = \alpha_{dipH} + \alpha_i + \alpha_e + \alpha_t \tag{5}$$

where $\alpha_t$ is polarizability due to the transfer of electrons/holes charge carriers. These carriers are produced under the effect of $H_2O$ molecules. The water molecule plays the role of dopants. If concentration of $H_2O$ molecules, charge carriers and total polarizability ($\alpha_h$) are humidity dependent then above discussion can be considered as a correct.

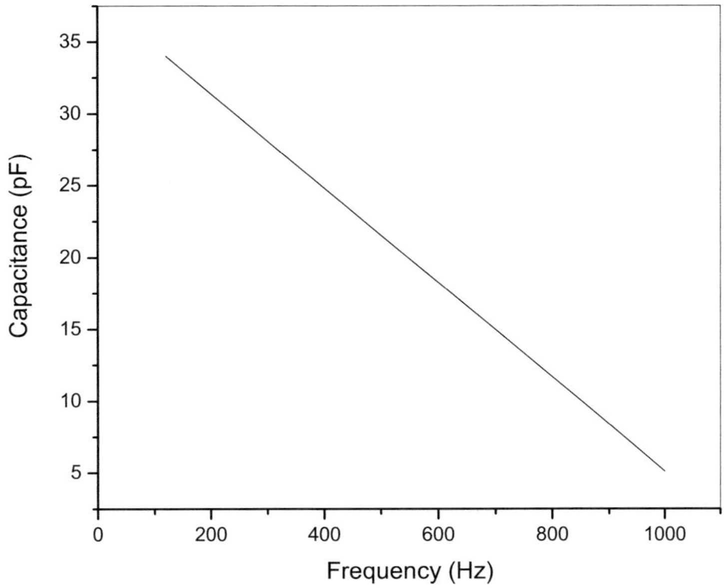


**Figure 8:** Capacitance-frequency relationships for the Ag/$Cu_2O$-PEPC/Ag device measured at relative humidity of 60%.

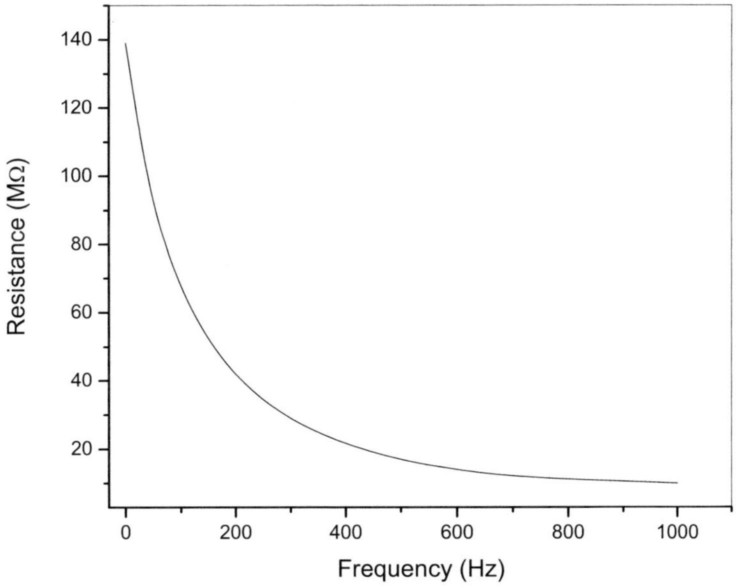


**Figure 9:** Resistance-frequency relationships for the Ag/$Cu_2O$-PEPC/Ag device measured at relative humidity of 46%.

Clausius-Mosotti relation given in [22] was presented in [16] for the simulation of capacitance-humidity relationships. For linearization of the experimental graphs, in particular, capacitance/resistance-humidity relationships, the op-amp logarithmic amplifiers [27] can be used as they allow linearizing the humidity response of this sample for practical applications in the instrumentation. From Figs. (**6**) and (**7**) it is seen that high sensitivity takes place at lower humidity for the resistance and at higher humidity for the capacitance. It means that the combination of resistance and capacitance measurement may be used to cover wider humidity range and the $Cu_2O$-PEPC composite may be used for the development of humidity sensors.

Thus the investigations of effect of humidity on the properties of $Cu_2O$-PEPC, nanocomposite films deposited from solution by drop-casting samples are observed sensitive to the effect of humidity. In particular, it was observed that the capacitance is increased and resistance is decreased with increase in humidity. From results of researchers following may be concluded: (i) dipolar polarization of the absorbed $H_2O$ molecules by the nanocomposite films and polarization due to transfer of charge carriers providing capacitive response of the sensor, (ii) the increase of $H_2O$ molecules concentration and displacement currents, and concentration of charge carriers doped by water molecules can decrease samples resistance.

## 2.2. Light Sensor

As a viable alternative to traditional sensors which are based on inorganic materials during the past decade, organic semiconductor devices based on conjugated polymers, oligomers and low molecular weight materials have been envisioned. An attracting extensive interest for potential applications in organic devices [28-31] is the minimum material and less fabrication cost of the organic semiconductors.

For some kinds of applications the heterojunction structures may be optimal in order to improve devices' characteristics. For the fabrication of photoelectric cells especially, it may be beneficial. Phthalocyanines [32-34, 37] and in particular, copper phthalocyanine (CuPc) are amongst the most well-studied organic photosensitive semiconductors. In wide spectra it has high absorption coefficient and high photo-electromagnetic sensitivity at low intensities of radiation. At the same time by vacuum sublimation the deposition of thin CuPc films is easy. As the sublimation occurs at relatively low temperatures (400~ 600°C) the purification of CuPc is simple and economical. It can be emphasized that energy band gap of the CuPc is approximately 1.6 eV [34] to 2.0 eV [38], *i.e.* it is actually optimal for solar energy applications.

Due to its potential applications in the area of photovoltaics and gas sensing [39-50], in recent years, nickel phthalocyanine (NiPc) has received increasing attention. Higher mobility of charge carriers (0.1 $cm^2/V$ s and $10^{-4}$ $cm^2/V$ s respectively) is one of the major advantages of NiPc over CuPc [50]. For indirect and direct allowed transitions [42], the energy band gap of the NiPc is 2.24 eV and 3.2 eV.

Many authors have investigated CuPc [36-38, 51, 52] and NiPc [32, 33, 35-37] based organic-inorganic semiconductors heterojunctions. It would be interesting to study phthalocyanine based organic-organic semiconductors heterojunctions due to their molecular and structural similarities that were observed in many cases. For example, it was found that the values of for effective density of states in CuPc and NiPc are the same ($N_v = 10^{27}$ $m^{-3}$, *i.e.* one electronic state per molecule). There molecular size and crystalline structures are also same [40]. This is important to minimize surface state density and defects concentration in the heterojunction interface. Moreover both CuPc [36, 38] and NiPc [40] form Schottky-type metal-semiconductor heterojunctions with aluminum. In this study, opto-electronic properties of ITO/NiPc/CuPc/Al and ITO/CuPc/NiPc/Al double-junction cells are presented.

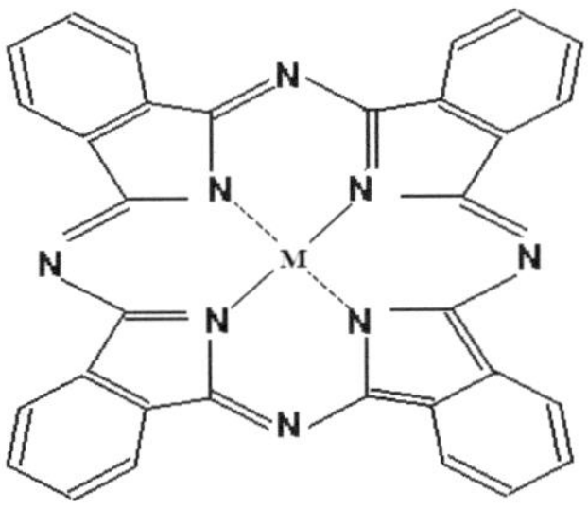


**Figure 10:** Molecular structure of CuPc and NiPc: M is Cu or Ni.

From Sigma-Aldrich, commercially available CuPc and NiPc were obtained. The structure of the CuPc and NiPc molecules [41] that were used as *p*-type organic semiconductors is shown in Fig. (**10**). From investigations done in normal terrestrial conditions and in the space it is known that there exist seven crystalline polymorph states of CuPc: $\alpha, \beta, \gamma$, $R, \delta$, $\varepsilon$ *etc.* [52, 53]. The metastable at T = 165 °C form is $\alpha$-CuPc that can be converted into the $\beta$-form thermally or with solution. Mostly investigated and frequently encountered states of CuPc are $\alpha$ and $\beta$-forms. In current investigation, X-ray diffraction data showed that the deposited CuPc films were in $\alpha$-form [54]. Particular crystalline orientations in the CuPc films were not found by investigation of scanning electron micrographs [54]. It was observed that NiPc films were in short-whisker like α-phase. The whiskers grow longer as the annealing temperature increases to 200 °C. It is known that annealing at 300 °C converted α-phase into β-phase [43]. Therefore we can assume that NiPc films in this case were in α-phase.

At 400-450 °C and at ~$10^{-4}$ Pa in an Edwards AUTO 306 vacuum coater having a diffusion pumping system thin films of CuPc were thermally sublimed onto a conductive glass substrate (ITO). At this process the deposition rate of thin film growth was 2 nm/min and the temperature of the substrate was equal ~40 °C. Under the same conditions, CuPc or NiPc films were deposited on the conductive glass (ITO) substrates. For measurements of absorption spectra a Perkin Elmer Lambda 19 UV/VIS/NIR spectrometer was used and it was found that the spectrum of these materials covers the NIR- UV, *i.e.*, wavelengths between 200-1000 nm [33, 37]. Absorption spectra of the deposited CuPc and NiPc films in NIR-visible spectra shows that the maximum of the spectrum corresponds to energy of 1.5 eV (Fig. **11**). Finally Al films were deposited on CuPc and NiPc films. By using an Edwards FTM5 film thickness monitor the thickness of the CuPc and NiPc was obtained. With Ag usually phthalocyanines form ohmic contacts and Schottky-type junction is formed with Al that was found in the results of a number of investigations [33, 36, 38, 49].

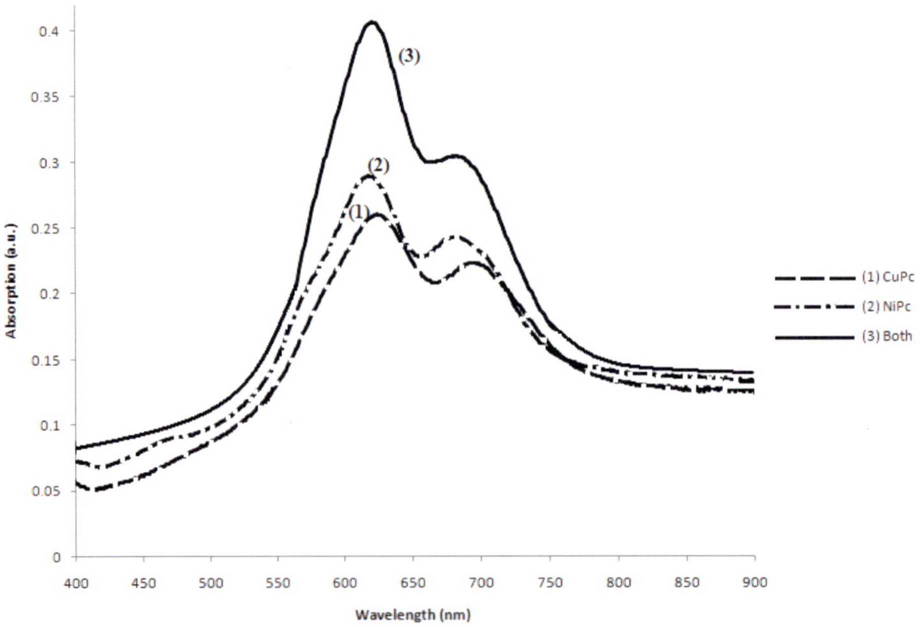


**Figure 11:** Absorption spectra of the CuPc/NiPc double layer film deposited on conductive glass (ITO) substrate by vacuum evaporation.

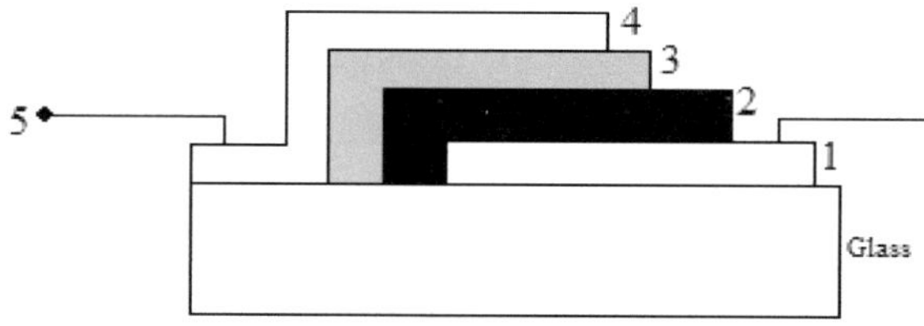


**Figure 12:** Cross sectional view of ITO/NiPc/CuPc/Al and ITO/CuPc/NiPc/Al double junction cells: 1-ITO, 2-CuPc or NiPc, 3-NiPc or CuPc, 4 – Al, 5 and 6 – terminals.

A cross-sectional view of the fabricated ITO/CuPc/NiPc/Al and ITO/NiPc/CuPc/Al cells is shown in Fig. (**12**). In three different cells the thicknesses of each organic semiconductor films were 25 nm, 50 nm and 75 nm respectively. Effective surface area of the samples was 80 $mm^2$ (8x10 $mm^2$). The cells were annealed at 150 °C for 2 hours and its effect on properties of the cells was studied. In experiments as the source of light a filament lamp and Red LED were used. The measurements of voltage and current were conducted at room temperature.

Fig. (**13**) shows dark and photo *I-V* curves of the ITO/CuPc/NiPc/Al cell at room temperature under illumination (340 lx) by filament lamp. The thickness of CuPc and NiPc layers was 150 nm (2x75 nm). It is seen that reverse saturation current of the cell increases under illumination approximately 5 times, as ratio of photo and dark currents ($I_{Ph}/I_D$) showing good response in the photo-diode mode of operation. Rectification ratio was around of 2 at an applied voltage of 5 V. The *I-V* characteristics of the ITO/NiPc/CuPc/Al cells were in principle similar, but the cells showed better photoresponse ($I_{Ph}/I_D$ was upto 5) and higher RR (up to 4). The forward bias *I-V* characteristics under dark and under illumination were similar for all cells.

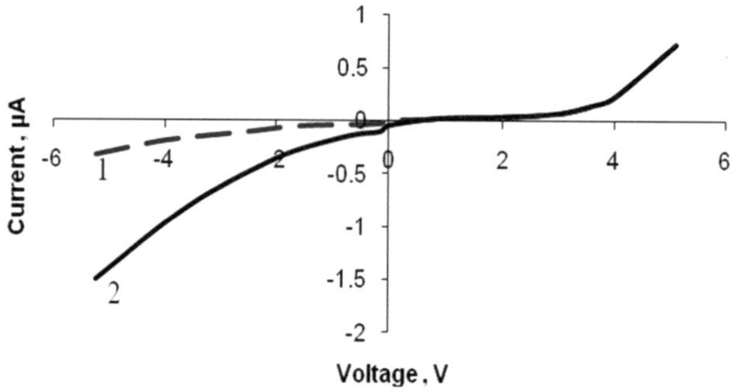


**Figure 13:** Dark (1) and photo (2) I-V curve of the ITO/CuPc/NiPc/Al. cell at room temperature.

For analysis of the dark *I-V* curves in the narrow potential range (0 - 0.5V), the modified Shockley equation may be used [52, 55]:

$$I = I_o \left[ Exp \frac{q(V - I R_s)}{nkT} - 1 \right] + \frac{(V - I R_s)}{R_{sh}} \tag{6}$$

where $I$ and $V$ are the terminal current and voltage, $R_s$ and $R_{sh}$ are the device series and shunt resistances, and $n$ is the diode ideality factor, $T$ is absolute temperature, $k$ is Boltzmann constant, and $I_o$ is the reverse saturation current given in ref. [56].

$$I_o = A^* T^2 \; Exp\,(-\frac{q\Phi}{kT}) \tag{7}$$

Where $q\Phi$ is the height of contact barrier (Schottky barrier [56]) and $A^*$ is the Richardson constant.

From *I-V* characteristics shown in Fig. (**13**), diode junction resistance can easily be determined as; $R_j = \frac{\delta V}{\delta I}$. As a maximum junction resistance () At reverse and forward bias, 9.1 MΩ and 2.2 MΩ was observed.

The value of $I_o$ may be determined and the $A^*$ of Eq. 7 may be calculated if at some approximation we assume that the forward bias voltage-current characteristics of the sensor may be described by Eq. 6 and processed in a similar manner to that of a heterojunction barrier diode [55, 56]. It means the current should be plotted on a log scale. Using this approach it can be found that the reverse saturation current calculated is $I_o$ = 0.007 μA, considering $q\Phi$ = 0.96 eV for Au/NiPc/Al cell [49]. Using Eq. 2, $A^* = 1.28\ 10^{\ 3}$ A/K$^2$ cm$^2$ is calculated.

Fig. (**14**) shows an open-circuit voltage – illumination relationships for the ITO/NiPc/CuPc/Al cells at different thickness of CuPc and NiPc films. As it was observed before in phthalocyanine cells these characteristics are nonlinear [55]. It is seen that with increase in thickness, $V_{OC}$ decreases that may be associated with absorption of light on the way to CuPc-Al Schottky junction.

The samples were illuminated with a filament lamp. The value of $V_{OC}$ on average for ITO/NiPc/CuPc/Al cells is higher by 20 % in comparison to $V_{OC}$ of ITO/CuPc/NiPc/Al cells. The polarity of the photo-induced voltage was negative on Al and positive on ITO for both ITO/NiPc/CuPc/Al and ITO/CuPc/NiPc/Al cells showing that Schottky barrier metal-semiconductor CuPc/Al and NiPc/Al junction plays a dominant role with respect to organic-organic semiconductor CuPc/NiPc (or NiPc/CuPc) junction. This means a buit-in potential barrier (equal to differences of work functions [56]), of Schottky junction in this case is higher than the barrier of organic-organic semiconductor junction. Therefore we can assume that thermionic emission theory may be used (Eq. (2)) [56] to describe current mechanism of majority charge carriers over the Schottky barrier. Short-circuit currents – illumination relationships of the same samples shows quasi-linear behavior.Under illumination the current increased up to 0.03 μA and a higher current was observed through the thin films cells (50 nm).

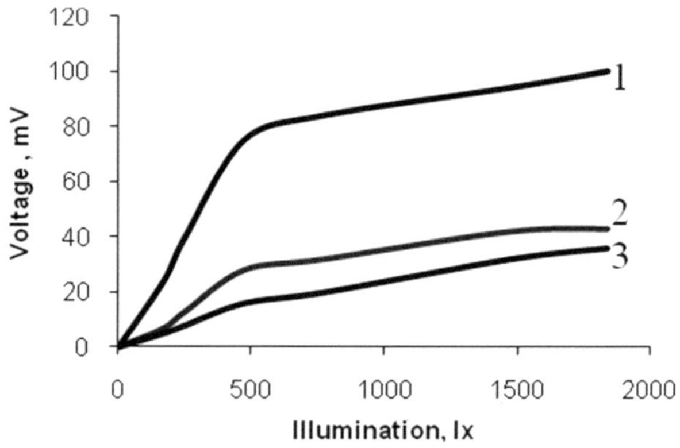


**Figure 14:** An open-circuit voltage – illumination relationships for the ITO/NiPc/CuPc/Al cells respectively at different thickness of CuPc and NiPc films: 1-50 nm (2x25 nm), 2- 100 nm (2x50 nm) and 3 – 150 nm (2 x 75 nm).

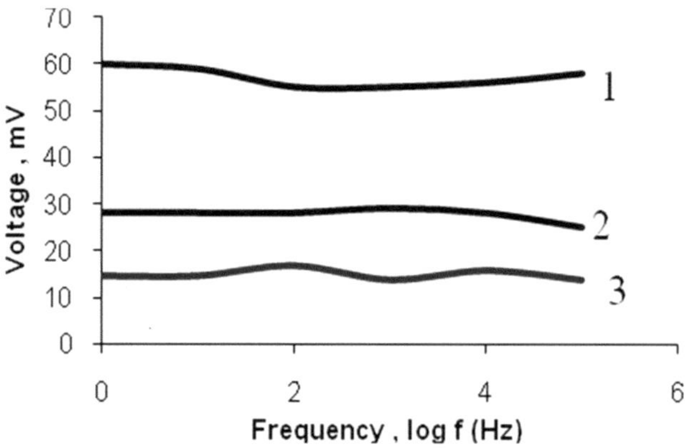


**Figure 15:** Dependences of ITO/NiPc/CuPc/Al cells' DC open-circuit voltage on Red.

Fig. (**15**) shows dependences of ITO/NiPc/CuPc/Al cells' DC open-circuit voltage ($V_{oc}$ (DC)) on Red LED modulated beam's frequency. It is seen that in the first approximation the DC open-circuit voltage is independent of the frequency of the modulation, whereas $V_{oc}$ (AC)'s response was up to 100 Hz respectively. The short-circuit currents responses were as $V_{oc}$ (DC) and $V_{oc}$ (AC) accordingly. The illumination of the Red LED was equal to 470 lx. The thickness dependence of the voltage is the same as was observed in Fig. (**14**).

LED modulated beam's frequency at different thickness of CuPc and NiPc films: 1-50 nm (2x25 nm), 2- 100 nm (2x50 nm) and 3 – 150 nm (2 x 75 nm).

Fig. **16** shows capacitance-illumination relationship for the ITO/NiPc/CuPc/Al cells having different thickness of the NiPc and CuPc films. It is seen that the relative value of the capacitance $C_{ph}$ increases by 50 % under illumination of 2000 lx, unlike the ITO/CuPc/NiPc/Al cells where the increase was 30 % only. It is seen (Fig. **16**) that a higher photo-response is shown by thicker film (150 nm) and a lower by the thinner one (50 nm). It may be due to more absorption of light in the former case. In [24, 26, 57] the polarizability due to the transfer ($\alpha_{td}$) of charge carriers as electrons and holes that are present at normal, including dark conditions were investigated. In this case we assume that photo-induced charge carriers are responsible for observed photo-capacitive effect in the ITO/NiPc/CuPc/Al and ITO/CuPc/NiPc/Al cells. Detail analysis of this effect based on Clausius-Mosotti retaion [22] with respect to poly-N-epoxypropylcarbazole was done earlier [16].

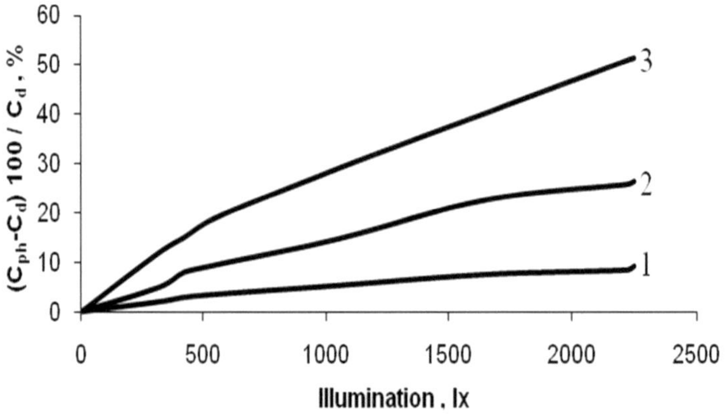


**Figure 16:** Capacitance-illumination relationship for the ITO/NiPc/CuPc/Al cells having different thickness of the NiPc and CuPc: 1-50 nm (2x25 nm), 2- 100 nm (2x50 nm) and 3 – 150 nm (2x75 nm).

It was found that annealing increased RR up to 13 and $V_{OC}$ up to 400 mV, that may be due to the decrease of total concentration of traps as was observed in Au/NiPc/Al cell [40]. At the same time a visible effect of the annealing to $I_{SC}$ was not observed, that may be caused due to oxidation of Al film in the annealing process and a decrease in its conductivity.

Taking into account the data presented above and the concept related to heterojunction properties [56], an energy-band diagram of the ITO/NiPc/CuPc/Al heterojunctions were developed (Fig. **17**). In Fig. (**17**) $\Delta E_c$ and $\Delta E_v$ are discontinuities in the conduction and valence bands, *i.e.* the differences between the two conduction bands energies and valence bands energies respectively. Accordingly the $eV_{bi1}$ and $eV_{bi2}$ are built-in potential barriers seen from holes in CuPc/Al and NiPc/CuPc heterojunctions interfaces respectively. In Fig. (**17**), it is shown that $eV_{bi1} > eV_{bi2}$. There are localized states on the CuPc and NiPc side of the junctions because in these organic semiconductors the mobility of charge carriers is much less than 1 $cm^2/Vs$ [34, 35] that is unlike to the delocalized states in Al or ITO sides.

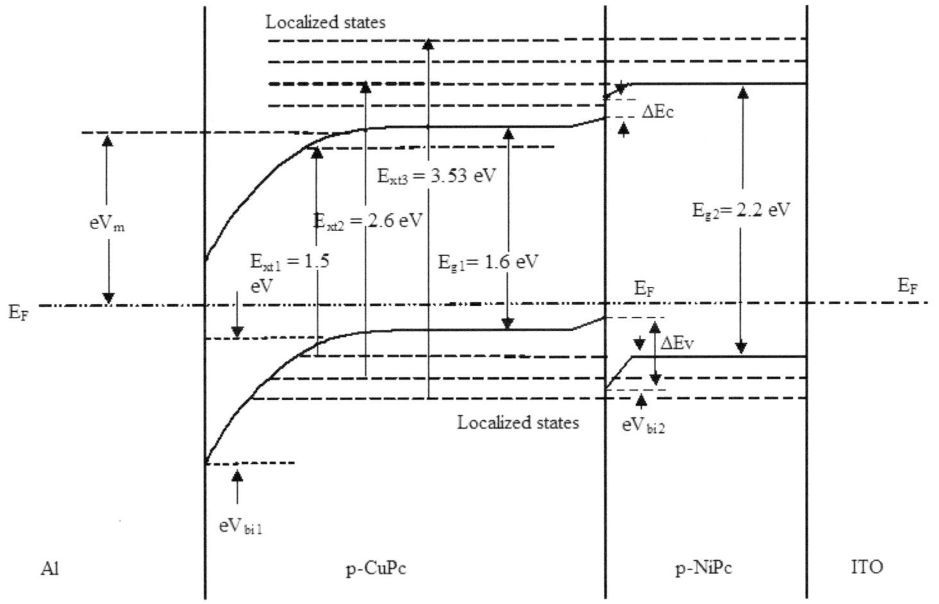


**Figure 17:** Energy-band diagram of the ITO/NiPc/CuPc/Al heterojunctions.

Fig. (**18**) shows an equivalent circuit of the organic photo-electric cell [52] developed for the ITO/NiPc/CuPc/Al double junction cell and current sources $I_1$ and $I_2$ represent CuPc/Al and NiPc/CuPc photo-electric cells. In the case of the

ITO/CuPc/NiPc/Al cell, the equivalent circuit is considered the same: only direction of $I_2$ will be opposite. From the dark *I-V* curves, $R_{sh}$ and $R_s$, *i.e.* cell's shunt and series resistances are determined, $C_1$ is the sum of the transition and diffusion capacitances of the heterojunction; $C_2$ is inter-terminal capacitance of the cell. In the first approximation it may be assumed that $C_1 = C_2$.

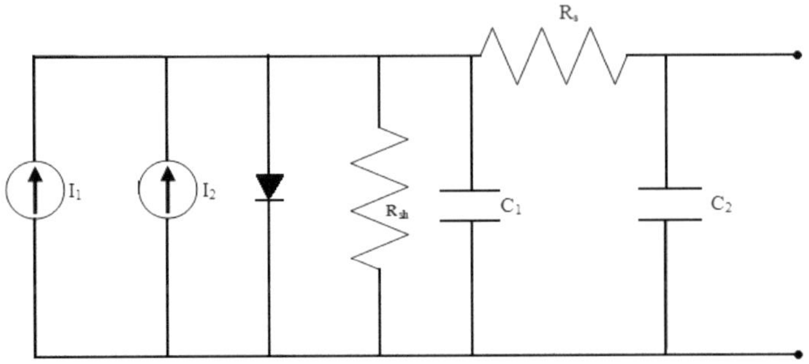


**Figure 18:** Equivalent circuit of the ITO/NiPc/CuPc/Al cell.

As a result of this work, ITO/CuPc/NiPc/Al and ITO/NiPc/CuPc/Al double junction cells were fabricated, investigated and it was found that I-V characteristics show rectification behavior and exhibit an increase in current under illumination under reverse bias. The cells show photo –electric response in photo-voltaic mode of operation: on average the response of ITO/NiPc/CuPc/Al cells is 20 % higher than the ITO/CuPc/NiPc/Al cell, revealing that Schottky barrier metal-semiconductor CuPc/Al or NiPc/Al junction plays a dominant role in comparison to organic-organic semiconductor CuPc/NiPc (or NiPc/CuPc) junctions. The cells show high photo capacitive response: in ITO/NiPc/CuPc/Al cells, capacitance increase under illumination by 50 % under illumination of 2000 lx. The energy-band diagram and an equivalent circuit of the double junction cells were developed that may be used for the simulation of their opto-electronic properties.

## 2.3. Pressure Sensor

Pressure transducers are mostly used with resistance, capacitance, inductance and piezoelectric sensors and with devices such as diaphragm and bellows [27, 58]. Organic semiconductor pentacene based transistors with network of pressure sensors fabricated is described in ref. [59]. It was observed that the drain-source current was increased from 15 nA to 6.7 μA under an applied pressure of 30 kPa. Due to the presence of non-uniform spatial electric field distribution in the junction and softness of organic semiconductors in the organic semiconductor

Schottky junctions, unusual electromechanical effects related to the piezoelectricity and electrostriction was observed. For the fabrication of electromechanical transducers potentially these effects can be used [60].

During the last years on the basis of low molecular organic semiconductors tetracyanoquinodimethane (TCNQ) ion-radical salts crystals, a number of resistance strain gauges have been fabricated and investigated [61]. It was found that resistance strain gauges based on poly-N-epoxypropylcarbazole (PEPC) showed high sensitivity [62].

Carbon nanotubes (CNTs) are observed interesting due to their unique electronic and mechanical properties. Regarding electronic properties, CNTs can be at different phases as metallic, semiconducting, or small-gap semiconducting (SGS), depending on the orientation of the graphene lattice with respect to the axis of the tube [63]. Therefore on the basis of CNTs, different kinds of sensors have been fabricated and investigated. Recently properties of humidity sensors based on single-walled carbon nanotubes (SWNTs), double-walled carbon nanotubes (DWNTs) and multi-walled carbon nanotubes (MWNTs) are studied [64-69]. Based on interesting electromechanical properties of the CNTs, these are used in practice as piezoresistors in mechanical sensors such as strain gauges, pressure sensors and accelerometers [63, 70-74]. Piezoresistance of CNTs on deformable thin-film silicon nitride membranes was investigated and it was found that the gauge factors ($\Delta R/R_\varepsilon$) of the samples were 400 and 850 for the semiconducting and SGS tubes, respectively, whereas the maximum value of gauge factor in silicon was 200 [7]. Investigations of small band-gap semiconducting (or quasimetallic) nanotubes showed that they exhibit the piezoresistive gauge factors from 600 to 1000 under axial strains, which are much larger than in metallic nanotubes [70]. The fabrication of SWNT thin-film transistors on plastic substrates were realized and it was shown that the resistance of the SWNT thin film decreases 38.2% and 47.1% with increase of the bending of the elastic substrate [71]. The thin films here contained 14 and 16 SWNT layers which are more than 10 time higher than the silicon. Piezoresistive effect was also investigated in the pristine CNTs films. It was found that at room temperature, the gauge factor under 500 microstrains was 65 and was demonstrated in [72]. It was observed that the gauge factor increased with temperature. In ref. [73], mechanical deformation-conductivity relationships of free-standing membranes of SWNTs are investigated. The nanotubes gauge factor of piezoresistivity is observed 2.3 to 2.5 times larger than that of the silicon substrate. The theoretical and experimental investigations of nano electromechanical piezoresistance transducers based on

SWNTs showed that ballistically conducting SWNTs show nonlinear piezoresistive gauge factors up to 1500 (applied strain was 1%) [74].

From practical point of view and for deepening of the knowledge about the physical properties of the CNTs the fabrication of the pressure transducers and investigations of squeezing effect to the CNTs layer would be useful. Concerning this it would be reasonable to investigate the resistance-pressure relationships. In this chapter the design, fabrication and investigation of the surface-type pressure resistance sensor based on CNT are presented.

Commercially produced (Sun Nanotech Co Ltd., China) CNTs powder was deposited on the adhesive elastic polymer tape of thickness of 35 μm, with built-in Al foil electrodes. After that the samples were covered with the same adhesive elastic tape. At the same time the tapes also played the role of the elastic casing (Fig. **19**). The diameter of the multiwalled nanotubes (MWNTs) varied between 10-30 nm. The thickness of the CNTs layers was in the range of 300-430 μm. The interelectrodes distance (length between electrodes) and width of the surface-type samples layers was in the interval of 4~6 mm and 3~4 mm, respectively. Using ESCORT ELC-132A meter, DC resistance of the sensors was measured. Earlier in ref. [75], a setup for the measurement of the influence of the pressure on the resistance of the sensors is described.

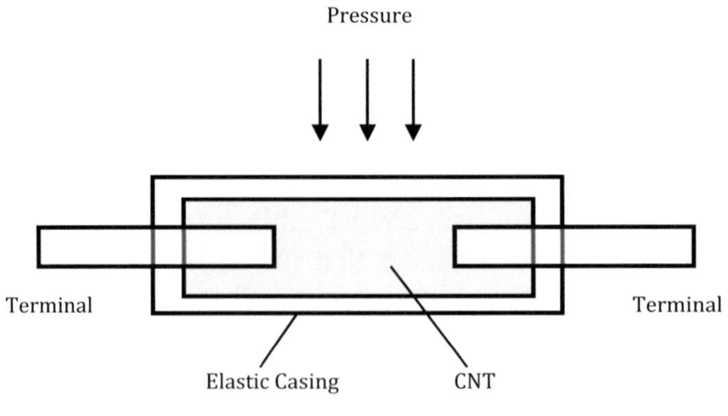


**Figure 19:** Al/CNT/Al resistance pressure sensor.

Fig. (**19**) shows resistance-pressure relationships for two Al/CNT/Al sensors. From experimental results it was found that the DC resistances of the sensors were decreased by 3~4 times, respectively, as the pressure was increased upto 17 kN/m$^2$. It was found that the effects of pressure on the sensors' resistances were more at thicker CNTs films (~430 μm) and less at thinner films (~300 μm). The sensor's resistance (*R*) was calculated using following expression [22]:

$$R = \frac{d\rho}{A} = \frac{d}{\sigma A} \tag{8}$$

where *d* is the length or inter-electrode distance and *A* is the cross-section of the aluminum foil electrode or CNT layer, and $\rho$ is the resistivity. Here $\rho = \frac{1}{\sigma}$, where $\sigma$ is conductivity of the CNT layer. From the observed resistance-pressure relationship, it was considered that the increase of the conductivity of the CNT film is due to the densification of the CNT under the effect of pressure. using exponential function $f(x) = e^{-x}$, Eq. 8 can be represented as [76]:

$$\frac{R}{R_o} = e^{-pK} \tag{9}$$

where *p* is pressure, *K* is resistance-pressure factor.

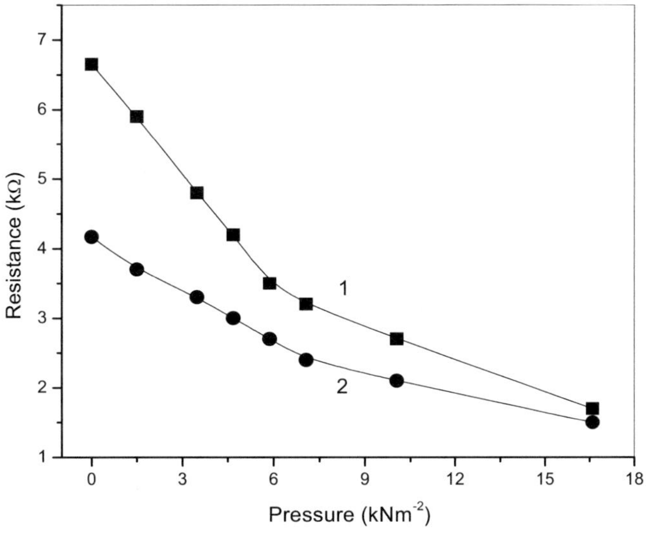


**Figure 20:** Resistance-pressure relationships of the two Al/CNT/Al sensors for; (1) thicker CNT ~430 μm, (2) thinner CNT ~300 μm films.

The average pressure factor $K$ =0.0706 $m^2$/kN was determined by using experimental data shown in Fig. (**19**). Experimental and simulated results are plotted in Fig. (**20**) using Eq. 9 and are observed in good agreement.

It can be considered that the mechanism of conductivity in CNT samples includes the transitions between spatially separated sites, or particles that can be attributed to the Percolation Theory [77, 78]. According to Percolation Theory the effective conductivity ($\sigma$) of the CNT samples can be calculated as:

$$\sigma = \frac{1}{LZ} \tag{10}$$

where $L$ is a characteristic length, depending on the concentration of the sites, or particles, $Z$ is the resistance of the path with the lowest average resistance. As pressure increases, the CNT layer which is between metallic electrodes, will be squeezed that may cause (i) decrease in $L$ and (ii)

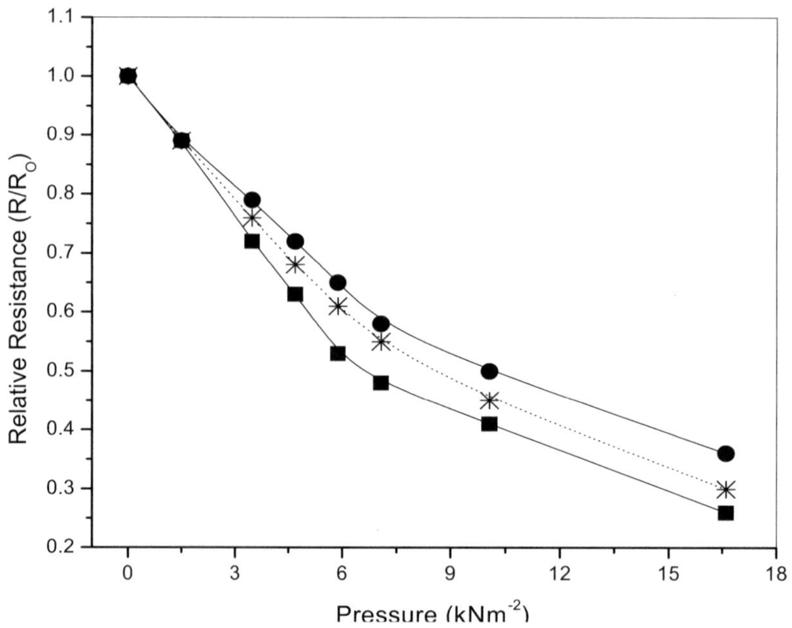


**Figure 21:** Experimental (solid lines for two sensors shown in Fig. (**2**)) and simulated (dotted line) relative resistance-pressure relationships of the Al/CNT/Al sensors ($R_o$ and $R$ are initial resistance and resistance under pressure, respectively).

reduction of $Z$. As observed experimentally from Fig. (**19**), conductivity increases and resistance of the sensor decrease with increase in pressure, accordingly. In ref. [63] the change of the conductivity of CNT under strain was also observed.

Obviously applied pressure (Fig. **18**) will result in decreases of the cross-section of the CNT layer (Eq. 8) that definitely will increase the resistance of the samples. At the same time in the experiments it was observed that the resistance decreases under the effect of pressure. It means that intrinsic properties, *i.e.* conductivity of the samples are changed much more than the geometrical parameters of the samples. Therefore it can be considered that the conductivity is increased, firstly, due to the densification of the samples, secondly, due to the increase of the conductivity of the nanoparticles.

The structure of the investigated sensors formally is the same as presented in [71]. Actually flexible substrates were used for the deposition of the CNT layer in both the cases. Only in [71], flexible substrate was fixed on silicon wafer by gluing. To prevent the effect of humidity on the resistance in our case the CNT layer was encapsulated in plastic casing as CNT based layers are very sensitive to humidity [64-69], *i.e.* the influence of the humidity on the electric properties of the sensors is eliminated.

Experimental results can be linearized by nonlinear op-amps [79]. It can be done if experimental resistance-pressure relationships for the Al/CNT/Al sensors (Fig. **19**) are slightly non-linear. In this work the surface-type Al/CNT/Al pressure sensors were fabricated and investigated. As is shown resistance of the sensors decreased with an increase in pressure. Finally the resistance-pressure relationships of the sensors were simulated. Percolation theory was used for the explanation of the conduction mechanism. It was assumed that the CNT system is a bulk hetero system that results in high sensitivity of the CNT layers at the squeezing effect. Here densification of the nanopowder and the increase of the inter-particles contact areas and intrinsic conductivity of the nanoparticles as well takes place. It results in increase of conductivity and decrease of the resistance of the samples with increase in pressure. It was found that the pressure sensors show good performance. As CNT layer is encapsulated in flexible elastic casing, the effect of humidity on the electric properties of the sensors was eliminated.

## 2.4. Displacement Sensors

Displacement sensors are used wherever it is necessary to acquire the exact position of two points that can be moved relative to one another. There are several types of displacement sensors currently in use. The displacement range of these sensors usually varies from microns to centimeters [80]. Practically displacement sensors are used with different kinds of transducers as resistive, capacitive, eddy-current, linear variable differential transformer, optical instruments and photoelectric sensors [27, 58]. Some types of transducers are contacting as linear variable differential transformer (LVDT), multiple resistor, resistance potentiometers and some are non-contacting, as capacitive, eddy-current, photoelectric *etc* [58].

Using diffused silicon strain gauges based on two kinds of alloys, Mitsuo *et al.* [81] developed a cantilever-type displacement sensor. Ferris *et al.* [82] examined the resistance-displacement characteristic of a magneto resistor moving in a magnetic field. In ref. [83] a new device was designed and a prototype was built and tested, that can simultaneously measure displacements and/or the components of a force in three orthogonal directions. This "3D sensor" consists of four or eight strain gauges attached to the four arms of a single cross-shaped force-measuring cantilever spring.

On the basis of low molecular organic semiconductors tetracyanoquinodimethane (TCNQ) ion-radical salts crystals [61], a number of resistance type linear and angular displacements transducers have been fabricated and investigated. Akhmedov [62] fabricated and investigated high sensitive resistance strain gauges based on poly-N-epoxypropylcarbazole (PEPC) that could be used for the measurements of displacements. Carbon nanotubes (CNT) are observed interesting due to their unique electronic and mechanical properties. It was found that electronically, CNTs can be metallic, semiconducting, or small-gap semiconducting (SGS), depending on the orientation of the lattice with respect to the axis of the tube [63]. Recently on the basis of CNTs, different kind of sensors have been fabricated and investigated: for example, humidity sensors based on single-walled nanotubes (SWNTs) [64-67], double-walled carbon nanotubes (DWNTs) [68], and multi-walled carbon nanotubes (MWNTs) [69] were studied.

CNT based electromechanical devices have interesting properties and could lead to their use as piezoresistors in mechanical sensors such as strain gauges, pressure sensors, and accelerometers [63, 70-74]. In the result of investigation of the piezoresistance of CNTs on deformable thin-film silicon nitride membranes [63] it was found that the gauge factor ($\Delta R/R_{\varepsilon}$) are of up to 400 and 850 for the semiconducting and SGS tubes, respectively. As is known the maximum value of gauge factor in silicon is only 200. Investigations of the small band-gap semiconducting (or quasimetallic) nanotubes showed that they exhibit the piezoresistive gauge factors from 600 to 1000 under axial strains. These effects are much larger than in the metallic nanotubes [70]. The fabrication of SWNT thin-film transistors on plastic substrates was described in [72]. It was observed that its resistance decreases 38.2% and 47.1% with increase in the bending of the elastic substrate for the thin films containing 14 and 16 SWNT layers respectively and was 10 order of magnitude higher than in silicon. In the pristine CNT films the piezoresistive effect was demonstrated at room temperature and its gauge factor under 500 micro strains was 65 [71]. It was found that the gauge factor increased with temperature. A mechanical deformation-conductivity relationship of free-standing membranes of SWCNTs was investigated [73]. In result it was found that the nanotubes gauge factor of piezoresistivity is 2.3 to 2.5 times more than that of the Si substrate. The nano electromechanical piezoresistance transducers based on SWNTs were investigated. Theoretically and experimentally it was shown that ballistically conducting SWNTs show nonlinear piezoresistive gauge factors up to 1500 (applied strain was 1%) [74].

From practical point of view and for deepening of the knowledge about the physical properties of the materials fabrication of the displacement transducers and investigations of squeezing effect on the CNT layer or its composites would be useful. In ref. [84] electric properties of the vanadium complex ($VO_2$ (3-fl)) were investigated. The authors have presented design, fabrication and investigation of the displacement resistance sensors based on composite film of vanadium complex ($VO_2$ (3-fl)) and CNT.

Fig. (**21**) shows the molecular structure of the vanadium complex ($VO_2$ (3-fl)) that was obtained from Aldric. The glass slides that were commercially available were used as substrates which were primarily cleaned and in the first go aluminum electrodes were deposited in a co-planar structure on the glass substrates using Edwards AUTO 306 vacuum evaporation technique. The pressure inside the chamber was maintained at $10^{-5}$ mbar. The thickness of the electrodes was 150 nm. The diameter of multi walled nanotubes (MWNTs) varied between 10-30 *nm*. 2.5 wt.% of CNT and 2.5 wt.% of ($VO_2$ (3-fl)) were mixed in benzol and deposited by drop casting on the pre-patterned Ag surface-type electrodes. In order to let the moisture evaporate from the films the device was kept at room temperature for 10 hrs. The composites' films thickness was in the range of 20-40 μm. The diameter of the samples was 6-8 mm. For SEM measurements the composite films were deposited on glass substrate and SEM micrographs of thin film of the composite were obtained from Hitachi SU-1500 scanning electron microscope. SEM micrographs have been obtained at 4K magnification (Fig. **22**). For the morphological study through SEM, initially thin film of the composite was made conductive by depositing a thin film of Platinum-Palladium (Pt-Pd) (thickness ~20 nm) on it. Hitachi E-1010 ion sputter was used for Pt-Pd thin film deposition. Morphological study showed that the sizes of CNT particles were ~ 100nm.

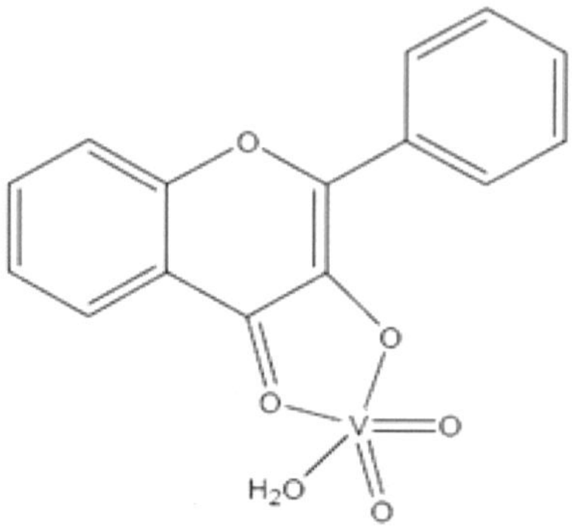


$VO_2$(3-fl)

**Figure 22:** Molecular structure of the vanadium complex $VO_2$ (3-fl).

**Figure 23:** SEM image of the $VO_2$ (3-fl)-CNT film.

A transducer for the investigation of the displacement is shown in Fig. (**23**). The transducer consists of the following elements as metallic support (1), sensor (2), thin plastic film (3) of thickness of 0.01 mm, elastic rubber (4) of height 7 mm and diameter of 6 mm, metallic disc (5) of thickness of 0.5 mm and diameter of 6 mm and micrometer (6). As micrometer moves down in the vertical direction the value of the displacement increases. The range of the displacement directly depends upon the heights of the elastic rubber and its elastic properties (Fig. **23**). By using digital HiTester 3256 meter at DC the resistance measurements were carried out. Dally *et al.* [58] calculated experimental errors for the measurement of displacement and it was found almost ±5 μm and the accuracy of electric resistance measurement was almost ±2%.

Fig. **24** shows resistance *versus* displacement relationship for two Ag/$VO_2$ (3-fl)-CNT/Ag resistive sensors. The resistance of the displacement sensors decreases 30-40% quasi-linear with 0-400 μm increase in displacement. The average resistance-displacement sensitivity of the sensor (*S*) can be calculated using [58] as;

$$S = \Delta R \, 100\% / (R_o \, \Delta D) \quad (11)$$

where $R_o$, $\Delta R$ and $\Delta D$ are initial value of the sensor's resistance, and changes in the resistance and displacement respectively. It was found that the sensitivity of the sensors in average was almost -0.1 %/μm. Using linear function given in [76], the relative resistance can be represented as:

$$R/R_o = kD + b \quad (12)$$

where $R$, $R_o$ and $K$ are resistance, initial resistance and displacement coefficient respectively.

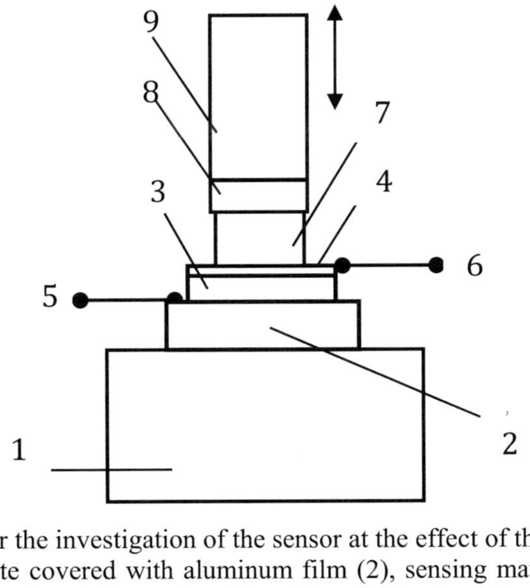


**Figure 24:** Transducer for the investigation of the sensor at the effect of the displacement: metallic support (1), glass substrate covered with aluminum film (2), sensing material (3), aluminum foil (4), terminals of the sensor (5 and 6), elastic rubber (7), metallic disc (8) and micrometer (9).

Experimental (1 and 3) and simulated (2) using Eq. 12, results are plotted in Fig. (**25**) and are observed in good agreement at *k* and *b* in Eq. 12 *i.e.* $-0.925\times10^{-3}$ 1/μm and 1, respectively.

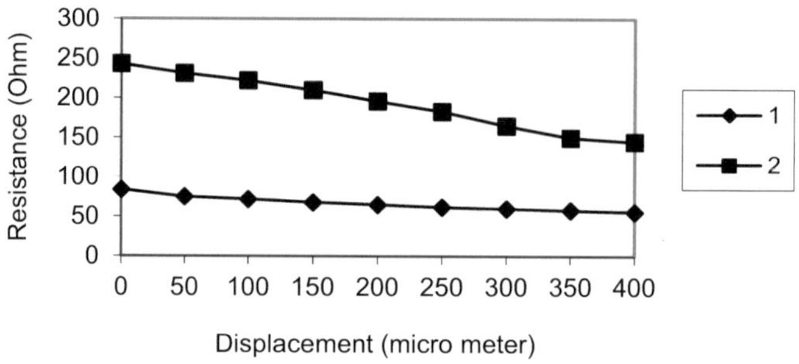


**Figure 25:** Resistance *versus* displacement relationships for two Ag/$VO_2$(3-fl)-CNT/Ag displacement sensors.

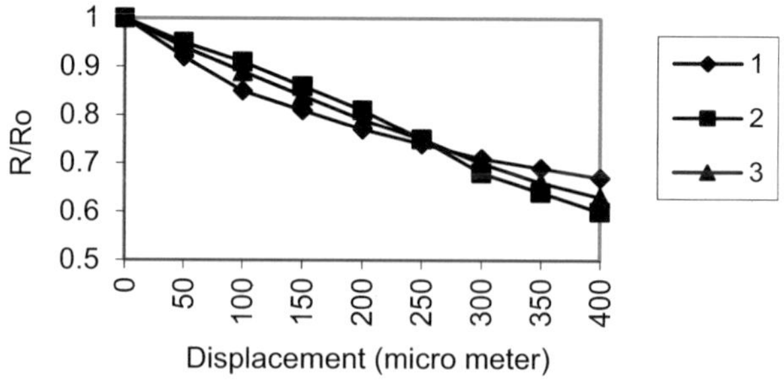


**Figure 26:** Experimental (1 and 2) and simulated (3) relative resistance *versus* displacement relationships for Ag/$VO_2$ (3-fl)-CNT/Ag displacement sensors.

The Percolation theory [77, 78] can be used for the explanation of the mechanism of conductivity in the $VO_2$(3-fl)-CNT composite samples that can be considered as transitions between spatially separated sites. Effective conductivity ($\sigma$) of the samples can be calculated according to Percolation Theory by use of the following equation:

$$\sigma = \frac{1}{LZ} \tag{13}$$

where *L* is a characteristic length, which depends on the concentration of sites, *Z* is the resistance of the path with the lowest average resistance. An increase in displacement may cause reduction of $L$ and $Z$ respectively. It may be due to increase of the concentration and mobility of charge carriers. As is observed experimentally (Fig. **24**) conductivity of the $VO_2$(3-fl)-CNT samples increases and the resistance decreases with increase of the displacement accordingly. By nonlinear op-amps [79] the obtained curves can be easily linearized as experimental resistance-displacement (Figs. **24** and **25**) relationships for the Ag/$VO_2$(3-fl)-CNT/Ag sensors are quasi-linear.

Comparison of properties of $VO_2$ (3-fl)-CNT with $VO_2$(3-fl) [85] and CNT [86] based sensors shows that the composite sensors have lower values of the resistances that are possible to change by changing ratio of $VO_2$(3-fl) and CNT in the composite. These properties of the $VO_2$ (3-fl)-CNT composite based sensors are very important for practical applications. The properties of the sensor studied in this work, make it beneficial potentially to be used in the instrumentation.

Thus the resistive-type displacement sensor was fabricated by drop-casting from solution based on composite of vanadium complex ($VO_2$ (3-fl)) and CNT. It was observed that with increase in the displacement, the resistance of the sensor is decreased and the resistive-displacement relationships of the sensors showed quasi-linear behavior. The mechanism of change of conductivity with change of the displacement in the $VO_2$ (3-fl)-CNT composite was considered as transitions between spatially separated sites that can be attributed to the Percolation Theory: with an increase in displacement, an increase of the concentration and mobility of charge carriers take place. Fabrication of the $VO_2$ (3-fl)-CNT composites allows to decrease the value of the sensors resistance due to higher conductivity of the CNT, that are important for the practical applications of the displacement sensors.

### 2.5. Strain Sensor

Usually different kinds of the electromechanical transducers, are mainly fabricated on the base of piezoresistive effect [87-89]. Actually these transducers are produced in large amount: significantly more than temperature, electrical and magnetic transducers. For the development of this field of technology it is reasonable to fabricate the new piezoresistive materials. Piezoresistive properties of number of organic semiconductors as press-tablets of tetracyanoquinodimethane (TCNQ) ion-radical salts as Li(TCNQ), Na(TCNQ), K(TCNQ) and $Cs_2(TCNQ)_3$ [90] was investigated. From investigations it was found that the resistance-strain relationships were linear and with an increase in pressure that was used during of fabrication of sample, the piezoresistive sensitivity was increased up to 20 to 50 for $Cs_2(TCNQ)_3$.

One of the first pressure transducers fabricated on the base of the TCNQ ion-radical salt strain sensors was described in Ref. [91]. As an active material for the strain sensors, ion-radical salts of TCNQ were used with diquinolinium and phenantrolinium. The piezoresistive sensitivity of these TCNQ salts based sensors was in the range of 150-300. In ref. [92], characteristics of the pressure sensors based on TCNQ salts with dimethyldipiridilium and methyltiuronium were described. It was observed that the sensors were sensitive to strain and showed low sensitivity to temperature changes. A probe for the measurement of the pressure in the stomach, strain sensor was fabricated. In the sensor, as an active material the composite of polymer ion-radical salt of TCNQ and trimethyl tiuronium TCNQ salt was used [93]. During the fabrication of the sensor, the composite was deposited on the polymer membrane. It was found that conductivity of the composite was 5 S/m, the strain sensitivity of the sensor was 10 and temperature resistance coefficient was less than 0.1 % $°C^{-1}$. Long time experiments showed that this composite showed good stability. Diameter of the sensor was 4 *mm* and the maximum sensitivity was 2 mm of water column.

CNTs unique structure resulted impressive properties [63]. Recent investigations showed that the electromechanical properties of the CNT are interesting and could lead to their use as piezoresistors in mechanical transducers *e.g.* as pressure sensors, and accelerometers [70-74]. In ref. [63] the piezoresistance of CNTs on deformable thin-film silicon nitride membranes were investigated. It was found that the sensitivity of the piezoresistance ($\Delta R/R\varepsilon$) is up to 400 and 850 for the semiconducting and SGS (small-gap semiconducting) tubes respectively, whereas the maximum value of sensitivity in silicon is 200. Small band-gap semiconducting (or quasimetallic) nanotubes was investigated and it was found that they exhibit the piezoresistive sensitivity from 600 to 1000 under axial strains that is much larger than in metallic nanotubes [9]. In ref. [71], fabrication of SWNT thin-film transistors on plastic substrates was described and it was shown that the resistance of the SWNT thin film decreases by 38.2% and 47.1% with increase in the bending of the elastic substrate for the thin films containing 14 and 16 SWNTs layers. This result is higher in order of magnitude than in silicon. In ref. [73] the piezoresistive effect was demonstrated in the pristine CNT films: at room temperature, the sensitivity under 500 microstrains was 65. It was observed that the sensitivity increased with temperature. Of free-standing membranes of SWCNTs mechanical deformation-conductivity relationships has been investigated [74]: it was shown that the nanotubes sensitivity of piezoresistivity is 2.3 to 2.5 times larger than that of the Si substrate. The nano electromechanical piezoresistance sensors based on SWNTs were investigated. In the result theoretically and experimentally it was shown that ballistically conducting SWNTs show nonlinear piezoresistive sensitivities of up to 1500 (applied strain was equal to 1%) [72].

As is known [20, 21] copper oxide ($Cu_2O$) is a photosensitive p-type semiconductor with a band gap of 2.0 eV. At the same time $Cu_2O$ is nontoxic and its fabrication process is simple, due to an abundance of copper in nature. In [100, 101] fabrication of nanodots and nanostructure thin films of $Cu_2O$ are reported.

The change in the composition *i.e.* the ratio of the composites components can bring changes in the properties is well known. Fabrication of materials based on composites allows by changing their composition to change such parameters, in particular, specific and sample resistances as well. From practical point of view and for deepening of the knowledge about the physical properties of the CNT and $Cu_2O$ the fabrication of the resistive strain sensors and investigations of their properties would be useful. For example, it would be reasonable to investigate the resistance-strain relationships. As this work may have practical application here we describe the design, fabrication and investigation of the properties of the resistive strain sensors based on composites of commercially produced CNTs and $Cu_2O$.

For the making of press-tablets of the CNT (25 wt.%) and $Cu_2O$ (75 wt.%) composites under pressure of 200 MPa commercially produced (Sun Nanotech Co Ltd., China) CNTs nanopowder and purchased from WINLAB UK $Cu_2O$ micropowder were used. The diameter of multiwalled nanotubes (MWNTs) variations were in the interval of 10 to 30 *nm*. Sizes of $Cu_2O$ micro particles were in the range of 3 to 4 microns. The nominal thickness of the fabricated press-tablets was in the range of 0.5 to 1.5 mm. For experimental purposes the samples were installed using glue on the polymer elastic beam. Interelectrodes distance (length) and width of the surface-type samples was measured in the range of 6 to 8 mm and 5 to 7 mm respectively. Silver paste was used to make the electric contacts to the samples.

A lever system for the creation of uniaxial tension, mechanisms for the creation of pressure with needles and devices with bending elastic beams were used for the investigation of the tensity resistive effect and calibration of the sensors [58, 94, 95]. The use of bending elastic beams as substrates for the deformation of the samples is the most acceptable for the investigation of organic semiconductor thin and fragile crystals or thin films. In ref. [96, 97] the devices with bending elastic beam and with elastic membrane are described respectively.

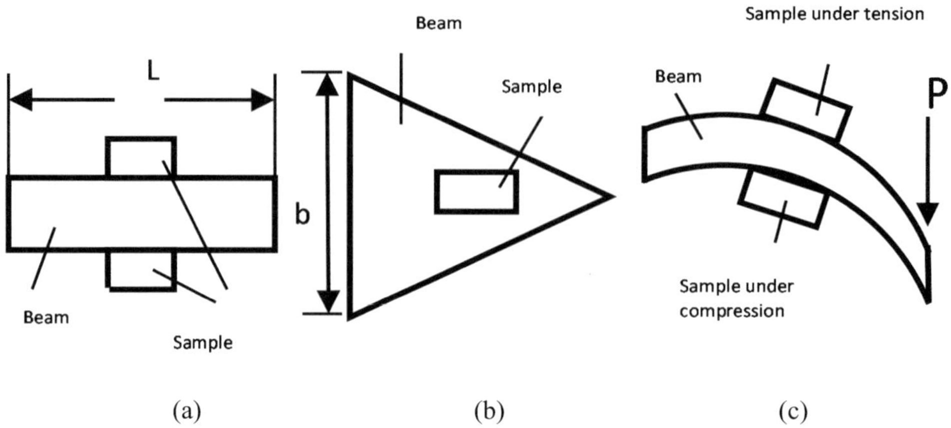


**Figure 27:** Simplified schematic diagram of the elastic beam of constant resistance to bending without load; (a) side view, (b) top view, and under load (c) side view.

Fig. (**26**) shows a simplified schematic diagram of the elastic beam of uniform bending with and without load. Its concrete setup is given in [61]. By using expression given in [61, 94] the longitudinal strain (or deformation) can be determined:

$$\varepsilon = 6\,L\,P/b_o\,h^2\,E \qquad (14)$$

where $L$ is length (150 mm), $b_o$ is width at foot of the beam (30 mm), $h$ is thickness of the beam (2 mm), $E$ is modulus of elasticity, $P$ is load.

Fig. (**27**) shows the resistive strain sensor installed on the elastic beam *i.e.* CNT-$Cu_2O$ press-tablets pasted on the surface of the elastic beam. The beam was fabricated from poly (methyl methacrylate) (PMMA). The average modulus of elasticity of the beam was 2.71 GPa.

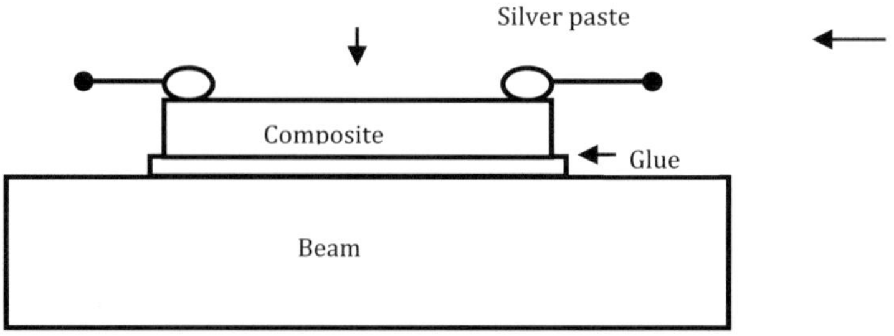


**Figure 28:** CNT based resistive strain sensor installed on the elastic beam.

For one of the samples fabricated under pressure of 300 MPa, Fig. (**28**) shows the resistance-strain relationships under tension and compression. According to Fig.

(**28**), the DC resistance of the strain sensor was increased at tension and decreased at compression. The strain sensitivity (S) was determined as given in ref. [58]:

$$S = \Delta R/R\ \varepsilon \qquad (15)$$

where $\Delta R$ and $R$ are change of the sensor's resistance and initial resistance (at $\varepsilon=0$) respectively. From investigations it was found, that the strain sensitivities were in average in the range of 44 to 46 at tension and 24 to 28 at compression. Initial resistances of the samples were in the range of 11 to 13 Ω for different sensors. The thickness of these samples was 1 mm.

Preliminary investigations showed that for CNT based strain sensors the value of strain sensitivity is dependent on pressure that was used for the fabrication of the samples. The sensitivity was higher for the samples that were fabricated at higher pressures, and lower for the samples that were fabricated at lower pressures. Usually the thinner samples showed higher strain sensitivity. Fig. (**29**) shows simplified structure of the investigated samples (only three particles are shown) at initial state (a), under tension (b) and under compression (c).

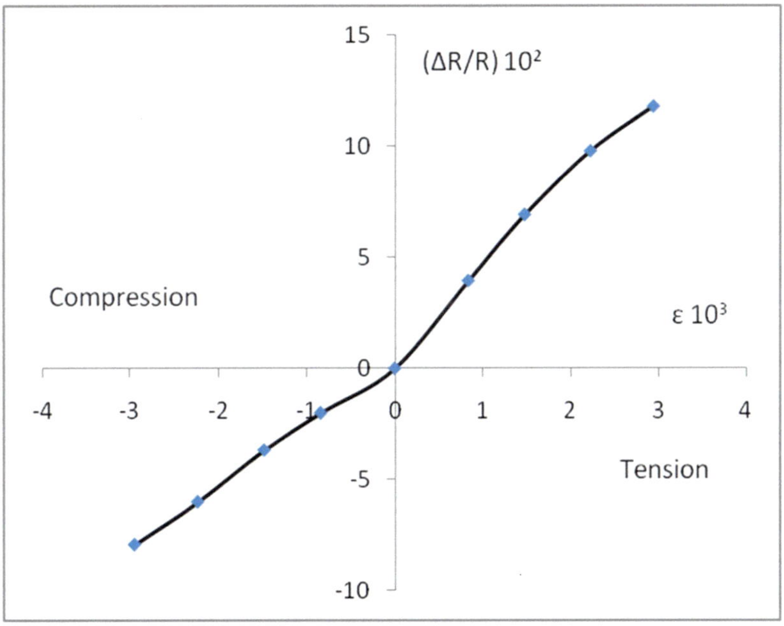


**Figure 29:** Resistance-strain relationships at tension and compression for the CNT-$Cu_2O$ composite based strain sensors.

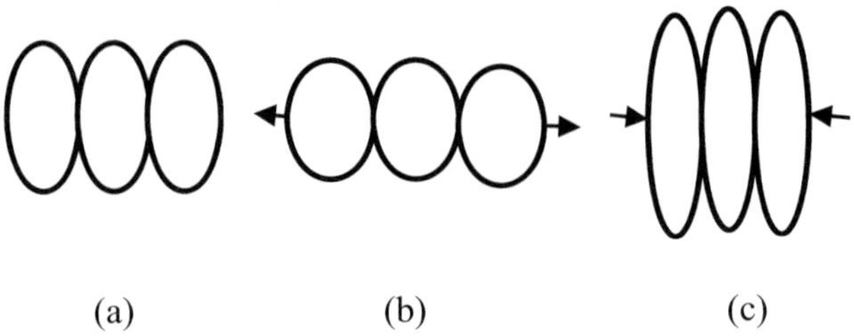


**Figure 30:** Simplified structure of the investigated CNT-$Cu_2O$ composite (only three particles are shown) at: (a) initial state, (b) under tension and (c) compression.

Using Fig. (**29**) we can explain the change of the sample's resistance ($R$), firstly, due to the change in geometrical parameters like inter-electrode distance ($L$) and cross-section area ($A$). Secondly, the change of the resistance may be considered is due to change in the intrinsic parameters as resistivity ($\rho$):

$$R = \rho\, L/A \tag{16}$$

The change of the geometrical parameters only, as in metal, provides, as a rule, strain sensitivity in average around of 2 [58], but the change of the resistivity, for example in Si strain sensors, results to the sensitivity around of 200 [58, 61]. Preliminary investigations of resistance-temperature relationships of the CNT done shows that the resistance increases with temperature. It means the CNT in this case has semiconductive behavior. The $Cu_2O$ on the other hand is semiconductor with gap in energy band diagram. Therefore it can be assumed that as a result of tension the charge concentration decreases due to the increase of the energy gap. At the same time mobility decreases due to increase of potential barriers and, probably the hopping length.

For practical utilization of these strain sensors the temperature compensation circuits should be used as the resistance of CNT-$Cu_2O$ based strain sensors can be changed due to change in temperature as well. Usually for these purposes the Wheatstone bridge can be used for the utilization of strain sensors [58]. As an example in Fig. (**30**), Wheatstone bridge circuit is shown that can realize four kinds of connection [58]. For example, in case-1 $R_1$ is active resistance strain sensor, $R_2$, $R_3$ and $R_4$ are ordinary resistors. This circuit usually is used if temperature compensation is not required. In case-2 and case-3, $R_1$ is active resistance strain sensor, $R_2$ or $R_3$ is dummy, $R_3$ or $R_2$ and $R_4$ are ordinary resistors. The active and dummy sensors should be pasted by the same adhesive material on the beams and be placed in the same thermal environment. At the same time active sensor be under strain and dummy sensor in stress-free region.

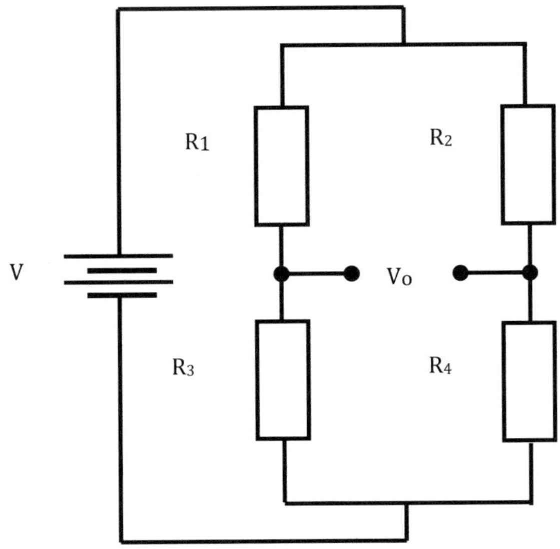


**Figure 31:** Schematic diagram of the strain sensors arrangement in the Wheatstone bridge.

In case-4, $R_1$ and $R_4$ should be under tension, $R_2$ and $R_3$ under compression, *i.e.* the circuit contains only active strain sensors. In this case the circuit has double sensitivity with respect to circuits 1, 2 and 3, and provides the temperature compensation as well [58]. $V_o$ is output voltage. Detail analysis of these four cases is given in Ref. [58].

Investigation of the CNT (25 wt.%)-$Cu_2O$ (75 wt.%) composite based strain sensors showed that these sensors have sufficient sensitivity for practical application. By decrease in the CNT amount in the composite, and accordingly increase in the amount of $Cu_2O$, for example, up to the range of 5-15 wt.% of CNT and 85-95 wt.% of $Cu_2O$ at the same time the initial resistances of the sensors can be increased. This work is very important and will be done by the authors in future. It will allow decreasing the influence of metallic wires resistances and hence increasing the sensitivity of the strain sensors [58].

Thus it was fabricated by making the press-tablets of CNT under pressure of 200 MPa and investigated the carbon nanotubes (CNT) and $Cu_2O$ composite based strain sensors. It was observed that the DC resistance of the strain sensors was increased under tension and decreased under compression. From experimental results it was found that resistance-strain relationships show quasi-linear behavior. Physical mechanism involves change of resistance under the effect of strain. The circuits based on Wheatstone bridge for practical utilization of the strain sensors and compensation of the effect of temperature is presented.

# 3. ORGANIC FIELD-EFFECT TRANSISTORS (OFET)

## 3.1. Thin-Film Field Effect Transistors

The mobility of organic semiconductors usually is observed much lower than of inorganic semiconductors like to Si and Ge. Therefore organic field-effect transistors (OFET) can be used only in such applications where high cutoff frequency is not required [102]. At the same time OFETs potentially are cheaper and can be fabricated at relatively low temperatures on large-area flexible substrates, for example, "electronic paper" displays [102, 103]. Pentacene crystals showed one of the highest (2.7 $cm^2$ $V^{-1}s^{-1}$) mobility among of organic semiconductors that can be used for OLEDs. The value of mobility at about of 10 $cm^2V^{-1}s^{-1}$ is considered as fundamental limits, that can be improved, firstly, by increase, by proper selection of organic semiconductors, of intermolecular forces and interactions, and secondly, by decreasing the gap between drain and source of OFET up to the size of single macromolecules where instead of intermolecular conduction the intramolecular conduction mechanism will be dominating [102]. It is known that in the conjugated conducting and semiconducting polymers the mobility of charge carriers can be equal to 1000 $cm^2$ $V^{-1}s^{-1}$.

*I-V* characteristics of the OFETs can be the same models that were developed for inorganic semiconductor field-effect transistors [102, 104]. Fig. (**31**) shows schematic diagram of two commonly used OLEDs configurations.

From Fig. (**31**) it is seen that the presented OFETs configurations are similar to the Si based MOSFET (metal-oxide semiconductor field-effect transistors) [56]. Detailed information about modeling of OFETs and their characteristics is presented in [102]. Experimentally it was found that performance of the top-contact OFETs is better than the bottom-contact devices [102].

Organic layers in the OFETs can be deposited by vacuum evaporation and from solution. Solution deposition process of organic layers is cheaper than the deposited at vacuum condition. But usually mobility of charge carriers in the solution deposited layers is lower than the vacuum deposited films [102]. At present OFETs are devices that are potential candidates to be implemented in practice. OFETs can be used not only for the amplification of the voltage, but as a different kind of sensors as well, such as humidity [105], light [106, 107], pressure [108], displacement [109] and temperature [110] sensors. All these OFETs have the similar configuration, like to MESFET (metal-semiconductor FET) structure [56], *i.e.* Schottky junction OFET. In this chapter details of the OFETs based temperature sensor is presented [110].

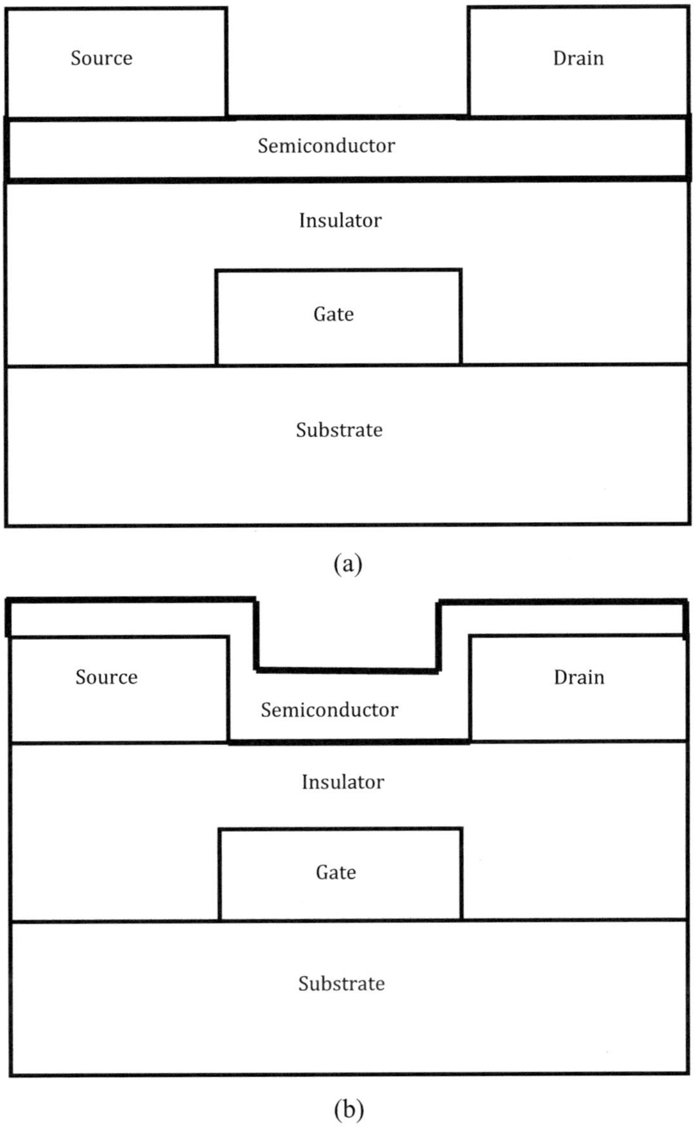


**Figure 32:** OFETs configurations: (a) Top-contact device with source and drain electrodes deposited onto semiconductor film by vacuum evaporation, (b) Bottom-contact device with the semiconductor film deposited onto prefabricated source and drain electrodes.

## 3.2. Temperature Sensitive Field Effect Transistor

During the past decade, field effect transistors based on conjugated polymers, oligomers and low molecular weight organic semiconductors have been investigated widely. The attracting extensive interest for potential applications of organic field-effect transistors (OFETs) as organic devices is due to their less material and fabrication cost with respect to the inorganic semiconductors based devices that are important in practice [102, 111-114]. Based on a biphenyl end-capped fused bithiophene oligomer, highly photosensitive organic phototransistor (OPT) was fabricated [115]. Similar to the absorption spectrum of the organic semiconductor the device showed photocurrent response under 380 nm UV light. It is considered that this phototransistor potentially can be used for the fabrication of UV sensors. The effect of UV light irradiation on the characteristics of OPTs containing sexithiophene (6-T) and pentacene was examined similar to the previous case in ref. [116].

From photoconductive and photovoltaic effects it was found that these devices showed fast and slow responses. One of the potential applications of these devices is in light-addressable field-effect transistor memory devices. Organic semiconductors pentacene, thiophene oligomers and regioregular polythiophene were investigated and found that these have potential application due to their good performance for use especially in OFET [102]. However, further improvements are still required for their properties and characteristics. Organic copper phthalocyanine/inorganic ferroelectric $PbZr_{0.2}Ti_{0.8}O_3$ heterojunction gate and $La_{0.87}Ba_{0.13}MnO_3$ ferromagnetic semiconductor oxide channel devices were investigated and presence of non-volatile and non-destructive photomemory operation was demonstrated [117]. The light information could be written by the device at the effects of light irradiation and the negative gate bias. The information could be deleted with the positive gate bias.

The FET with a floating gate was fabricated for the detection of gas species [118]. Nanoscale organic and polymeric FETs as chemical sensors were fabricated and investigated showing high sensitivity [119]. In ref. [120], gas sensors based on conducting polymers were also fabricated.

Organic semiconductors, phthalocyanines and especially CuPc are well-studied organic photosensitive semiconductors [36]. High absorption coefficient in wide spectrum and high photo-electromagnetic sensitivity at low intensities of radiation was observed. The deposition of thin CuPc films by vacuum sublimation is easy and this technology is used widely. As the sublimation of CuPc occurs at relatively low temperatures (400-600 °C), its purification is simple and economical. In ref. [51, 52], fabrication of heterojunction based organic-on-inorganic Ag/*p*-CuPc/*n*-GaAs/Ag photoelectric sensors was fabricated. Properties of Ag/*p*-CuPc/*n*-GaAs/Ag sensor were investigated under filament lamp illumination at room and elevated temperatures in photovoltaic and photoconduction mode of operation. Through investigations of photocurrent and photo-voltage spectra, cell was found sensitive in the large spectral range of wavelengths from UV to visible and NIR spectrum (200-1000 nm).

Electric field of molecules can be detected by OFETs and recently fabricated OFET was able to detect charged/uncharged chemical species in aqueous media through field effect. In particular, chemical sensitivity of the transistor was illustrated for protons and glucose [121]. To improve stability and Nernstian behavior in ion-sensitive field effect transistor (ISFET), gate oxide was covered by $Si_3N_4$ or $Al_2O_3$ [122]. It was found that phthalocyanine based Langmuir-Blodgett films conductivity is highly sensitive to the concentration of NO and $NO_2$ gases [123]. It was observed that copper phthalocyanine films were sensitive to concentration of chlorine ($Cl_2$) [124]. High sensitivity to $NO_2$ gas was observed using CuPc based FET with suspended gate [125].

Effect of humidity on the electric properties of gateless FET was presented in ref. [126]. When a pentacene based FET was exposed to the humidity due to which decrease in its saturation current and FET mobility was reported [127, 128]. It was observed that the sensitivity of this FET decreases when the thickness of the pentacene increases and humidity is more than 30 %. It was found memory effect at humidity less than 30% and thicker film.

Along with FETs, humidity sensing resistors based on phthalocyanines [127] was fabricated and investigated as well. At the same time it is well known that unlike to resistors the transistors have several advantages over resistors [127]. For example, due to the presence of the gate, very thin films (20-50 nm) may be used, and the value and polarity of gate potential can control absorption process of species on sensor's surface.

Semiconductor thermistors as temperature sensors, fabricated from metal oxides are used in very wide scale in practice [58]. Therefore investigation of the effect of temperature on phthalocyanine based FET would be reasonable, first of all, for the investigation of potentialities to use transistor as temperature sensor. Secondly is to investigate the interval of temperature where the transistors' parameters will be stable for practical application of the devices. Temperature sensitive OFET (TSFET) is reported with CuPc and Schottky junction (Al-CuPc) [110] that have the structure of MESFET. For the fabrication of TSFET, CuPc was obtained from Sigma Aldrich. The molecular structure of CuPc is presented in [34, 35]. The investigations showed that there are at least, seven crystalline polymorph states of CuPc exist: $\alpha$, $\beta$, $\gamma$ *etc.* [53]: some of form ($\alpha$-CuPc) is a metastable at T = 165 °C and can be converted thermally or with solution to the other $\beta$-form. The $\alpha$ and $\beta$-forms are concluded the most frequently observed states of CuPc. Investigation of the deposited CuPc films by X-ray diffraction showed that CuPc is in $\beta$-form [129]. In the CuPc films, by investigations with SEM, any particular crystal orientations [36] were not found. The band gap of CuPc is in the range of (1.6-2.0) eV and conductivity is equal to (1.2 $10^{-8}$ - $5\times10^{-13}$ ) $\Omega^{-1}$ $cm^{-1}$ at T = 300 K [34-36]. One of the most important parameters at thin film deposition of CuPc is the sublimation temperature that varies from 400 °C at a pressure of $10^{-4}$ Pa to 580 °C at $10^{-4}$ Pa [36].

Thin films of CuPc were thermally sublimed at 400 - 450 °C and ~$10^{-4}$ Pa onto the glass substrate (with pre deposited 50 nm thick silver electrodes) in Edwards AUTO 306 vacuum coater with diffusion pumping system. The CuPc film deposition rate was 0.2 nm/s. The substrate's temperature in this process was almost 40°C. By Edwards FTM5 film thickness monitor the thickness of the CuPc film was obtained. After deposition of CuPc film the semi-transparent 20 nm Al film (transparency was 10 -15%) was deposited. From literature it is known that CuPc forms ohmic contacts with Ag and Schottky type rectifying contacts with Al [51]. Usually the CuPc films show p-type conductivity [36, 52, 121] due to doping of the films by oxygen.

Fig. (**32**) shows cross-sectional view of the fabricated OFET where source and drain terminals are connected with Ag films and gate terminal with Al film. The geometrical parameters of the OFET as length, width and thickness of the CuPc channel were 100 μm, 20 mm and 40 nm respectively. The measurements of voltage, current and temperature were conducted with standard digital meters that were available in the laboratory. The sample was placed in special laboratory chamber where the temperature could be changed in the range of 25 °C-125 °C.

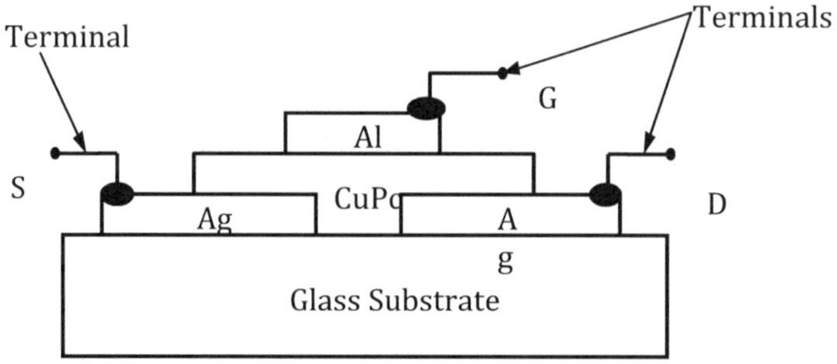


**Figure 33:** Cross-sectional view of the fabricated OFET: S, G and D are source, gate and drain respectively.

At different temperature 25°C, 60 °C, 90 °C and 125 °C, Fig. (**33**) shows the output (drain current-drain voltage) characteristics of the CuPc based OFET. In these experiments the gate of the OFET was suspended. The output characteristics of the OFET are shown in Fig. (**33**): the characteristics show quasi-linear behavior at low temperatures and slightly super linear deviation at elevated temperatures and these characteristics donot cover saturation region. It is noted that at voltage of 5 V, current approximately increases to 1.5 times with increase in temperature.

The output (drain current-drain voltage) characteristics of the CuPc based OFET at different temperature 25oC, 60 °C, 90 °C and 125 °C show non-linear behavior when Gate and Drain of the OFET was grounded are shown in Fig. (**34**). At particular voltage, for example 5 V, the current increased approximately 7.5 times with increase in temperature. It was calculated that the ratio of the resistances at T=125 °C measured at 1 V and 5 V is approximately 10. This behavior shows "switching effect" from low conductance to high conductance state that was observed in a number of materials. For example, the "switching effect" was found in some inorganic semiconductor films [130], organic films of poly-N-epoxypropylcarbazole (PEPC) [131]. At Zener diodes this behavior is well-known and was observed at reverse bias [56]. Unlike to this in the case of OFET it was observed at the forward bias.

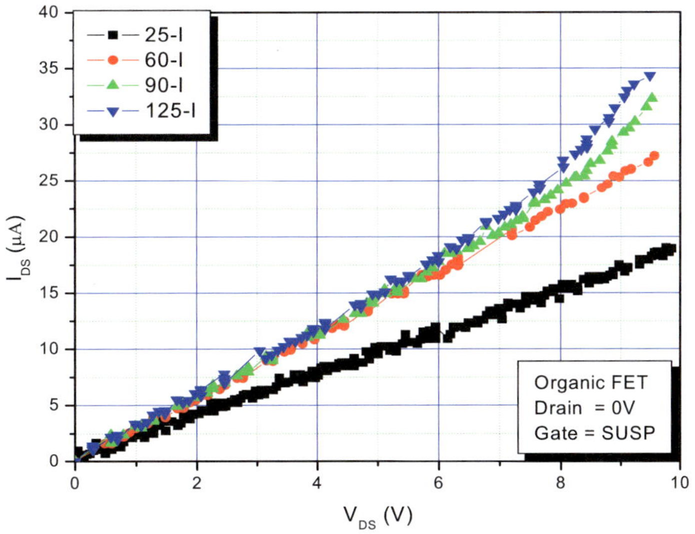


**Figure 34:** The output (drain current-drain voltage) characteristics of the CuPc based OFET at different temperature 25 °C, 60 °C, 90 °C and 125 °C when Gate of the OFET was suspended.

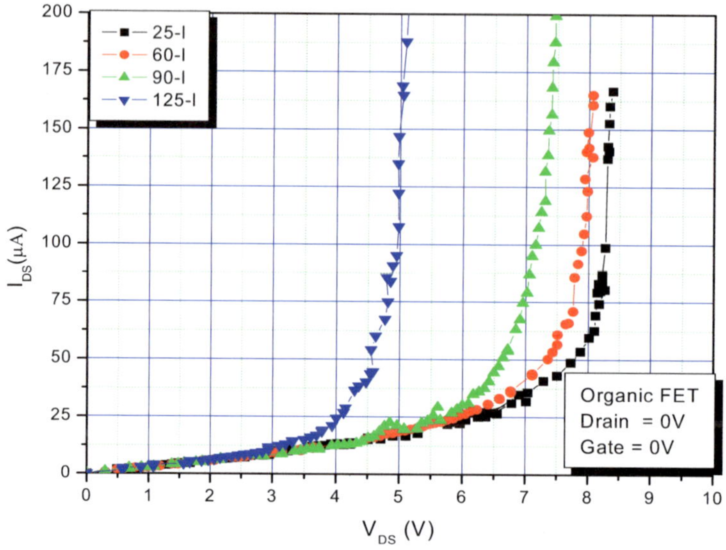


**Figure 35:** The drain current-drain voltage characteristics of the CuPc based OFET at different temperature 25°C, 60 °C, 90 °C and 125 °C when Gate and Drain of the OFET was grounded.

Fig. (**35**) shows threshold voltage ($V_{Th}$) -temperature relationship *i.e.* with increase in temperature $V_{Th}$ decreases that may be due to the fact that with decrease with the temperature of the effective width of the depletion region at Al-CuPc Shottky junction, that have the character of the avalanche process, of intensive generation of charge carriers due to the collisions of the charge carriers with CuPc molecules.

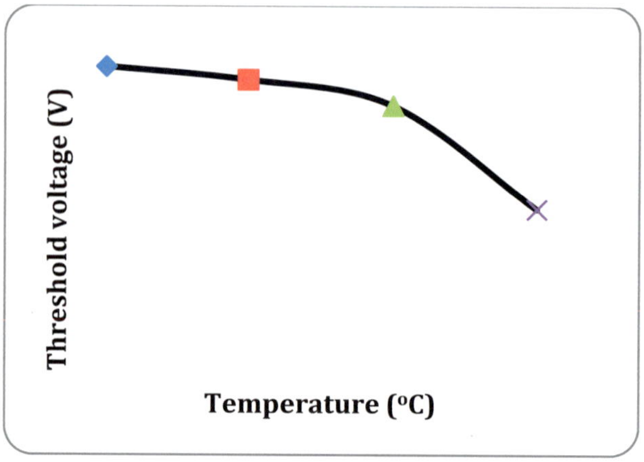


**Figure 36:** The threshold voltage-temperature relationship obtained using data presented in Fig. (**5**).

It would be reasonable to estimate temperature sensitivity of the drain-source resistance ($R=V_D/I_D$) for some practical applications of the OFET as temperature sensitive FET (TSFET). This relationships obtained using data presented in Fig. (**34**) is shown in Fig. (**36**) at constant voltage ($V=5\ V$) and constant current ($I=150\ \mu A$). It is seen that in both the cases resistance decreases with temperature approximately 9.4 and 1.7 times respectively.

Temperature sensitivity ($S$) of the TSFET in average can be evaluated as per ref. [58]:

$$S = \Delta R/R\ \Delta T \qquad (17)$$

where $\Delta R$ and $\Delta T$ are resistance and temperature changes. At constant voltage measurement it was found that $S$ = 1 %/K. Made on the base of metal oxides, conventional semiconductor thermistors have sensitivity in the interval of (3-4) %/K [32], but, firstly, fabrication of these devices is energy consuming as these sensors are processed at temperatures of around of 1000 °C. Moreover as TSFET is multifunctional device it can be used potentially for the measurement not only of temperature but humidity and light as well [106, 132]. Fig. (**34**) shows that TSFET can also be used for the stabilization of the voltage as Zener diodes and the value of threshold voltage can be controlled by temperature.

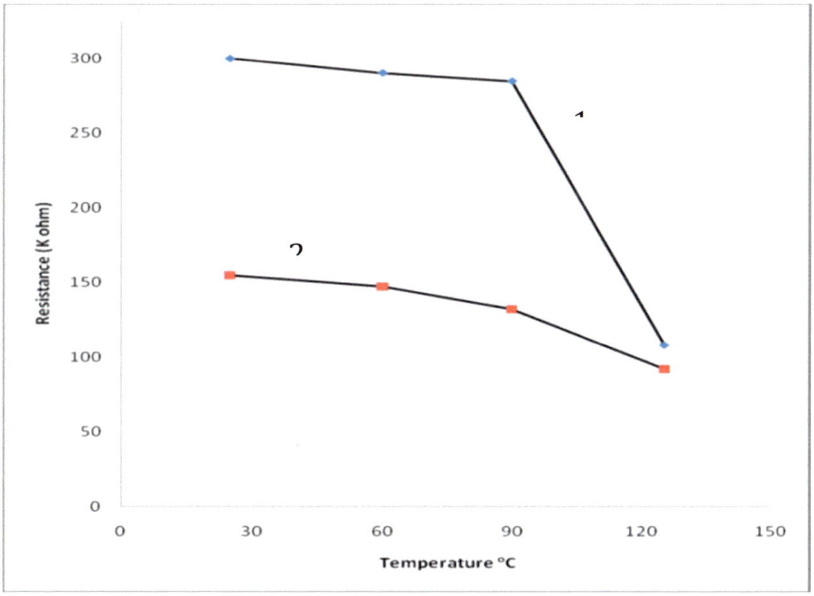


**Figure 37:** Resistance ($R=V_{DS}/I_D$) – temperature relationships obtained by use of data presented in Fig.34 at; (1) constant voltage (V=4.5 V), (2) constant current (I=μA).

Fig. (**37a** and **b**), shows suspended Gate (current pass only through channel) and grounded Gate and Drain conditions (currents $I_1$ and $I_2$ pass through forward biased junction and channel respectively) of the OFET.

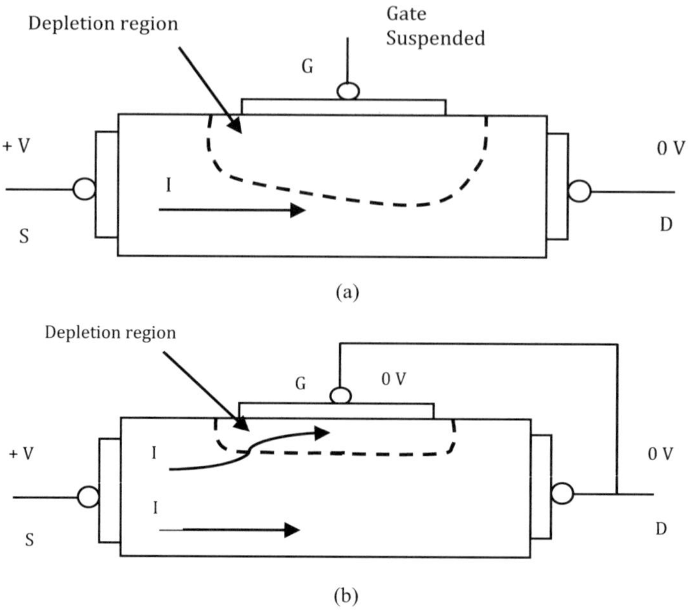


**Figure 38:** Schematic diagrams of the OFET at (a) suspended Gate (b) grounded Gate and Drain.

Using following approaches the simplified conduction mechanism in the organic semiconductors can be explained. In disordered system in multi crystalline thin films, mostly hopping mechanism of conduction was observed [34, 35] due to the phonon assisted hopping of carriers from one localized state to another. Usually mobility increases with temperature, but it depends on actual contribution from scattering phenomenon. Experimentally it was found that the mobility may increases, decreases or be independent of temperature in different organic semiconductors [34, 35]. A value of mobility around 1 $cm^2 V^{-1}s^{-1}$ is considered as boundary between band transport and hopping mechanisms.

Ioffe and Regel in ref. [130] has shown that at mobility less than 1 $cm^2 V^{-1}s^{-1}$, the mean free path of carriers calculated from band transport approach is less than the inter atomic distances in semiconductors that would be incorrect from the physics point of view.

The most universal phenomenon is tunneling effect, but it may contribute mostly at very low temperatures where band and hopping mechanism have less contribution. It is believed that in tunneling mechanism the values of mobility are very low ($\mu << 1$ cm$^2$ V$^{-1}$s$^{-1}$) and it shows temperature independent behavior.

The conductivity mechanism in CuPc can be considered as thermally activated hopping transitions between spatially separated sites of molecules [34, 35]. As temperature increases, the concentration and mobility of the charges in the CuPc film also increase. It is observed that experimentally conductivity increases and the resistance of the sample decreases.

Structure and properties of organic semiconductors and in particular, thin films are depend not only on the molecular structure or concentration of impurities but are also dependent on the technology of deposition [34, 35, 132]. Therefore further improvement of properties of OFET would be expected in this area in future *i.e.* simulation and parameters optimization of the OFET.

> *The electrical characteristics of CuPc based temperature sensitive organic FET were investigated in the temperature interval of 25 °C – 125 °C and it was shown that the drain current of this organic field effect transistor (OFET) is increased with temperature. Presence of switching effect from low to high conduction states in this device was observed. With increase in temperature the threshold voltage of the switching effect decreased is observed. A average ntemperature sensitivity of the FET was -1 %/K was also observed. The OFET can be used as multifunctional sensor as it is sensitive to the temperature, light and humidity.*

## 4. SOLAR CELLS

### 4.1. Organic Solar Cells

#### *4.1.1. Schottky Junction and p-n Heterojunction Solar Cells*

Research works in the area of organic solar cell (OSC) started in the 1950's that finally bring efficiencies of about 1% in the 1970s [134]. Majority of the solar cells was fabricated on the base of Schottky junction or organic-on-organic p-n heterojunction [135, 136]. For practical application it was considered that efficiency of OSCs should be at least around of 5%. Calculation made by G.A.Chamberlain [134] says that efficiency of about of 10% can be realized in Schottky barrier solar cells. Later organic-on-inorganic structures (Ag/n-GaAs/p-CuPc/Ag) was fabricated, and an efficiency of around of 4% was observed [38] (Fig. **38**).

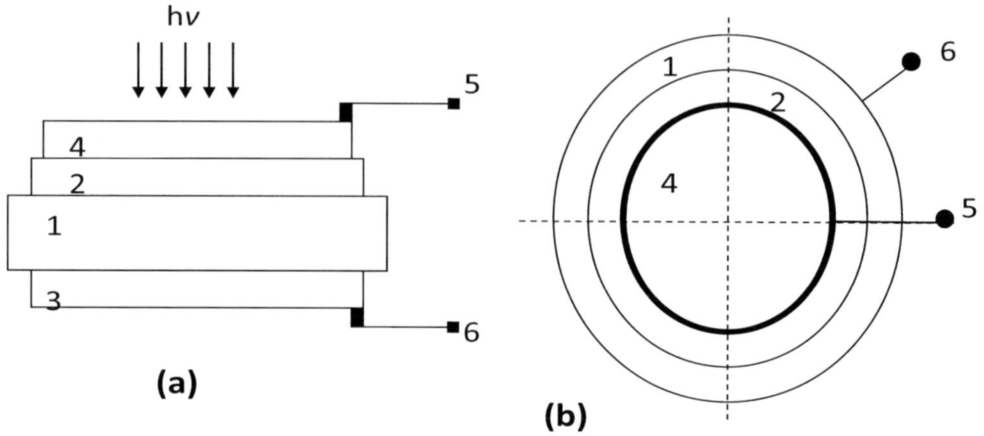


**Figure 39:** Schematic diagram of organic–on-inorganic heterojunction solar cell fabricated by using *n*-type GaAs substrate and a thin *p*-CuPc film: 1-GaAs, 2-CuPc, 3-Ag, 4-Ag (semi-transparent), 5 and 6- terminals: side view (a), top view (b).

### *4.1.2. Bulk Heterojunction Solar Cells*

In 1990s researches started the investigations of bulk heterojunction solar cells [77, 136-141]. Peculiarity of these OSC was in the presence of the blend of two materials donors and acceptors, unlike to two layer devices, and presence of electrodes with different work function, high work function metal and low work functional metal, for creation of built-in electric field in the bulk of the cell. The metals-blend contact is considered as ohmic. Fig. **39** shows the simplified schematic diagram of the bulk heterojunction solar cell [77]. Table **1** shows photovoltaic parameters of number of bulk heterojunction solar cells [77]: open-circuit voltage ($V_{oc}$), short-circuit current ($I_{sc}$), fill-factor (*FF*).

One of the most important points was that in these solar cells organic low molecular and high molecular semiconductors could be used. It was found that the efficiency of the solar cell was up to 3% [77] and even 5% [133]. It was probably due to the photoinduced charge transfer from donors into acceptors that takes place at the absorption of light by the blend and transportation of electrons through acceptors and holes through donors. As in bulk hetrojunction solar cell it is considered the presence of uniform electric field that accelerate the charges, the separate transfer of electrons and holes may results to relative decrease of the electron-hole recombination rate, that finally increases the efficiency. On the other hand it is obvious that recombination and trapping mechanism definitely effect to the performance of the cells [77].

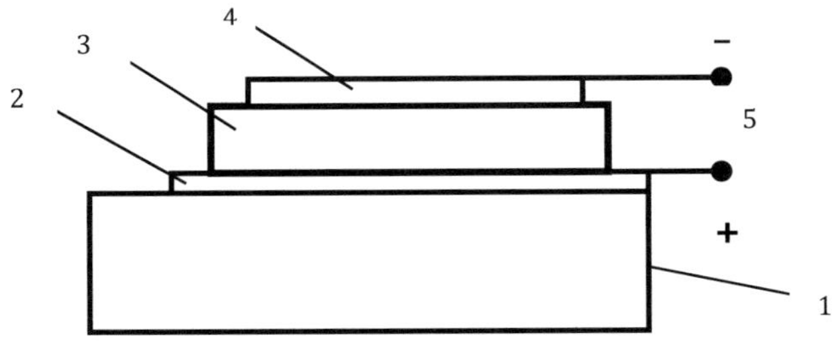


**Figure 40:** Schematic diagram of the bulk heterojunction solar cells: glass substrate (1), ITO/PEDOT (PANI) (2), blend (p-type/n-type: MDMO-PPV/PCBM) (3), Ca, Ba, Mg, Al or LiF-Al (4), terminals (5).

**Table 1. Photovoltaic parameters of bulk heterojunction solar cells at AM 1.5 conditions.**

| **Active layer** | **$V_{oc}$(v)** | **$I_{sc}$ (mA cm$^{-2}$)** | **FF** | **η (%)** |
|---|---|---|---|---|
| PTPTB/PCBM 1/3 | 0.72 | 2.95 | 0.37 | 1.00 |
| PTPTB/PCBM + 10% Nile red 1/3 | 0.53 | 2.25 | 0.32 | 0.48 |
| PTPTB/PPV/PCBM 0.5/0.5/4 | 0.51 | 1.00 | 0.30 | 0.20 |
| PPV/PCBM ¼ | 0.81 | 4.90 | 0.60 | 3.00 |

PPV is poly(para-phenylenevinylene), PTPTB is p-type low band gap polymer, consisting of N-dodecul-2,5-bis(2'-thienyl) pyrrole and electron-deficient 2,1,3-benzothiadiazole units, PCBM is {6,6}-phenyl $C_{61}$-butyric acid methyl ester [77]. Unlike to inorganic solar cells, bulk heterojunction solar cells show practically very interesting efficiency-temperature relationship (Fig. **40**). As seen from Fig. (**3**) as temperature increases up to around of 50 °C, the efficiency also increases.

For the improvement of the properties of bulk heterojunction solar cells proper electrodes that form ohmic contacts with blend, and, donor and acceptor pairs should be selected metallic [77]. It is very important if mobility of the charges and absorption of light by the blend components will be higher as possible [77].

### *4.1.3. Dye-Sensitized Solar Cells*

At present dye-sensitized solar cells (DSSC) are very popular among researchers in different countries due to the potential high efficiency and cheapness. Though the photogalvanic effect was discovered in 1839 in liquid state device by Edmond Becquerel, the fundamental contribution in the development of photoelectrochemical cell' technology was done by Michael Gratzel (1991) [142-145]. Fig. (**41**) shows simplified schematic diagram of the DSSC [77].

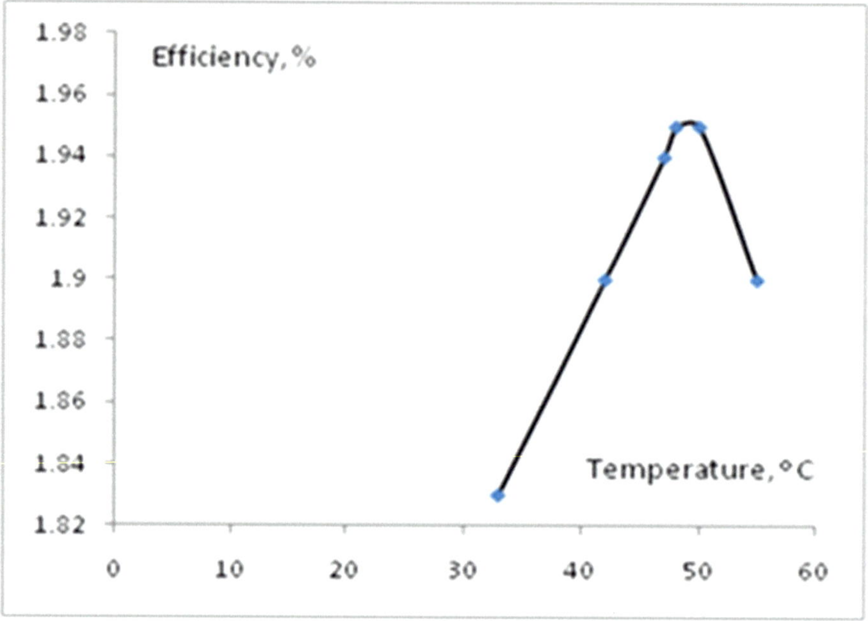


**Figure 41:** Efficiency-temperature relationship for bulk heterojunction solar cell.

As TCO the $SnO_2$:F is used, TCO coated by $TiO_2$ and dye is photo-electrode. Open-circuit voltage of the DSSC is around of 0.7 V (in Si solar cells $V_{oc}$=0.6 V) and short-circuit current is equal to 20 mA/cm$^2$ (in Si solar cells $I_{sc}$=35 mA/cm$^2$). In relatively large (1-5 cm$^2$) and small (0.25 cm$^2$) area cells the efficiencies were equal to 5-8% and 11% respectively [77] (Modules efficiencies were about of 7 % [Nat]). It was found [143] that the presence of the mesoporous $TiO_2$ with nanometer particles of size of 10-20 nm (Fig.42) plays crucial role in increasing of the incident-photon-to-current conversion efficiency to around of 600 times [143]. Definitely due to this nanocrystalline ($TiO_2$) effect the efficiency of the DSSCs were increased. More over the efficiency DSSC depends on the selection of the dye and electrolyte. Thickness of $TiO_2$ layer was in the range of 5-15 μm. The absorption of light by monolayer dye deposited on $TiO_2$ as observed was high due to the nanocrystalline structure of the $TiO_2$. As a dye, N3 ruthenium complex can be used [144]. The electrolyte contained a redox couple, for example, iodide/triiodide ($I^-/I^-_3$) in a non-viscous organic solvent [77].

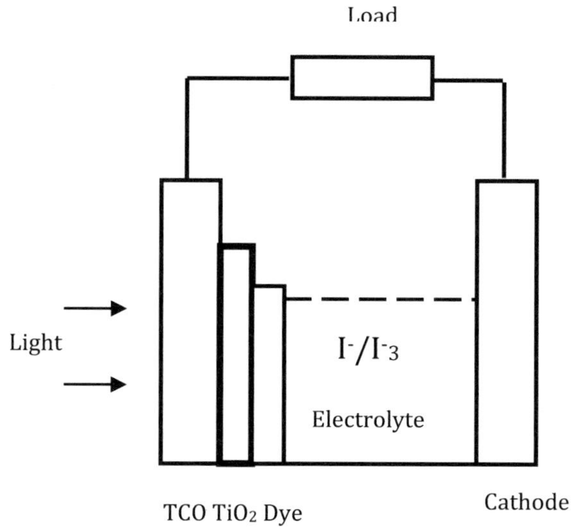


**Figure 42:** Schematic diagram of the DSSC: transparent conductive oxide (TCO), mesoporous TiO2, Dye (sensitizer), electrolyte (non-viscous organic solvent), $I^-/I^-_3$ (a redox couple: iodide/triiodide), cathode (counter-electrode: TCO coated by platinum).

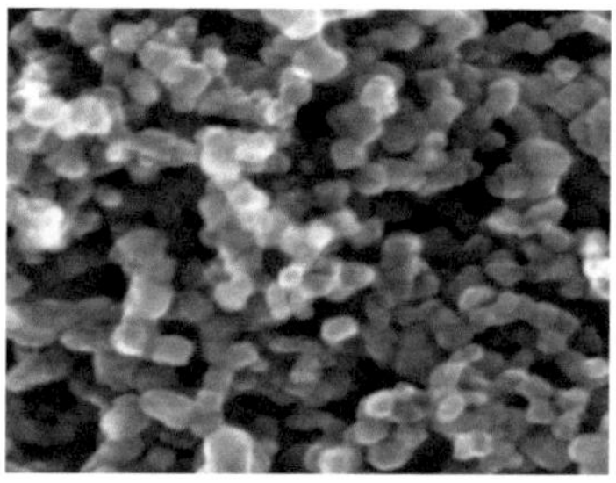

**Figure 43:** Scanning electron microscope picture of a nanocrystalline $TiO_2$ (anatase) film used in the DSSC [144].

Operation principles of the DSSC are described in Ref. [77] *i.e.* light penetrate through TCO and $TiO_2$ layer and is absorbed by photo sensitive layer of the dye (Fig. **41**). In the result the molecule of the dye (*S*) is excited ($S^*$) and electron (*e*) is injected into the conduction band (*CB*) of the $TiO_2$ [77]:

$$S^* \longrightarrow S^+ + e_{CB} \qquad (18)$$

where $S^+$ is oxidized sensitizer molecule. Through $TiO_2$ and TCO the electrons can reach external load and finally counter electrode. The oxidized dye is reduced in electrolyte (by $I^-$) [77]:

$$2S^{+} + 3I^{-} \longrightarrow 2S + I^{-}_{3} \quad (19)$$

The triiodide ($I^{-}_{3}$) diffuses to the counter electrode and is reduced by electron from external circuit at presence of platinum:

$$I^{-}_{3} + 2e^{-} \longrightarrow 3I^{-} \quad (20)$$

Generated by light, the open-circuit voltage of the DSSC depends on the differences of thc Fermi energy level in $TiO_2$ and of the electrochemical potential of the electrolyte [77]. The recombination processes in the dye and $TiO_2$ decreases the efficiency of the DSSCs. At the manufacturing of the DSSC the $TiO_2$ and platinum films are deposited by screen-printing and sintered at 400-500 °C [144].

The performance of DSSCs was investigated during one year [77]. It was found that for two test cells with active area of 4 $cm^2$ the efficiency was actually constant: the efficiency of the cells was measured under irradiation by indoor solar simulator, but cells were exposed mostly at outdoor conditions.

Further development of the DSSC will be focused to solve some problems that were observed at fabrication and investigation of the cells as deposition of large area uniform $TiO_2$ layer, internal interconnection of the cells, and especially the sealing of the cells and modules [77]. Table **2** shows comparison of the parameters of an amorphous silicon module and five-cell DSSCs module [77]. It is seen from Table **2** that performances of the amorphous silicon and DSSCs modules approximately are equal.

**Table 2. Comparison of the parameters of an amorphous silicon module and five-cell DSSCs module.**

| **Kind of module** | **Light intensity (lux)** | **$I_{SC}$ (μA/cm²)** | **$V_{OC}$ (mV/cell)** | **FF** | **$P_{max}$ (μW)** |
|---|---|---|---|---|---|
| Three-laye | 50 | 4.9 | 502 | 0.56 | 12.6 |
| Monolithic | 250 | 26.1 | 574 | 0.52 | 69.8 |
| Amorphous | 50 | 3.9 | 612 | 0.63 | 13.9 |
| Silicon | 250 | 18.5 | 677 | 0.62 | 70.1 |

The liquid electrolyte in principle can be replaces by gels. At the same time during the last years, solid-state dye-sensitized solar cells were also developed

[77] where instead of the electrolyte the solid p-type semiconductor (spirobifluorene) was used. The efficiency of the cells was almost 3.2%. Later [145] in solid-state DSSCs as a hole transporting conductor organic 2,2',7,7'-tetrakis-(N,N-di-p-methoxyphenyl-amine)9,9'-spirobifluorene (spiro-OMeTAD) and as a dye organic metal-free indoline were used.

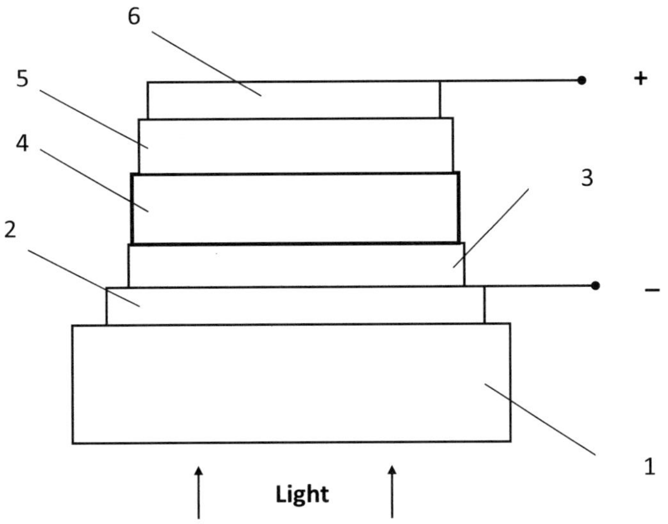


**Figure 44:** Schematic diagram of the solid-state DSSC: glass substrate (1), $SnO_2$ doped by F (2), 100 nm dense $TiO_2$ (3), nanoporous $TiO_2$ covered by dye and filled with spiro-OMcTAD (4), 30 nm Au electrode (5).

In this case the efficiency of the cells was almost 4 %. Fig. (**43**) shows simplified schematic diagram of the solid-state DSSC [145].

## 4.2. Rectennas

Unlike to conventional solar cells that convert the energy of light into electric power, it may be done by physically small antennas with built-in rectifiers or so called rectennas [146, 147] *i.e.* rectifying antennas. In the conventional solar cells actually quantum nature of light is used, whereas in antenna concept the electromagnetic nature of light is working. It was observed that 90% conversion efficiency of microwave radiation (at 2.45GHz) into DC current [146, 148] take place. In photovoltaic cell, the conversion is limited by band gap of the semiconductor material. It is considered that antenna approach, in principle, will provide opportunity to fabricate high efficiency converters of visible and IR radiations into electricity. This idea was first patented by J.C.Fletcher of NASA and R.L.Bailey of the university of Florida in the late 1960s [146, 149]. This antenna was tested at microwave frequency of (100-1000MHz). Using modern semiconductor technology it may be fabricated for visible and IR spectrum. For the antenna array as pyramidal dielectric the organic materials as carbon nanotubes, algae cells (biological antenna) or Langmuir-Blodgett films may be used [146]. The dielectric pyramidal antenna is connected with load *via* the p-n junction or metal-semiconductor Schottky type junction as shown in Fig. (**44**). Another type of rectenna described in [149] is shown in Fig. (**45**). The efficiency of the antenna as expected would be in the range of 75-80%.

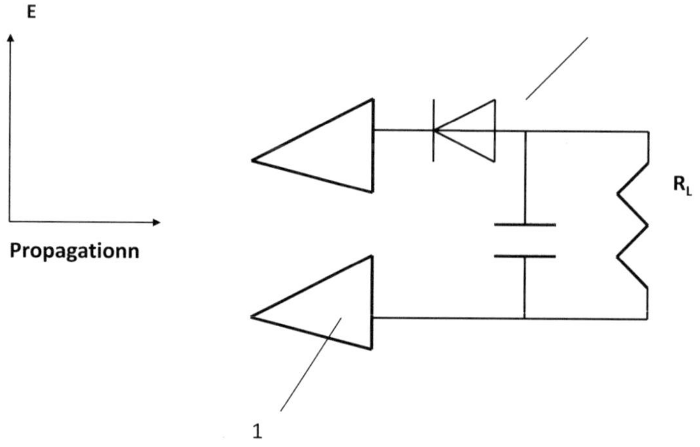


**Figure 45:** Pyramidal rectenna: (1) broadband pyramidal antenna, (2) rectification arrangement, $R_L$ –load resistance.

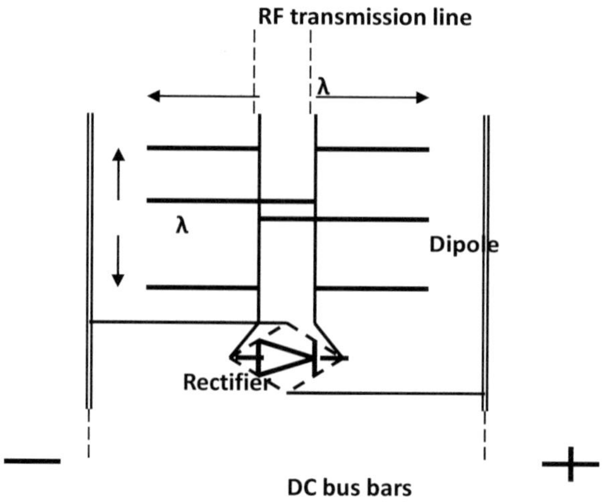


**Figure 46:** Dipole array for direct conversion of light to DC power. The wavelength λ=600 nm.

The diodes that will be used in this case will need several requirements such as small switching time, ability to work at high frequencies and small sizes *etc*. It means for fabrication of the rectennas new materials and nanotechnology will be used. Moreover between antenna and rectifier the impedance matching should be implemented that is important for the maximum power transfer. But in any case this technology is considered as emerging one and in future may bring a revolution in electromagnetic waves (solar, infrared *etc*.) energy utilization. To achieve this aim, a very well-focused innovative research work should be done by interdisciplinary team of researchers.

## 5. LIGHT-EMITTING DIODES (LED)

Light-emitting diodes (LED) technology is one of the fast developing areas of molecular electronics. Fig. (**46**) shows single layer organic light-emitting diode (OLED) fabricated on the base of Poly(p-phenylene vinylene (PPV) that was reported in 1990 [150] and emitted green light. The threshold voltage was equal to 2.3. V. Thickness of PPV layer was equal to 100 nm. The luminance of the polymer OLED was around of 650 cd/m$^2$ at 4 V. The Luminance efficiency was equal to 2 lm/W. Unlike to inorganic semiconductor LEDs in OLEDs polymer is not doped. In OLEDs metallic contacts should be fabricated from different metals: one metallic contact injects into polymer electrons only, and another one only holes.

Fig. (**47**) shows schematic diagram of typical organic light-emitting diode (OLED) which uses $Alq_3$ (tris(8-0hydroxyquinolinato)aluminum, as a electron transport and emitting layer, and NPB (N,N'-di(naphthalene-1-yl)-N,N'diphenyl-benzidine, as the hole transport layer [2].

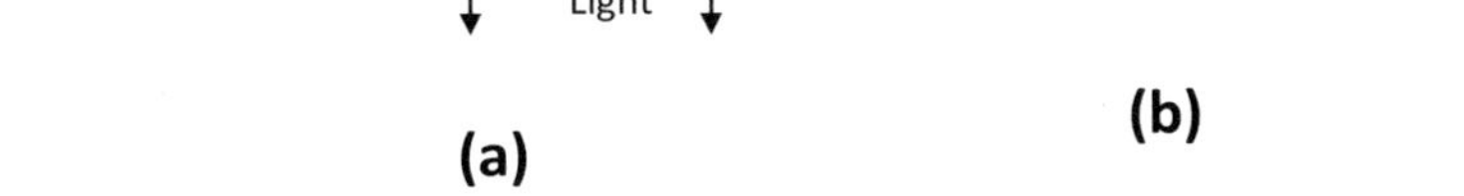


**Figure 47:** Single layer organic light-emitting diode (OLED) (a) fabricated on the base of Poly(p-phenylene vinylene (PPV) (b): ITO is anode and Al is cathode.

If forward bias voltage (a few volts) is applied to LED, the motion of injected from cathode electrons (e) and from anode holes (h) respectively takes place (Fig. **47**). In the interface of n-type and p-type semiconductors the recombination of electron and holes is observed, in the result of that the light is emitted.

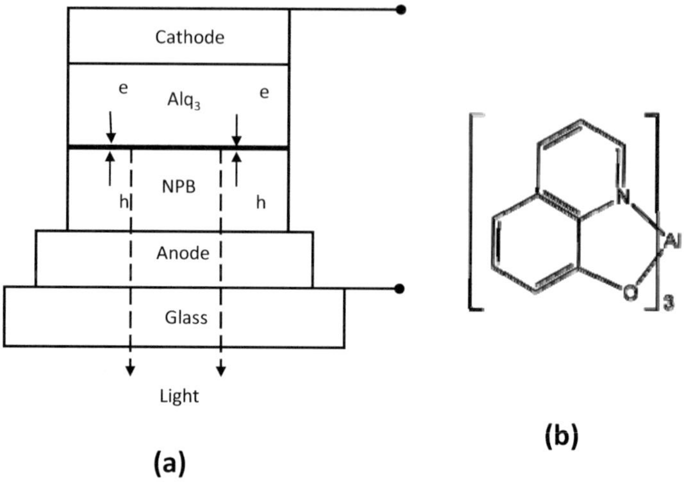


**Figure 48:** Schematic diagram of typical two-layer organic light-emitting diode (a) which uses $Alq_3$ (tris(8-hydroxyquinolinato)aluminum (b) and NPB (N,N'-di(naphthalene-1-yl)-N,N'diphenyl-benzidine (e-electrons, h- holes).

Table **3** shows data of LED/OLED performances [151]. It is seen that during the last years' LEDs and OLEDs performances were much improved. Actually the performance of light-emitting diodes became better or comparable with the performance of traditional light sources as filament and fluorescent lamps.

**Table 3. LED/OLED performance.**

| LED/OLED (year of publication) | Performance (lumens/watt) |
|---|---|
| GaAsP:N (1980) | 1 |
| Thomas Edison's bulb (filament lamp) | 1.4 |
| $Alq_3$ (1986) | 1 |
| AlInGaP/GaAs (1994) | 10 |
| $Alq_3$ (1994) | 15 |
| InGaN (1995) | 18 |
| Fluorescent lamp | 80 |
| AlGaInP/GaP (2000) | 100 |

At present, through the optimization of the properties and parameters of the devices, a great progress in fabrication and investigation of OLEDs and flat-panel displays is observed, *e.g.* amorphous silicon thin-film transistors, 20-inch size full-color active matrix OLED display [133, 152]. Further improvement of the properties of OLEDs was achieved by increasing number of layers using different materials [153].

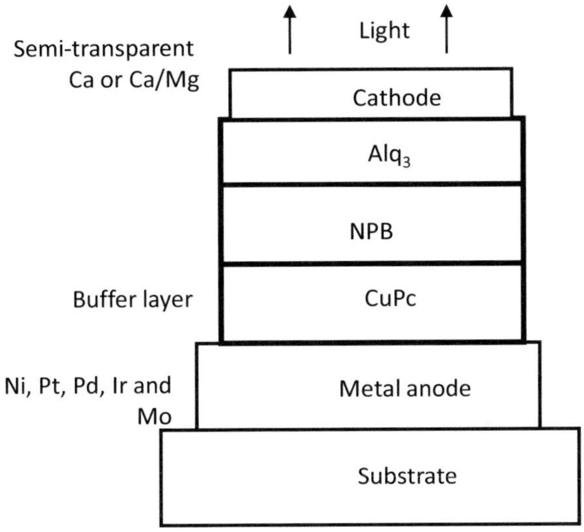


**Figure 49:** Schematic diagrams of the planar top-emitting OLEDs with three CuPc/NPB/$Alq_3$ layers.

Figs. (**48**) and (**49**) show schematic diagrams of the planar top-emitting OLEDs with three and five layers [133].

In Fig. (**49**), BCP is 2,9-dimethyl-4,7-diphenyl 1,10-phenanthrolin, that was used as hole-blocking layer. NPB is hole-transporting layer. CuPc (copper phthalocyanine) is buffer layer. $Alq_3$ is electron transporting layer. CBP is 4,4'-N,N' –dicarbazole-1,1'-biphenyl, that was doped with 6% tris(2-phenylpyridine)iridium ($Ir(ppy)_3$), was used as emission layer [133].

On the glass substrates (50x50 $mm^2$) the OLEDs (2x3 $mm^2$) were fabricated [133]. Thickness of the CuPc was equal to 20 nm, NPB - 40 nm, CBP:$Ir(ppy)_3$ – 20 nm, BCP – 10 nm and $Alq_3$ – 40 nm [133, 154]. Cathodes (semi-transparent) films thickness was 20 nm (Ca) or 12 nm (Ca/Mg). The evaporation rates of dopants and host materials were 0.05 and 1.0 Å/s respectively [133]. It was found that a dielectric capping layer (ZnSe) improve the performance of OLEDs [133].

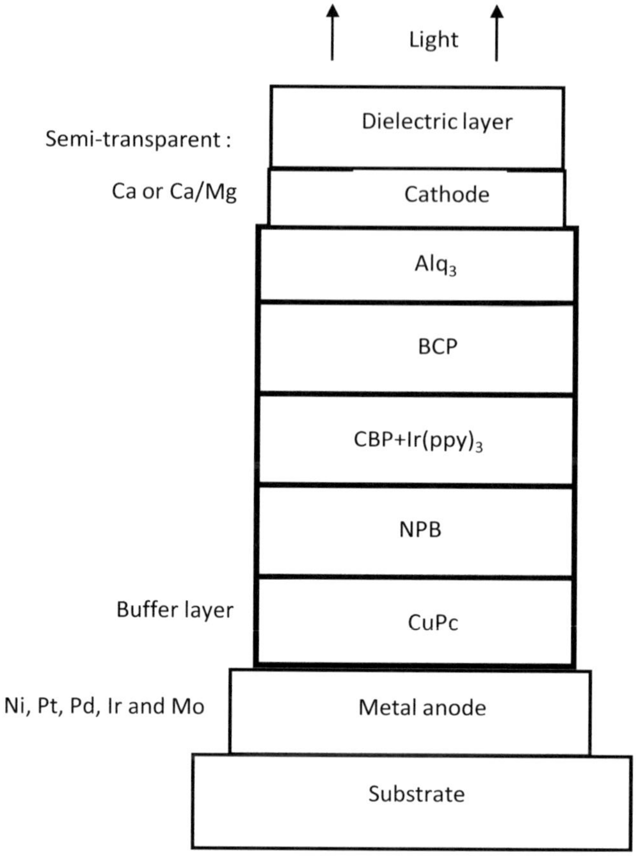


**Figure 50:** Schematic diagrams of the planar top-emitting OLEDs with phosphorescent five CuPc/NPB/CBP+Ir(ppy)$_3$/BCP/Alq$_3$ layers.

For the improvement of the OLEDs performance, at presen,t research works are foxed not only to use of different technologies and materials or their combinations but to deep investigations of electrophysical properties of the devices [155-158]. In [155] the properties of ITO/PEDOT:PSS/TPD:Alq3:C60/Al with different $C_{60}$ concentrations (0-6.0 wt.%) were investigated. It was found that at high applied voltages in the I-V characteristics of the diodes the field-emission tunneling injection of charges dominates. It was observed that at concentration of $C_{60}$ of 3 wt.% the current efficiency is optimum. In ref. [156], electrophosphorescence and its mechanism in poly(N-vinylcarbazole)(PVK):N,N'-diphenyl-N,N'-bis(3-methylphenyl)-[1,1-biphenyl]-4,40-diamine(TPD)/2-(4-biphenylyl)-5-(4-tert-butylphenyl)-1,3,4-oxadiazole(PBD):fac-tris(2-phenylpyridine) iridium[Ir(ppy)3] concentration graded bilayer electroluminescence devices were investigated. It was found that in comparison to the conventional blended devices, polymer phosphorescent devices prepared by adapting a bilayer system of PVK:TPD as the bottom layer and Ir(ppy)3:PBD as the upper layer showed decreased turn-on voltage of 5V and an operating voltage of 9.1V. Moreover an efficiency of 10.16cd/A was obtained which showed that the properties were improved [156].

When a simple cross-patterning technique was used for the fabrication of phosphorescent white organic light-emitting diodes (WOLEDs) with red, green and blue (RGB) emitters [157], it was observed that the OLEDs has relatively low driving voltage and high color stability. It was found that the driving voltage was decreased up to of 0.5–2.0V, while the luminous efficiency was maintained. Also investigation was done for a new blue emitting materials composed of 9,10-di(2-naphthyl) anthracene and highly twisted ortho-methylated naphthyl or ortho-phenylated phenyl on the 2-position of anthracene [158]. The following devices ITO/DNTPD/NPD/(ADN host/MNAn or BIPAn 3%)/Alq3/LiF OLED showed a maximum current efficiency of 3.55 cd/A (maximum luminance efficiency of 2.11 lm/W and maximum external quantum efficiency of 3.38.

## CONCLUSION

It was found that humidity sensing properties of $Cu_2O$-PEPC nanocomposite films, deposited from solution by drop-casting showed [159] increase in capacitance and decrease in resistance with increase in humidity level. It was considered that the capacitive response of the sensor is due to the contribution of dipolar polarization of the absorbed $H_2O$ molecules by the nanocomposite films and polarization due to the transfer of charge carriers. The decrease in resistance probably is the result of the increase of $H_2O$ molecules concentration and displacement currents, and increase in the concentration of charge carriers due to doping of the composite by water molecules.

In the result of investigations of the ITO/CuPc/NiPc/Al and ITO/NiPc/CuPc/Al double junction cells w [160], the rectification behavior in the *I-V* characteristics was found. The cells show photo –electric response in both photo-voltaic and photo-diode modes of operation. On average the response of ITO/NiPc/CuPc/Al cells is 20 % higher than the ITO/CuPc/NiPc/Al cell, revealing that Schottky barrier metal-semiconductor CuPc/Al or NiPc/Al junction plays a dominant role in comparison to organic-organic semiconductor CuPc/NiPc (or NiPc/CuPc) junctions. In the ITO/NiPc/CuPc/Al cells it was observed that the capacitance increase under illumination of 2000 lx by 50 %. As a result, the energy-band diagram and an equivalent circuit of the double junction cells were developed that may be used for simulation of their opto-electronic properties.

In the surface-type Al/CNT/Al pressure sensors, the decrease in resistance with increase in pressure [161] was observed. Experimental resistance-pressure relationships of the sensors were simulated. The Percolation theory was used for explanation of the conduction mechanism. In this case the CNT system is assumed as a bulk hetero system that results high sensitivity of the CNT layers for squeezing effect where densification of the nanopowder and the increase of the inter-particles contact areas and intrinsic conductivity of the nanoparticles takes place; resulting in increase of the conductivity and decrease of the resistance of the samples with increase in pressure. To eliminate the effect of humidity on the electric properties of the sensors the CNT layer with electrodes was encapsulated in flexible elastic casing.

The decrease in the resistance with increase in the displacement was observed in the resistive-type displacement sensor based on composite of vanadium complex ($VO_2$ (3-fl)) and CNT [162]. Quasi-linear behavior of the resistive-displacement relationships of the sensors was observed. Transitions between spatially separated sites that can be attributed to the Percolation Theory; the mechanism of change of conductivity with change of the displacement in the $VO_2$ (3-fl)-CNT composite was considered. Fabrication of the $VO_2$ (3-fl)-CNT composites allows to decrease the value of the sensors resistance due to higher conductivity of the CNT. It is especially important for the practical applications of the displacement sensors.

An increase in resistance of the strain sensors under tension and decrease under compression [163] in the carbon nanotubes (CNT) and $Cu_2O$ composite based strain sensors was observed. It was found that resistance-strain relationships show quasi-linear behavior. Physical mechanism is observed for the change of resistance under the effect of strain. Wheatstone bridge circuit for practical utilization of the strain sensors and compensation of the effect of temperature is presented.

*A basic structure and properties of organic field-effect transistors (OFET), including the use of the transistors as sensors was discussed. In the temperature interval of 25 °C – 125 °C [110], the electrical characteristics of CuPc based temperature sensitive organic FET were investigated. Experimentally it was found that the drain current of this organic field effect transistor (OFET) is increased with temperature. The presence of switching effect from low to high conduction states was observed in organic polymer semiconductor. The threshold voltage of the switching effect decreased with increase of the temperature. The OFETs can be used as the multifunctional sensors as it is sensitive to temperature, light, humidity, pressure and displacement.*

During the last years, bulk heterojunction solar cells were fabricated and investigated. It was found that the efficiency of the bulk heterojunction solar cell was up to 3% [77] and even 5% [133]. Unlike to inorganic solar cells, bulk heterojunction solar cells show practically very interesting efficiency-temperature relationship: as temperature increases up to around of 50 °C, the efficiency also increases.

At present dye-sensitized solar cells (DSSC) based on $TiO_2$ nanostructure is very popular among of researchers in different countries due to potential high efficiency and cheapness. Though the photogalvanic effect was discovered in 1839 in liquid state device by Edmond Becquerel the fundamental contribution in the development of photoelectrochemical cells' technology was done by Michael Gratzel (1991) [142-145]. Open-circuit voltage of the DSSC is around of 0.7 V (in Si solar cells $V_{oc}=0.6$ V) and short-circuit current is equal of 20 mA/cm$^2$ (in Si solar cells $I_{sc}$=35 mA/cm$^2$). In relatively large (1-5 cm$^2$) and small (0.25 cm$^2$) area cells the efficiencies were equal to 5-8% and 11% respectively Modules efficiencies were about of 7 %. It was found that performances of the amorphous silicon and DSSCs modules approximately are equal.

Rectenna or rectifying antenna, is one of the most promising device for solar energy conversion into electric power that can be realized only by use of nanotechnoly. This idea was first patented by J.C.Fletcher of NASA and R.L.Bailey of the university of Florida in the late 1960s [146, 149]. In the conventional solar cells actually quantum nature of light is used, in antenna concept the electromagnetic nature of light is working. Expected efficiency of the rectenna based solar cells will be almost 90 %.

*At present organic light-emitting diodes (OLEDs) are one of the commercialized devices of organic electronics* [133, 152]. *It was discussed the single layer OLED fabricated on the base of Poly(p-phenylene vinylene (PPV), two-layer organic light-emitting diode which uses $Alq_3$ (tris(8-hydroxyquinolinato)aluminum and NPB (N,N'-di(naphthalene-1-yl)-N,N'diphenyl-benzidine, three CuPc/NPB/$Alq_3$ layers OLED, and device with five CuPc/NPB/CBP+Ir(ppy)$_3$/BCP/$Alq_3$ layers. For improvement of the OLEDs performance at present research works are foxed not only to use of different technologies and materials* *or their combinations but to deep investigations of electrophysical properties of the devices* [155-158].

*In* [156] *it was investigated bilayer OLED based on poly(N-vinylcarbazole)(PVK):N,N'-diphenyl-N,N'-bis(3-methylphenyl)-[1,1-biphenyl]-4,40-diamine(TPD)/2-(4-biphenylyl)-5-(4-tert-butylphenyl)-1,3,4-oxadiazole(PBD): fac-tris(2-phenylpyridine)iridium[Ir(ppy)3]. It was found that the OLED showed decreased turn-on voltage of 5V and an operating voltage of 9.1V, in addition to an efficiency of 10.16cd/A.*

## ACKNOWLEDGEMENTS

Authors thank GIK Institute of Pakistan for the support of this work.

## CONFLICT OF INTEREST

The author confirms that he has no conflict of interest to declare for this publication.

## REFERENCES

[1] W. Qihao, Y. Shihong, "Urban air pollution patterns, land use and thermal landscape: an examination of the linkage using GIS", *Environ. Monit. Assess.,* 117, pp. 463-489, 2006.
[2] Y. Anjaneyulu, I. Jayakumar, V.H. Bindu, P.V.M. Rao, G. Sagareswar, K.V. Ramani, T. H. Rao, "Real time remote monitoring of air pollutants and their online transmission to the web using internet protocol", *Environ. Monit. Assess.,* DOI: 10.1007/s10661-006-9233-6, 1-11, 2006.
[3] A. Joyce, J. Adamson, B. Huntley, T. Parr, R. Baxter, "Standardization of temperature observed by automatic weather stations", *Environ. Monit. Assess.*, vol. 68, pp. 127-136, 2001.
[4] Z. M. Rittersma, Recent achievements in miniaturised humidity sensors. "A review of transduction techniques", *Sens. Actuators A, Phys.*, vol. 96, pp. 196 -210, 2002.
[5] C. Y. Lee and G. B. Lee, "Humidity sensors: A review", *Sens. Lett.* Vol. 3, no. 1, pp. 1–15, 2005.
[6] Z.Chen, C. Lu., "Humidity sensors: a review of materials and mechanisms", *Sens. Lett.,* vol. 3, pp. 274–295, 2005.
[7] J.G. Korvink, L. Chandran, T. Boltshauser, H. Baltes, "Accurate 3D capacitance evaluation in integrated capacitive humidity sensors", *Sens. Mater.,* vol. 4. pp. 323-328, 1993
[8] Z.M. Rittersma, "Recent achievements in miniaturised humidity sensors - a review of transduction techniques", *Sens. Actuators A.,* vol. 96. pp. 196–210, 2002.
[9] G. M. O'Halloran, P. M. Sarro, J. Groeneweg, P. J. Trimp, and P. J. French, "A bulk micromachined humidity sensor based on porous silicon", *International conference on Solid-State Sensors and Actuators*., pp. 563-566, 1997.
[10] S. Park, J.H. Kang, J. Park, S. Mun, *Sensor Actuat. B, Chem*., vol. 76, pp. 322–326, 2001.
[11] K. Carr-Brion, "Sensors in Process Control". Elsevier Applied Science Publishers, London and New York, 1986.
[12] H. Grange, G. Delapierre, T. Seiyama (ed.), "Chemical Sensor Technology", 2/154, Kodansha, Tokyo/Elsevier, Amsterdam, 1989.
[13] H. Grange, G. Delapierre, T. Seiyama (ed.), "Chemical Sensor Technology", Kodansha, Tokyo/Elsevier, Amsterdam, 2/155, 1989.
[14] M. Matsugnchi, Y. Sadaoka, N. Nosaka, M. Ishibashi, Y. Sakai,T. Kuroiwa, A. Yto, *J. Electrochem. Soc.,* vol. 140, pp. 825-830, 1993.

[15] Kh.S. Karimov, I. Qazi, T.A. Khan, P.H. Draper, F.A. Khalid, M.M. Tahir, "Humidity and illumination organic semiconductor copper phthalocyanine sensor for environmental monitoring", *Environ. Monit. Assess.*, vol.141, pp.323-328, 2008.

[16] Kh.S. Karimov, K. Akhmedov, I. Qazi, T.A. Khan, "Fabrication and investigation of the charge/discharge characteristics of zinc/PVA-KOH-H2O-Iodine/carbon cell", *J. Optoelect. Adv. Mater.*, vol. 9, pp.477-479, 2007.

[17] A. O. Musa, T. Akomolafe, and M. J. Carter, "Production of cuprous oxide, a solar cell material, by thermal oxidation and a study of its physical and electrical properties", *Sol. Energ. Mater. Sol. Cells.*, vol. 51/3-4, pp. 305–316, 1998.

[18] M.T.S. Nair, L. Guerrero, O. L. Arenas, P.K. Nair, "Chemically deposited copper oxide thin films: structural, optical and electrical characteristics", *Appl. Surf. Sci.,* vol. 150, pp. 143-151, 1999.

[19] H. Zhang, D.M. Goodner, M.J. Bedzyk, T.J. Marks, R.P.H. Chang, "Formation and Kinetics Study of Cuprous Oxide Nanodots on LaAlO3(001)", *Chem. Phys. Lett.,* vol. 395, pp. 296-301, 2004.

[20] N.A. M, Shanidb, M. A. Khadar, "Evolution of nanostructure, phase transition and band gap tailoring in oxidized Cu thin films", *Thin. Sol. Films.*, vol. 516/18. pp. 6245–6252, 2008.

[21] J.D. Irwin, "Basic Engineering Circuit Analysis", Sixth Edition. John Wiley & Sons, New York, 1999.

[22] M.A. Oma, "Elementary Solid State Physics", Pearson Education Inc. Singapore, 1999.

[23] D.B. Hibbert, "Introduction to Electrochemistry", *Mackays of Chatham PLC Chatham*, Kent, UK, 1993.

[24] V.A. Kargin, "Organic Semiconductors", Nauka, Moscow, 1968.

[25] M. Iwamoto and T. Manaka, "Experimental Evidence of Surface Polarization in Organic Films and Control of Current-Voltage. Characteristics by the Surface Polarization", *Proc. Int. Symp. Tokyo, Japan. IPAP Conf. Series.,* vol. 6, pp. 63-68, 2005.

[26] F. Amy, C. Chan and A. Kahn, *Org. Elect.,* vol. 6, pp. 85-89,2005.

[27] C.D. Simpson, "Industrial Electronics", *Prentice Hall*, Inc., USA, 1996.

[28] A. Al-Shawabken, H.A. Al-Wahab, Y.A.Shahab, "Optoelectronics and Advanced Materials", *Rapid. Communications*. Vol. 1/4, pp. 186-188, 2007.

[29] T.N. Murakami, N. Kawashima and T. Miyasaka, "A high-voltage dye-sensitized photocapacitor of a three-electrode system", *Chem. Commun,* vol. 10, No.11, pp.3346-3348, 2005. DOI: 10.1039/b503122b.

[30] W. Riess, H. Riel, T. Beierlein, W. Brutting, P. Muller, P.F. Seidler, "Influence of trapped and interfacial charges in organic multilayer light-emitting devices", *IBM J. of Res. & Dev.* Vol.45/1, pp.77-88, 2001.

[31] L. Wang, D. Fine, D. Sharma, L. Torsi, A. Dodabalapur, "Nanoscale organic and polymeric field-effect transistors as chemical sensors", *Analytical and bioanalytical chemistry.* Vol.384/3, pp.310-321, 2006.

[32] L. Valli, "Phthalocyanine-based Langmuir–Blodgett films as chemical sensors", *Adv. Colloid. Interface Sci.,* vol. 116, pp. 13-44, 2005.

[33] D.B. Mitzi, K. Chondroudis, C.R. Kagan, "Organic-inorganic electronics", *IBM. J. Res. & Dev.,* vol.45/1, pp.29-45, 2001.

[34] F. Gutman, L.E. Lyons, "Organic semiconductors", Part A, Robert E. Krieger Publishing Company, Malabar, Florida, 1980.

[35] F. Gutman, H. Keyzer, L.E. Lyons, R.B. Somoano. "Organic semiconductors", Part B, Robert E. Krieger Publishing Company, Malabar, Florida, 1983.

[36] M.I. Fedorov, "Influence of the doping on conductivity and photoconductivity of the phthalocynanine", *Ph.D. Thesis*. Chernogolovka, Moscow, Russia, 1973.

[37] A. Oprea, U. Weimar, E. Simon, M. Fleischer, H.P Frerichs, Ch. Wilbertz, M. Lehmann, "Copper phthalocyanine suspended gate field effect transistors for NO 2detection", *Sensors and Actuators B Chemical*, vol. 118/1, pp. 249-254, 2006.

[38] M.I. Fedorov, "Investigation of organic-inorganic semicondiuctors heterojunctions", *D.Sc., Thesis*, Ryazan, Russia. 2004.

[39] T.S. Shafai, T.D. Anthopoulos, "Junction properties of nickel phthalocyanine thin film devices utilising indium injecting electrodes ", *Thin. Sol. Film.*, vol. 398, pp. 361-367, 2001.

[40] T.D. Anhopoulos, T.S. Shafai, "Electrical properties of $\alpha$-nickel phthalocyanine/aluminium interfaces: effects of oxygen doping and thermal annealing", *J. Phys. Chem. Solids.*, vol. 64, pp. 1217-1223, 2003.

[41] R.B. Chaabane, A. Ltaief, C. Dridi, H. Rahmouni, H.B. Ouada, "Study of organic thin film transistors based on nickel phthalocyanine: effect of annealing", *Thin. Sol. Films.*, vol. 427, pp. 371-376, 2003.
[42] P.R. Binu, C.M. Joseph, K. Shreekrishnakumar, C.S. Menon. "Electrical conductivity and optical absorption studies on NiPc-substituted borate glass matrix", *Mater. Chem. Phys.,* vol. 80, pp. 591-594, 2003.
[43] C.J. Liu, J.J. Shih, Y.H. Ju, "Surface morphology and gas sensing characteristics of nickel phthalocyanine thin films", *Sensors and Actuators, B 99*, pp.344-349, 2004.
[44] T.D. Anthopoulos, T.S. Shafai, "Influence of oxygen doping on the electrical and photovoltaic properties of Schottky type solar cells based on α-nickel phthalocyanine", *J. Phys. Chem. Solids.*, vol. 65, pp. 1345-1348, 2004.
[45] M.M. El-Nahass, K.F. Abd-El-Rahman, A.A.M. Farag, A.A.A. Darwish. "Photovoltaic properties of NiPc/p-Si (organic/inorganic) heterojunctions", *Org. Elect*, vol. 6, pp. 129-136, 2005.
[46] M.M. El-Nahass, A.F. El-Deeb, F. Abd-El-Salam. "Influence of temperature and frequency on the electrical conductivity and the dielectric properties of nickel phthalocyanine", *Org. Elect,* vol. 7, pp. 261-270, 2006.
[47] M.M. El-Nahass, K.F. Abd-El-Rahman, A.A.A. Darwish, "Fabrication and electrical characterization of p-NiPc/n-Si heterojunction", *Microelectr. J.*, vol. 38, pp. 91-95, 2007.
[48] T. D. Anthopoulos and T. S. Shafai, " Influence of oxygen doping on the electrical and photovoltaic properties of Schottky type solar cells based on α-nickel phthalocyanine", *Thin. Sol. Films,* vol. 441, S. 207–213, 2003.
[49] M.M. El-Nahass, K.F. Abd-El-Rahman, "Investigation of electrical conductivity in Schottky-barrier devices based on nickel phthalocyanine thin films", *J. Alloy. Comp.*, vol. 430, pp. 194-199, 2007.
[50] F.Petraki, V.Papaefthimiou, S.Kennou, "The electronic structure of Ni-Phthalocyanine/metal interfaces studied by X-ray and Ultraviolet Photoelectron spectroscopy", *Org. Elect*, vol. 8, pp. 522-528, 2007.
[51] Kh.S. Karimov, Kh.M. Akhmedov, A.A. Dzhuraev, M. N. Khan, S. M. Abrarov, M. I. Fiodorov, "Resistance, Capacitance and Impedance Based Humidity Sensors", *Eurasian Chem. Tech. J.*, pp.251-259, 2000.
[52] Kh.S. Karimov, M.M. Ahmed, S.A. Moiz, M.I. Fedorov. "Temperature-dependent properties of organic-on-inorganic Ag/p-CuPc/n-GaAs/Ag photoelectric cell". *Sol. Ener. Mater. Sol. Cells,* vol. 87, pp. 61-75, 2005.
[53] M. K. Debe, R. I. Poirer, D. D. Erickson, T. N. Tommet, D. R. Field, K. M. White, "Homogeneity of Copper Phthalocyanine", *Thin Sol. Films,* vol. 186, pp. 257-272, 1990.
[54] Kh.S. Karimov, S. Bellingeri, Y. Abe, "Physical Vapor Transport of CuPc Organic Thin Filmsin High Gravity. Processing by Centrifugation", Kluwer Academic/Plenum Publishers, New York. pp.99-106, 2001.
[55] R. C. Loutfy, J. H. Sharp, C. K. Hsiao and R. Ho, "Phthalocyanine organic solar cells: Indium/x-metal free phthalocyanine Schottky barriers", *J. Appl. Phys.,* vol. 52/8, pp. 5218-523, 1981.
[56] D.A. Neamen, "Semiconductor Physics and Devices", Richard D. Irwin, Inc., USA, 1992.
[57] M. Iwamoto, T. Manaka, "Experimental Evidence of Surface Polarization in Organic Films and Control of Current-Voltage Characteristics by the Surface Polarization", Proc. Int.Symp. Super-Functionality Organic Devices, *IPAP. Conf. Series,* vol. 6, pp. 63-68, 2005.
[58] J.M. Dally, W.F. Riley and K.G. McConnell, "Instrumentation for Engineering Measurements", 2nd Ed. Wiley, New York, 1993.
[59] T. Someya, Y. Kato, T. Sekitani, S. Iba, Y. Noguchi, Y. Murase, H. Kawaguchi and T. Sakurai, "Conformable, flexible, large-area networks of pressure and thermal sensors with organic transistor active matrixes", *Proc. Natl Acad. Sci., USA.,* vol. 102, pp. 12321–5, 2005.
[60] G. Dennler, C. Lungenschmied, N.S. Sariciftci, R. Schwodiauer, S. Bauer and H. Reiss, "Unusual electromechanical effects in organic semiconductor Schottky contacts: between piezoelectricity and electrostriction", *Appl. Phys. Lett.,* vol. 87, pp. 163501-3, 2005.
[61] Kh.S. Karimov. "Electrophysical properties of low-dimensional organic materials at deformation", *DSc Thesis.* Academy of Sciences, Tashkent, Uzbekistan, 1993.
[62] Kh. Akhmedov. Synthesis, "Properties and application of carbazolile containing Polymers", *DSc Thesis*. Academy of Sciences, Dushanbe, Tajikistan, 1998.
[63] R.J. Grow, Q. Wang, J. Cao, D. Wang, H. Dai, *Appl. Phys. Lett.*, vol. 86, 093104, 2005.

[64] R.J. Grow, Q. Wang, J. Cao, D. Wang, H. Dai, "Piezoresistance of carbon nanotubes on deformable thin-film membranes", *Appl. Phys. Lett.*, 86/9, Article ID 093104, 2005.

[65] M. Saleem, Kh.S. Karimov, Z.M. Karieva, A. Mateen, "Humidity sensing properties of CNT–OD–VETP nanocomposite films", *Physica. E.*, vol. 43, pp. 28-32, 2010.

[66] C. Cantalini, L. Valentini, I. Armentano, L. Lozzi, J.M. Kenny, S. Santucci, "Sensitivity to $NO_2$ and cross-sensitivity analysis to $NH_3$, ethanol and humidity of carbon nanotubes thin film prepared by PECVD", *Sensors and Actuators, B 95*, pp. 195-202, 2003.

[67] P.S. Na, H. Kim, H.M. So, K.J. Kong, H. Chang, B.H. Ryu, Y. Choi, J.O. Lee, B.K. Kim, J.J. Kim and J. Kim, "Investigation of the humidity effect on the electrical properties of single-walled carbon nanotube transistors". *Appl. Phys. Lett.*, vol. 87, pp. 093101-3, 2005.

[68] M. Rinkiö, M.Y. Zavodchikova, P. Törmä, and A. Johansson, "Effect of humidity on the hysteresis of single walled carbon nanotube field-effect transistors", *Phys. Stat. Sol. B*, vol. 245/10, pp. 2315–2318, 2008.

[69] D. Tang, L. Ci, W.Zhou, S. Xie, "Effect of H2O adsorption on the electrical transport properties of double-walled carbon nanotubes", *Carbon,* 44/11, 2155–2159, 2006.

[70] O.K. Varghese, P.D. Kichambre, D. Gong, K.G. Ong, E.C. Dickey and C.A. Grimes, "Gas sensing characteristics of multi-wall carbon nanotubes", *Sensors. Actuat., B 81,* pp. 32-41, 2001.

[71] J. Cao, Q. Wang, and H. Dai, "Electromechanical properties of metallic, quasimetallic, and semiconducting carbon nanotubes under stretching". *Phys. Rev. Lett.,* vol. 90/15, pp. 157601-1–157601-4, 2003.

[72] T. Fang, A. Konar, H. Xing, and D. Jena, "Carrier statistics and quantum capacitance of graphene sheets and ribbons". *Appl. Phys. Lett.*, vol. 91/9, 092109, 2007.

[73] W. Xue, T. Cui, "Electrical and electromechanical characteristics of nano-assembled carbon nanotube thin film resistors on flexible substrate", *Proc. 14th International Conf. on Solid State Sensors, Actuators and Microsystems,* Lyon, France, June 10–14, pp. 1047–1050, 2007.

[74] L. Yong, W. Waulu, L. Kejun, H. Chenguo, H. H.Zhi and F. Oing, "Nanotube Films", *Chinese. Sci. Bull.*, vol. 48, pp. 125-127, 2003.

[75] P. Regoliosi, A. Reale, A. Di Carlo, S. Orlanducci, M. L. Terranova, and P. Lugli, "Piezoresistive behaviour of single wall carbon nanotubes", *Nanotechnology, 4th IEEE Conf.,* pp. 149–151, 2004.

[76] C. Stampfer, T. Helbling, A. Jungen, and C. Hierold, "Piezoresistance of Single-Walled Carbon Nanotubes", *14th Intl. Conf. on Solid-State Sensors, Actuators and Microsystems*, T*ansducers and Eurosensors XXI*, June 10-14, Lyon, France, 2, pp. 1565-1568, 2007.

[77] Kh.S. Karimov, M. Saleem, M. Mahroof-Tahir, T.A. Khan, A. Khan, "Pressure Sensitive Organic Field Effect Transistor", *Phys. E.,* vol. 43, pp. 547-551, 2010.

[78] A. Croft, R. Davison and M. Hargreaves, "Engineering Mathematics", A Modern Foundation for Electronic, Electrical and Control Engineers, Addison-Wesley, London, 1993.

[79] C.J.Brabec, V.Dyakonov, J.Parisi and N.S.Sariciftci, "Organic Photovoltaics Concepts and Realization", *Springer*, Heidelberg, 2003.

[80] H. Bottger and V.V. Bryksin, "Hopping Conduction in Solids", *Akademie,* Berlin, 1985.

[81] R.G. Irvine, "Operational Amplifiers Characteristics and Applications", 3rd Ed. Prentice Hall, Englewood Cliffs, NJ, 1994.

[82] A. Chiba, T. Fukao, O. Ichikawa, M. Oshima, M. Takemoto, D.G. Dorrell, "Magnetic Bearings and Bearingless Drives", *Elsevier*, Linacre House, Jordan Hill, Oxford, UK, 2006.

[83] M. Ai, M. Shimazoe, K. Soeno, M. Nishihara, A. Yasukawa, Y. Kanda, "Cantilever-type displacement sensor using diffused silicon strain gauges", *Sensors and Actuators.*, vol. 2, pp. 297-301, 1981.

[84] S.A. Ferris, J.M. Ivison, D. Walker, "The magnetoresistor as a displacement transducer element", *J. Phys. E. Sci. Instrum.,* vol. 3, pp. 639-644, 1970.

[85] K. Kristiansen, P. McGuiggan, G. Carver, C. Meinhart, J. Israelachvili, Langmuir, "3D force and displacement sensor for SFA and AFM measurements", *Am. Chem. Soc.*, vol. 24, pp. 1541-1546, 2008.

[86] M. Mahroof-Tahir, Kh.S. Karimov, "Fabrication of heterojunctions and current-voltage characteristics of vanadium coordination compounds", *Optoelectron. Adv. Mat.*, vol. 11/1, pp. 83-88, 2009.

[87] Kh. S. Karimov, M. Saleem, M. Mahroof-Tahir, R. Akram, M. T. Saeed Chanee, A. K. Niaz, "Resistive humidity sensor based on vanadium complex films", *J. Semicond*, vol. 35/9, 094001, 2014.

[88] Kh.S. Karimov, M.T.S. Chani, F.A. Khalid, "Carbon nanotubes film based temperature sensors", *Physica E, Low-dimensional systems and nanostructures.*, vol. 43/9, pp. 1701-1703, 2011.

[89] B.A. Atakulov, A.Ya. Afuzov, E.I. Bilalov, "Electromechanical properties of non-uniform semiconductors". *Fan, Tashkent*, 1977.

[90] M. Sikorsky, "Application of p-n junction semiconductor devices for measurement of pressure and deformations", *Physical Acoustics*, Edited by U.Mezon, 1 B, pp. 210-302, 1967.

[91] I.I. Farbshtein, S.A. Zaitsev, V.G. Krigel *et al*, "PTE, No.3", 30-232, 1976.

[92] A.A. Berlin, M.I. Brodzeli, R.M. Vlasova, *et al*, "Pisma v ZhT.F., 4", No.15, pp. 885-888, 1978.

[93] I.I. Krivonosov, N.A. Markova, G.E. Zaikov *et al.*, "Pressure sensor". Patent No. 301583 of USSR, 1971.

[94] G.G. Abashev, V.Sh. Lunegov, V.S. Pavlov *et al.*, "Semiconductor pressure sensor". Patent No. 985719 of USSR, 1982.

[95] V.I.Lunegov, "Pressure sensor with piezoresistive element on the base of composite of polymer ion-radical salt, 11 Universities Seminar on Organic Semiconductors", *Gorkiy*, USSR, 1987.

[96] C.S. Smith, "Piezoresistance Effect in Germanium and Silicon", *Phys. Rev.*, vol. 94/1, pp.42-49, 1954.

[97] S.M. Sze, "Semiconductor Sensors", *Wiley*, New York, 1994.

[98] V.I. Vaganov, "Integrated Tensity Strain Gauges", *Energoatomizdat*, Moscow, 1983.

[99] O.D. Sorokoletov, N.P. Marin, B.V. Landishev, "Thermostable sensors for investigation of thermo power parameters", *Pribori i Sistemi Upravlenia.*, vol. 12, pp. 6-9. 53, 1990.

[100] A.O. Musa, T. Akomolafe, M.J. Carter, "Production of cuprous oxide, a solar cell material, by thermal oxidation and a study of its physical and electrical properties", *Sol. Energ. Mater. Sol. Cells.*, vol. 51, pp. 305-316, 1998.

[101] M.T.S. Nair, L. Guerrero, O.L. Arenas, P.K. Nair, "Chemically deposited copper oxide thin films: structural, optical and electrical characteristics", *Appl. Surf. Sci.*, vol. 150, pp. 143-151, 1999.

[102] H. Zhang, D.M. Goodner, M.J. Bedzyk, T.J. Marks, R.P.H. Chang. "Formation and Kinetics Study of Cuprous Oxide Nanodots on LaAlO3(001)". *Chem. Phys. Lett.*, vol. 395, pp.296-301, 2004.

[103] N.A.M. Shahid, M.A. Khadar. "Cu2O-PEPC Composite Based Pressure Transducer". *Thin Sol. Films*. vol. 516, pp. 6245-6252, 2008.

[104] C.D. Dimitrakopoulos, D.J. Mascaro, "Organic Thin Film Transistors for Large Area Electronics", *IBM J. Res. & Dev*, vol. 45/1, pp.11-27, 2001.

[105] R. Wisnieff, "Printing Screens", *Nature*, vol. 394, pp. 225-232, 1998.

[106] S.M. Sze, "Physics of semiconductor devices", 2nd Edition, *Wiley-Interscience Press*, New York, 1981.

[107] I. Murtaza, Kh.S. Karimov, Z. Ahmad, I. Qazi, M. Mahroof-Tahir, T.A. Khan, "Humidity Sensitive Organic Field Effect Transistor". *J. Semicond.*, vol. 31/5, 05001, pp.1-5, 2010.

[108] Kh.S. Karimov, I. Qazi, M. Mahroof-Tahir, T.A. Khan, U. Shafique, "Photo organic field effect transistor's propertie", *Turk. J. Phys*. vol. 32, pp.1-7, 2008.

[109] M.H. Sayyad, Z. Ahmed, Kh.S. Karimov, M. Yaseen, M. Ali, "Photo organic field effect transistor based on a metallo-porphyrin D", *Appl. Phys.*, vol. 42, pp. 1-3, 2009.

[110] Kh.S. Karimov, M. Saleem, M. Mahroof-Tahir, T.A. Khan, A. Khan, "Pressure Sensitive Organic Field Effect Transistor", *Phys. E.*, vol. 43, pp. 547-551, 2010.

[111] Kh.S. Karimov, T.A. Qasuria, "The use of displacement sensitive organic field effect transistor for telemetry system applications", *Measurement*, 45/1. DOI: 10.1016/j.measurement.2011.10.006, 2012.

[112] Q. Zafar, R. Akram, Kh.S. Karimov, T.A. Khan, M. Farooq, M.M. Tahir, "Temperature Variation Effects on I-V Characteristics of Cu-Phthalocyanine based OFET", *J. Chem. Env. Eng.*, vol. 58, pp. 517-521, 2011.

[113] H.L. Gomes, P. Stallinga, "Ανάπτυξη τεχνολογίας κατασκευής τρανζίστορς λεπτών υμενίων βελτιστοποιημένων ως προς τη δομή των υμενίων πυριτίου που προκύπτει από την κρυσταλλοποίηση". *App. Phy. Lett.*, 84/16, 1-6, 2004.

[114] P. Stallinga, H.L. Gomes, "Electronic transport in field-effect transistors of sexithiophene", *J. App. Phys.*, vol. 96, pp. 5277-5281, 2004.

[115] H.L. Gomes, P. Stallinga, F. Dinelli, M. Murgia, F. Biscarini, D.M. de Leeuw, M. Muccini, K. Mullen, "Electrical characterization of organic based transistors: stability issues", *Polym. Adv. Technol.*, vol. 16, pp. 227-232, 2005.

[116] S.F. Nelson, Y-Y. Lin, D.J. Gundlach, T.N. Jackson, "Temperature-independent transport in high-mobility pentacene transistors", *App. Phy. Lett.,* vol. 72/15, pp. 1854-1856, 1998.

[117] J. Ghim, K. Jun. Baeg, Y.Y. Noh, S.J. Kang, J. Jo, D.Y. Kim, S. Cho, J. Yuen, K. Lee, A. J. Heeger, "Perfluorocyclobutane containing polymeric gate dielectric for organic thin film transistors with high on/off ratio". *App. Phy. Lett.,* vol. 89/20, 202516, 2006.

[118] Y.Y. Noh, J. Ghim, S.J. Kang, K.J. Baeg, D.Y. Kim, K. Yase, "Effect of light irradiation on the characteristics of organic field-effect transistors", *J. App. Phys.,* vol.100, pp.1-5, 2006.

[119] Y-G. Park, T. anki, H-Y. Lee, H. Tanaka, T. Kawai, "Temperature Variation Effects on I-V Characteristics of Cu-Phthalocyanine based OFET", *Sol. Stat. Electr.*, vol.47, pp.2221-2225, 2003.

[120] M. Burgmair, H.P. Frerichs, M. Zimmer, M. Lehmann, I. Eisele, "Field effect transducers for work function gas measurements: device improvements and comparison of performance", *Sensors and Actuators., B* 95, pp.183-188, 2003.

[121] L. Wang, D. Fine, D. Sharma, L. Torsi, A. Dodabalapur. *Anal. Bional. Chem.*, 384,310-315, 2006.

[122] H. Bai, G. Shi, "Gas Sensors Based on Conducting Polymers", *Sensors.,* vol.7, pp.267-271, 2007.

[123] C. Bartic, A. Campitelli, S. Borghs, "Field-effect detection of chemical species with hybrid organic/inorganic transistors". *App. Phy. Lett.,* vol.82, 3, pp.475-480, 2003.

[124] M.C. Petty. "An Introduction to Molecular Electronics", Published by Edward Arnold, Great Britain, 1995.

[125] L. Valli. "Phthalocyanine-based Langmuir-Blodgett films as chemical sensors". *Adv. Coll. Interf. Sci.*, vol.116, pp.13-18, 2005.

[126] T. Miyata, S. Kawaguchi, M. Ishii, T. Minami, "High sensitivity chlorine gas sensors using Cu-phthalocyanine thin films", *Thin. Sol. Films*, vol.425, pp.255-260, 2003.

[127] A. Oprea, U. Weimar, E. Simon, M. Fleischer, H.P. Frerichs, Ch. Wilbertz, M. Lehmann, "Copper phthalocyanine suspended gate field effect transistors for NO2 detection". *Sensors and Actuators, B 118*, pp.249-253, 2006.

[128] M. Burgmair, M. Zimmer, I. Eisele. "Field effect transducers for work function gas measurements: device improvements and comparison of performance", *Sensors and Actuators, B 93,* pp.271-277, 2003.

[129] M. Bouvet, A. Bioanal. "Phthalocyanine-based field-effect transistors as gas sensors", *Chem.,* vol.384, pp.366-371, 2006.

[130] Z.T. Zhu, J.T. Mason, R. Dieckmann, "Humidity sensors based on pentacene thin-film transistors". *Appl. Phys. Lett.,* vol.81, pp.4643-4648, 2002.

[131] Kh.S. Karimov, Kh.M. Akhmedov, A.M. Achourov, R.M. Gul, "Influence of Centrifugation on the Conductivity of Poly-N-epoxypropylcarbazole-Copper Phthalocyanine Films Deposited from Solution, Processing by Centrifugation", edited by L.L. Regel and W.R. Wilcox. Kluwer Academic/Plenum Press, New York. 93, 2001.

[132] N.F. Mott, E.A. Davis, "Electronic Processes in Noncrystalline Materials", *Clarendon Press, Oxford,* 1971.

[133] K. Akhmedov, M. Rahimova, Kh.S. Karimov, M.I. Cherkashin. *J. Acad. Sci. Tajikistan.,* vol.25, pp.24-29, 1982.

[134] I. Murtaza, Kh.S. Karimov, Z. Ahmad, I. Qazi, M. Mahroof-Tahir, T.A. Khan, "Impedance Hygrometer based on Cellulose and CuPc", *J. Semicon.*, vol.31/5, 05001. pp.1-5. 2010.

[135] W. Brutting. "Physics of organic semiconductors". Wiley-vch Verlag GmbH & Co. KGaA, Weinheim, Germany, 2005.

[136] G.A. Chamberlain, "Organic Solar Cells: A Review", *Sol. Cells.*, vol.8, pp.47-83, 1983.

[137] Kh.M. Akhmedov, Kh.S. Karimov, M.I. Fiodorov, "Organic Solar Cells", *Geliotekhnika*,vol.1-3, pp.178-183, 1995.

[138] C.W.Tang, "Two-layer organic photovoltaic cell", *Appl. Phys. Lett.,* 48, 183-185, 1986.

[139] P. Peumans, V. Bulovic, S.R. Forrest, "Efficient photon harvesting at high optical intensities in ultrathin organic double-heterostructure photovoltaic diodes". *Appl. Phys. Lett.*, vol.76, pp.2650-2652, 2000.

[140] J.H. Schon, Ch. Kloc, E. Bucher, B. Batlogg, "Efficient organic photovoltaic diodes based on doped pentacene", *Nature.*, vol.403, pp.408-410, 2000.

[141] J.F. Nierengarten, J.F. Eckert, J.F. Nicoud, L. Ouali, V.V. Krasnikov, G Hadziioannou, Krasnikov, "Synthesis of a C-60-oligophenylenevinylene hybrid and its incorporation in a photovoltaic device", *Chem. Commun.,* pp.617-618, 1999.

[142] K. Yoshino, K. Tada, A. Fujii, E.M. Conwell, A.A. Zakhidov, "Novel photovoltaic devices based on donor-acceptor molecular and conducting polymer systems", *IEEE. Trans. Electron. Devices,* vol.44, pp.1315-1324, 1997.

[143] M. Granström, K. Petritsch, A. Lux, M. R. Andersson and R. H. Friend, "Laminated fabrication of polymeric photovoltaic diodes", *Nature.,* vol.395, pp.257-260, 1998.

*[144]* B. O'Regan and M. Grätzel, *"A low-cost, high-efficiency solar cell based on dye-sensitized colloidal TiO2 films", Nature.,* vol.353, pp.737–740, 1991.

[145] M. Gratzel, *"Photoelectrochemical Cells", Nature.,* vol. 414, pp. 338-344, 2001.

[146] M. Gratzel, "Dye-sensitized solar cells", *J. Photochem. Photobiol. C. Photochem. Rev.,* vol.4, pp.145-153, 2003.

[147] L. Schmidt-Mende, U. Bach, R. Humphry-Baker, T. Horiuchi, H. Miura, S. Ito, S. Uchida, and M. Gratzel, "Organic dye for highly efficient solidstate dye-sensitized solar cells", *Adv. Mater,* vol.17/7, pp.813-815, 2005.

[148] D.Y. Goswami, S. Vijayaraghavan, S. Lu, G. Tamm, "New and emerging developments in solar energy", *Sol. Energ,* vol.76, pp.33-43, 2004.

[149] J.D. Kraus, "Antennas", 2nd Edition, McGraw-Hill International Editions, New York, 1988.

[150] W.C. Brown, "The History of Power Transmission by Radio Waves", *Microwave Theory and Techniques. IEEE. Trans. MT32.* vol. 9, pp.1230-1242, 1984.

[151] R.L. Bailey, "A Proposed New Concept for a Solar-Energy Converter", *J. Eng. Power.,* vol.94, pp.73-77, 1972.

[152] G. Hadziioannou and P.F. van Hutten, "Semiconducting polymers, Chemistry, physics and engineering". Edited by WILEY-VCH Verlag GmbH, Weinheim, Germany, 2000.

[153] M. Shaw, P.F. Seidler, "Organic Electronics: Introduction", *Ibm. J. Res. Dev.*, vol.45/1, pp.3-9, 2001.

[154] T. Tsujimura *et al*, "A 20-in. OLED display driven by Super-a-SiTechnology", *SID Symposium Digest,* vol.34/6, 2003.

[155] Z. Liu, J. Pinto, J. Soares, E. Pereira, "Efficient multilayer organic light emitting diode", *Synthetic. Metals.*, vol.122, pp.177-179, 2001.

[156] T.A. Beierlein, "Combinatorial Methods for Organic Light-Emitting Materials and Devices", *Ph.D. Thesis,* University of Bayreuth, Cuvillier Verlag, Gottingen, Germany, 2003.

[157] M. Neghabi, A. Behjat, "Current. Electrical and electroluminescence properties of ITO/PEDOT:PSS/TPD:Alq3:C60/Al organic light emitting diodes", *Appl. Phys.,* vol.12, pp.597-601, 2012.

[158] J.K. Kim, D-H. Lee, D.H. Wang, S.M. Cho, J.Y. Lee, J.H. park, "Optimized composition for efficient and low potential operative host/guest bilayer polymer electrophosphorescence devices", *J. Luminescence,* vol.132, pp.870-874, 2012.

[159] W.J. Hyun, J.K. Park, O.O. Park, B.D. Chin, "Hysteresis-less inverted solar cells: Performance enhancement by Li-assited hole conduction", *J. Phys. D: Appl. Phys.* vol.45, pp.1-5, 025101, 2012.

[160] M.G. Shin, S.O. Kim, H.T. Park, S.J. Park, H.S. Yu, Y-H. Kim, "Synthesis and characterization of ortho-twisted asymmetric anthracene derivatives for blue organic light emitting diodes (OLEDs)", *Dyes. Pigmen.,* vol.92, pp.1075-1082, 2012.

[161] Kh.S. Karimov, M. Saleem, Z.M. Karieva, A. Mateen, M.T.S. Chani and Q. Zafar, "Humidity Sensing Properties of Cu2O-PEPC Nanocomposite Films", *J. Semicond.*, vol.33/7. doi:10.1088/1674-4926/33/7/073001, 2012.

[162] Kh.S. Karimov, I. Qazi, M. Mahroof-Tahir, I. Murtaza, T. Amin, S.Z. Abbas, "Opto-electronic properties of ITO/CuPc/NiPc/Al and ITO/NiPc/CuPc/Al double junction cells", *Proc. 6th Intl. Conf. on Elect. Eng.,* Cairo, Egypt. Pp.1014-1027. 27-29 May 2008.

[163] Kh.S. Karimov, M. Saleem, Z.M. Karieva, A. Khan, T.A. Qasuria, "A carbon nanotube-based pressure sensor", *Phys. Scr.* vol.83/6. doi:10.1088/0031-8949/83/06/065703, 2011.

[164] Kh.S. Karimov, M. Saleem, Z.M. Karieva, A. Khan, A. Mateen, "CNT Based Displacement Sensor", *Intl. J. Mat. Res.*, vol.103, pp.897-900, 2012.

[165] Kh.S. Karimov, F.A. Khalid, M.T.S. Chani, "Carbon nanotubes based strain sensors", *Measurement.*, vol.45, 5, pp.918–921, 2012.

CHAPTER 4

# Nanocomposites and Importance of Nanofiller in Nanocomposites

**Sher Bahadar Khan[1,2,*], Kalsoom Akhtar[3] and Abdullah M. Asiri[1,2]**

*[1]Center of Excellence for Advanced Materials Research, King Abdulaziz University, Jeddah 21589, Saudi Arabia; [2]Chemistry Department, Faculty of Science, King Abdulaziz University, Jeddah 21589, Saudi Arabia and [3]Division of Nano Sciences and Department of Chemistry, Ewha Womans University, Seoul 120-750, Korea*

**Abstract:** The progress of nanostructured materials have untied a new standard whereas the nanoreinforcement and matrix resin can be modified to optimize the physical and mechanical properties of various composite materials. This resulted in appearance of new frontiers in mechanical and electronic structures centering on multi functionalized, self-assembling, self-healing materials and have shown outstanding applications in many fields. This chapter emphasizes about nanocomposites, their classification and characterization of different types of nanocomposites.

**Keywords:** Clay, nanocomposites, nanofillers, nanotechnology, synthetic methods, XRD.

## 1. INTRODUCTION

It is clear from the heading that this book is related to nanomaterials which are the materials having minimum one aspect (dimension) below 100 nm. In nanotechnology, nano indicate too small. The prefix ''Nano'' symbolizes nanometer, which equals one billionth of a meter ($10^{-9}$ meter) [1]. The nanometer size ranges from 1 to 100 nm. If a nanostructure magnify over 10 million times then its features possibly can see with the naked eye. To realize the nanometer dimension, a nanometer is equal to 1/80,000 of human hair diameter and 1/2,000 of human red blood cell length. Nanotechnology refers to knowledge on nano-level that has function in the real globe [2]. Nanotechnology generally employed to describe science and technology related to the control and exploitation of materials and devices on nanometer scale *i.e.,* a size below 100 nm in dimension [3].

***Corresponding author Sher Bahadar Khan:** Center of Excellence for Advanced Materials Research (CEAMR) and department of Chemistry King Abdulaziz University, Jeddah, Saudi Arabia 21589; Chemistry Department and Center of Excellence for Advanced Materials Research (CEAMR), King Abdulaziz University, Jeddah 21589, Saudi Arabia; Email: sbkhan@kau.edu.sa

On the basis of dimensions, nanomaterials have been divided into various categories *i.e.,* zero-dimensional nanomaterials, one-dimensional nanomaterials and two-dimensional nanomaterials. Zero-dimensional nanomaterials are those compounds which have zero dimensions. Nanoparticles are the example of zero-dimensional nanomaterials. Lamellar (layered) structures are one-dimensional nanostructures having much bigger length and width as compared to nanoscale thickness. Thin films used for electronic devices are examples of lamellar (layered) structure. Two-dimensional nanostructures have much bigger length as compared to nanoscale width or diameter. Nanowires and nanotubes are examples of two-dimensional nanostructure. The nanostructures are three-dimensional and encompass of crystallites, or in some cases quasicrystals and/or amorphous substance of nanoscale dimension. "Nanocrystalline" refers to crystalline materials possessing a size <100 nm, but this classification is not always precise.

Even though nanotechnology has been around since the beginning of life, the innovation of nanotechnology has not been endorsed until lately by Richard Feynman in 1959. Since then nanotechnology has advanced into a multidisciplinary field of applied science and technology in the design, production, characterization [1, 3], and application in different fields. In the early 21$^{st}$ century, nanotechnology is expected to have an insightful and useful effect on economy and society, similar to other technologies. Nanotechnology is considered as the next industrial revolution as it assures step forward in many areas of research [2].

Generally, two main approaches have been used in order to enter the nano-world recognized as 'bottom-up' and 'top-down' approaches. In the former approach, assembling of nanostructure materials and devices takes place from atoms into molecules, while the latter involves preparing nanostructures and devices from bigger units to low level. Just in recent years, the progression in both procedures has been hastened with the availability and development of highly susceptible apparatus like atomic force microscope (AFM), scanning tunnelling microscope (STM), electron beam lithography, molecular beam epitaxy, *etc.* These instruments can be used for the inspection and exploitation of new nano-materials. Most research being carried worldwide in developing nanotechnology and their emerging applications have focused on manufacturing nanotechnology-based objects at high rate, low cost and reliability [3].

## 2. NANOMATERIALS

Although nanotechnology is quite new, nanomaterials have been used in applications that already affecting our daily lives, from drinking-water purification using nanofilters to stain-resistant ‘‘nanopants’’ made from fibers processed with fluorinated nanopolymers to food packing using the antibacterial and deodorizing properties of nanoparticles. Nanotechnology have been extensively used in electronics, magnetics, optoelectronics, energy, catalytic, biomedicine, pharmaceuticals, clothing, cosmetics, and environmental detection and monitoring [1]. Depending on the desired application, nanomaterials can be engineered into different forms as particles, rods, tubes, wires, layers, films, composites of varying sizes, ranging from a few to several hundred nanometers [1, 4].

The tiny size of nano-material guides to a number of changes in their physical properties such as enhance in volume to surface area ratio and domination of quantum effects on the particles [5]. At a scale between 1 and 100 nm some unique and interesting properties can arise, possessing for example novel mechanical, thermal, catalytic, magnetic, optical, or electronic properties. It has been shown that the properties of matter are controlled at this range. It can be said for instance that the chemical properties get benefit of high surface to volume ratio which is essential in various application [3, 6].

Nanomaterials can be classified into three types of nanostructures associated with the term quantum as variations in properties take place from the quantum-mechanical nature of physics in the area of the ultra-small. These types include a quantum well, wire and dot. A quantum well referred to a three-dimensional nanostructure with only one length in the nano-dimension. If two sides of any three-dimensional material are in nanometer length, the material is denoted as quantum wire. If all three dimensions are in the range of nano, the term quantum dot is applied [2].

Various bulk materials, with either organic or inorganic components as their major constituents, can be used to produce different types of nanomaterials. Nanomaterials can be sorted into three groups: polymeric, semiconductor, and metallic. Organic polymeric nanomaterials embrace a great and diverse group of carbon nanotubes, emulsions, liposomes, dendrimers, *etc*. Metallic nanomaterials include several elemental metals, metal oxides, and magnetic materials. Fluorescent quantum dots (QDs) are considered as the most commonly used semiconductor nanomaterials today [1].

With the advent of nanomaterials research, synthesis of inorganic nanoparticles, that is, particles having minimum one dimension in nanometer range, is promptly attainable [7]. As we have already mentioned, nanomaterials may synthesized using either the molecular nanotechnology (bottom-up approach) or the top-down approach. The bottom-up approach could be further classified according to the temperature used into wet and dry synthesis. The dry synthesis include methods require using high temperature (> 250 °C) like gas condensation, wire explosion and liquid aerosol thermolysis. Meanwhile, the wet synthesis (< 250 °C) involves using mechanical attrition, replication method and chemical precipitation. The chemical precipitation involves hydrothermal synthesis, inverse micelle method and sol–gel synthesis (either hydrolytic or non-hydrolytic) [5].

## 3. IMPORTANCE OF NANOMATERIALS

Nanomaterials are material of elevated potency and rising with large applications. Nanotechnology already finds many applications in different areas such as new variety of foods, medical equipments, chemical coatings, human health testing kits, security system's sensors, water purification components for manned space craft, displays for hand-held computer games, and high-resolution cinema screens. Nanotechnology is most important expression for materials and appliances that work at the nanoscale. Nanomaterials show remarkable physical and chemical properties originated due to numerous aspects comprising amplification in surface area.

There are two major ways of access to the nano-world. Molecular development includes the manipulation of unit atoms (working from the bottom up); and from larger size to smaller (building from the top down). The progress of nanotechnology as the most recent mega propensity in science and engineering which will carry a wave of radical innovation because of its prospectively wide impact, spark new industrial rebellion in different application regions. It has its impact on all materials: ceramics, metals, polymers, and biomaterials.

Some parameters have significance importance in the synthesis of nanomaterials. These are as follows: 1) size-controlled synthesis and 2) shape-controlled synthesis. There are two main classifications of nanoscale material based on path-way *i.e.* used to build them. This building of nanostructures or devices can be achieved either by top down approach or by bottom up approach. "Top-down" is defined as the building something by starting with a large component and carving away material. "Bottom-up" is stated as building something by assembling smaller components (like building a car engine).

The study of matter interactions at atomic level gives a wide scope in almost every discipline of science *i.e.,* physics, biotechnology, engineering, and medical. If we notice, nature also starts from nano scale *i.e.,* the basic building blocks of all biomolecules works at nano scale [1].

From the concise introduction mentioned above a question may come in mind that if nanotechnology has such a numerous applications then why it becomes common from just last 10 years. Actually, manipulation at atomic level of matter requires proper instruments and latest techniques which were not available in the previous times even now researchers have to face lots of difficulties to study the mechanisms at atomic scale [2].

## 4. NANOCOMPOSITES

Nanocomposites are nanostructure materials composed of two or more phases. Among different phases of nanocomposites, one phase has one or more dimensions below 100 nm. Nanocomposites are combination of bulk material and nano scale phase which differ in structure and chemistry.

Nanocomposites have shown outstanding performance in different areas, particularly in sensor fabrication [17-21]. Therefore, nanocomposites got remarkable attention due to their exceptional assets, characteristic behavior and broad range of applications in different areas particularly concern about human health and environmental application [22]. These widespread applications of nanocomposites are essentially considered due to meticulous shape and high surface area of the nanomaterials introduced into the polymer matrix [16]. Introduction of nanomaterial into the matrix of polymer is one of the major noteworthy means to adjust the texture of the nanostructures. Nanomaterials displayed prospective role as doped material and have improved the properties of polymer [23-26].

In nanocomposites, change in composition has high influence on the mechanical, chemical, electrical and optical properties. Nanocomposite differs from the usual composite because nanocomposites possess extraordinary lofty surface to volume ratio of the nano phase. The interface area among the matrix and nano phase is normally larger than for usual composite materials. By adding nanoclay, the electrical and thermal conductivity can be greatly enhanced. Similarly other nanoparticles can affect the optical, mechanical, thermal properties and resistance to bear damages. Basically the nano strengthening is discreted onto the matrix during its formation. The percentage by weight of nanoparticle introduce can stay very little because of tiny filler percolation threshold.

The field of material science and nanotechnology has motivated scientists all over the world in order to advance materials with innovative properties that cannot be met by the conventional materials [8]. Advancement of nanocomposites as one of the growing topics of composites research has increased markedly in recent years due to the great potential associated with this new generation of extraordinary materials [9-11].

Silica containing nanocomposites possess customized assets and incomparable behaviors appropriate for large variety of applications [7-11]. The key course for the development of these innovative materials involves the sol-gel technique [8]. This multipurpose synthetic procedure engages the hydrolysis and condensation of molecules (typically alkoxides) in mild conditions, which gives hybrid materials in which one of the components exist at nanosize scale [7]. These substances have extraordinary thermal, mechanical assets and capable of different applications in a variety of areas [12-16].

Polymer-inorganic nanocomposite based on nanoclay is another growing class of innovative nano-structures with generous range of useful applications [21-23]. The main method for the development of such advanced materials is simple intercalation reaction. The copious applications of these materials are principally due to nano size and huge surface area of the nano-fillers, the strong interfacial contact linking the nano-filler and the polymer [22-24]. These properties are accountable for grand progress in many assets of polymer.

Terms such as "hybrids" and "nanocomposites" have been widely used for many different systems in the material filed and therefore it is rather confusing to recognize whether materials are "nanocomposites" or not. Thus, it is essential to clarify the terms "hybrids" and "nanocomposites" prior to discuss nanocomposites. Actually, no apparent borderline exist between hybrids and nanocomposites. The hybrid materials are defined more broadly as materials that contain two substances mix together on the molecular level, whereas nanocomposite refers to structural components in a distinct size range of 1–100 nm. Thus normally the term nanocomposites is used if distinct structural units in the respective size system are utilized (*e.g.*, nanocomposites are using nanoparticles, nanorods, carbon nanotubes and clay minerals) and the term "hybrid" is commonly used when the inorganic particles are synthesized *in situ* by molecular precursors (*e.g.*, applying the sol-gel process) regardless to the possible exchanges between inorganic and organic species [13]. Yet, several reports showed a tendency to use the term "nanocomposites" even though the scale of all phases was above 100 nm [14].

Most of the composite materials contain two main phases: the continuous phase (the matrix), and the phase it surrounds (the dispersed phase, fillers or reinforcements) [8, 12]. As shown in Fig. (**1**), the interaction between these two phases causes the formation of the interfacial region, which extends well beyond the adsorption layer of the matrix bound to the filler surface. The interfacial region interconnects between the matrix and filler and ascribes structure and properties different than either the filler or matrix phases depending to its proximity to the bound surface [7, 15].

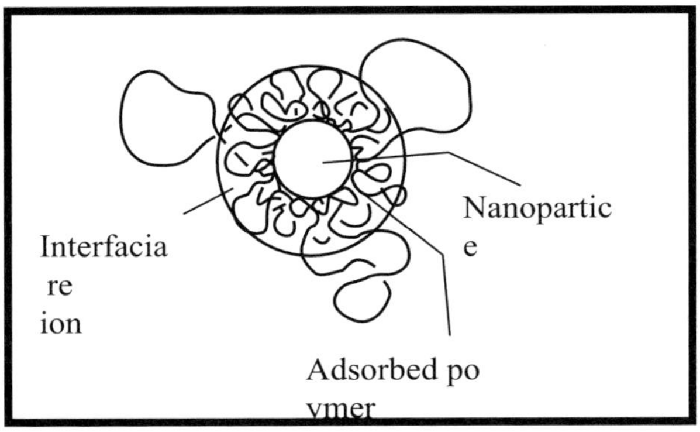


**Figure 1:** Schematic of interfacial region between a filler and polymer matrix [7].

Depending on the geometrical shape of the dispersed phase, nanocomposites can be divided into three groups: One-, two-, and three- dimensional nanocomposites. Each of these groups can be further divided into at least two subdivisions as shown in Fig. (**2**). The matrix in one-dimensional nanocomposites are reinforced with layered materials with a thickness in the nanometer range such as clays, graphite, layered silicate, and other layered minerals. For two dimensional nanocomposites, the matrix is fibrous reinforced (*i.e.*, with tubes or whiskers) with a diameter in the nanometer range; examples are carbon nanotubes, cellulose, and chitin whiskers. The dispersed phase in the three-dimensional nanocomposites has the geometry of spherical particles in the nanometer range. Nano-sized metals, metal oxides, silica, carbon and other organic and inorganic particles are examples of nanoreinforcement in this group [8, 9, 11].

In the field of composites research, different materials have been used in hopes of exploiting the unique properties of their conventional constituents, which are not obtained in one component by itself [8]. Polymers for example carbohydrates, lipids, and proteins along with many natural reagents form many strong composites in nature e.g bones, shells, and wood. These nanocomposites (PNC; Fig. **3**) are designed by merging particles, layers or fibers, in which one phase is in the order of a nanometer (less than 100 nm) [4, 6, 7, 9, 10]. However, comparing to metals and ceramics, polymer systems usually display less attractive mechanical properties such as low strength and low elastic modulus [7]. Even though, in the last decades, polymer nanocomposites are capable to make a great display in the media and throughout numerous industries [6, 7, 10]. They have been broadly exploited and anticipated to be one of the most beneficial applications of nanotechnology due to their light weight, design flexibility, processibility [7], dielectric [14], solubility and/or film-forming qualities. Moreover, they have been utilized to contribute additional properties such as improving protection from surrounding media (*e.g.* air or water vapor) or preventing of agglomeration and precipitation of the metal particles [6]. Thus, the major concern in PNC research is to enhance the mechanical properties of the polymeric components such as stiffness, toughness, and ambient and high-temperature strength by using molecular or nanoscale reinforcements [7, 8, 10].

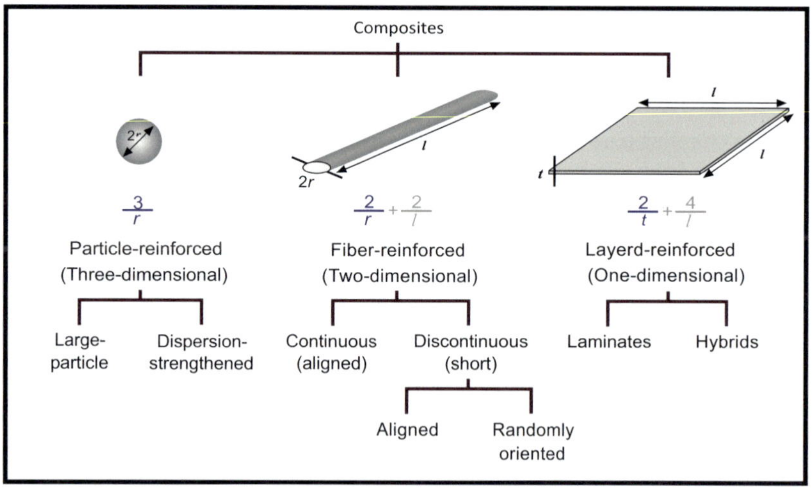


**Figure 2:** A classification scheme for the various composite types [8, 16].

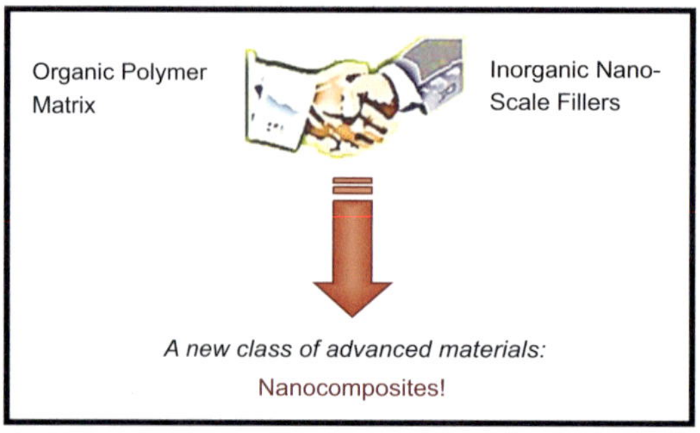


**Figure 3:** Illustration of nanocomposite.

Even though PNC embracing nanomaterials are commonly examined where fortification of the polymer matrix is attained [7, 17], several other variations are also in vigorous study and commercialization [17]. Depending on the nanomaterial used and the composition of the nanocomposite, it is possible to design materials with remarkable property enhancements. PNCs often can significantly enhance and influence many basic properties of the composite (*e.g.*, mechanical, thermal, rheological, optical, electrical, *etc.*) [9, 10, 18]. The advantages of incorporating nanoparticle can lead to sufficient property profile for countless application and utilization possibilities, which can't be met adequately by the incorporation of analogous larger particle.

Comparing with the conventional polymer microcomposites, polymer nanocomposites are demonstrated by separate fillers on the order of few nanometers, whereas the conventional composites are loaded with fillers on the order of microns. PNC exhibit a drastic alternative to conventional polymer based microcomposite [15]. The nanocomposite properties would be significantly affected by size "size effects" of its constituent phases [4].

In contrast to conventional composites, nanocomposites are notable for their use of relatively small amounts of fillers (less than 10%) to optimally achieve the desired properties [7, 14-16]. This is of considerable advantage; hence at high concentrations needed in the conventional composites, the fillers can adversely impact other benefits of the polymers such as processability, appearance [7], ductility [9] and toughness. However, in PNC these properties will obtain without sacrificing the polymer's inherent properties, or adding excessive weight [15]. The possibilities for nanocomposites improvements, often at such low amount, originate from the properties and the nanometric size of the fillers. Compared to micro or macro sized materials, the use of nano sized particles can significantly reduce the likelihood finding defects, such as grain boundaries, voids, dislocations and imperfections, which leads to considerably improved properties in the composite. In addition, the relatively low content and extremely small size of the nanoparticles can help preserving the optical clarity or transparency of the composite; since materials much smaller than the light wavelength do not scatter light [9].

In addition to the size effect of the constitutes, the properties of composite materials rely on several other parameters including the properties of the constituent phases, their volume fraction, matrix-inclusion interface and the geometry of the dispersed phase *i.e.*, their particle shape, dispersion, and orientation [7, 8, 10]. Additional variables such as interaction type and strength between the monomers and the nanoparticles, nanoparticle mobility, temperature, entanglement of the polymers, relative crystalline or amorphous nature of the polymer matrix, and degree of polymerization are also have significant effect on the properties of the composites [15, 18].

A morphological feature of great importance related to the structure–property association of nanocomposites is surface area of the fillers in either form (particles, fibrous or layers). The surface area is not directly dependent on material's diameter in case of particles and fibers, consequently, the lesser the diameter, larger the surface area. For the layered material, especially for nanomaterials, the surface area/volume is dominated by the thickness of the layers. Therefore, change in the diameter of a particle or fibrous, or the thickness of a layer, or a change in the range from micro to nano meter, will influence the surface area-to-volume ratio *via* three orders of magnitude. Nanostructured materials have an extremely high surface area to a given volume and thus the switch from micro to nano scale yields dramatic alterations in physical and chemical properties [4, 7, 10, 11].

Significant improvements in composite assets may attained reliant on the degree of mixing, the nature of the constituents used (layered or nanofiber, cation exchange capacity, and polymer matrix) and the process of synthesis [4]. For example, stacked silicate layers, may form three different types of composites during the nanocomposite formation process (phase separated microcomposite, and intercalated and exfoliated nanocomposites). In order to obtain polymer/clay nanocomposites, these layers must be opened up to increase the interfacial contact between the polymer and clay. The nanocomposite is said to be intercalated when the layers are begun to separate and single extensive polymer chain can enter into the silicate layers. In this intercalated structure a well-organized multilayer morphology outcomes by means of polymer and inorganic layers one above other. The more effective structure is referred to as exfoliated or delaminated. It occurs when the silicate are no longer stacked and are wholly and unvaryingly dispersed in the polymeric matrix. Yet, phase-separated microcomposites may obtain instead of the desired nanocomposites in case the polymer is incapable to intercalate among the silicate interlayers. In this case, the properties are same as those for traditional microcomposites. Thus, it can be said that intercalated structure is more desirable than traditional phase separated, but exfoliated nanocomposite is the most effective one [4, 16].

Similarly, in fibrous or particle-resistant PNCs, spreading of nanoparticle in polymer matrix and sticking together at the particle–matrix interface show decisive role in defining the assets of the nanocomposite. Good dispersion will present improved properties as compared to typical composites, whereas a inadequately dispersed nanomaterial may worsen these properties [4, 12]. The degree with which the nanoparticles could be uniformly spread in polymer matrix would considerably influence the thermal, mechanical and optoelectronic assets of the composite. However, achieving proper dispersion for nanoparticles in the composites is extremely difficult, since nanoparticles have a propensity to aggregate together during fabrication. Hence several parameters and techniques have been considered to control the aggregate size and dispersion. The parameters include controlling residence time, pressure, loading, nature of precursor, thermal treatment, densification, surface chemistry, structure modification and dispersing energy, while the techniques embrace methods like using mechanical agitation (*e.g.*, ball milling or magnetic stirring), ultrasonic vibration, shear mixing, non-contact mixing and using the dispersing agent [19].

Furthermore, optimizing the interfacial bond connecting the nano-filler and the polymer matrix can alter the assets of nanocomposite. For instance, proper attraction at the border line will enhance properties such as inter-laminar shear strength, delamination resistance, fatigue, and corrosion resistance. The level of contact among the nano-filler and the polymer is an imperative matter. The weak interaction, (*e.g.*, hydrogen bonding, van der Waals forces) may obtain during physical or simple mechanical mixings. On the other hand, using typical preparation process for the fabrication of organic–inorganic nanocomposites such as sol gel technique may introduce a strong chemical bond (covalent or ionic–covalent bond) linking the organic and filler [4, 12].

In fact, several approaches were employed to synthesis polymer inorganic nanocomposites containing sol gel process, solution mixing process, melt mixing process, in-situ polymerization and in-situ intercalative polymerization. However, all these approaches can be essentially divided into two classes "*in situ*" and "*ex situ*" depending on whether the inorganic component is developed in the presence of the polymer (monomer) matrix or pre-fabricated [10].

Several particles are utilized to develop polymer/particle nanocomposites, e.g metals (Al, Fe, Au, Ag, *etc.*), metal oxides (ZnO, $Al_2O_3$, $CaCO_3$, $TiO_2$, *etc.*), nonmetal oxides ($SiO_2$) and other materials (SiC). The assortment of nanoparticles relies on the anticipated thermal, mechanical, and electrical properties of the nanocomposites. For instance, aluminum nanoparticles are frequently preferred for their good conductivity; calcium carbonate particles are preferred due to their relative low cost, and silicon carbide (SiC) nanomaterials are chosen because of their high hardness, resistivity towards corrosion, and strength [4].

Nanocomposites are hard, lighter, stiffer, less fragile, and more dimple, scratch-resistant, eco-friendly, more flame retardant, less absorbent, improved conductors of electricity, and can be tinted more easily than typical plastics [20, 21]. Therefore, car companics are using nanocomposites in different car parts instead of plastics since last several years. In 2001, Toyota consumed nanocomposites for cars bumpers. In 2002, General Motors (GM) made nanocomposite "stepassists" – external running boards that help people get into and out of cars. In future, one will look at a car that in reality will looking at nanotechnology without even realizing it.

### 4.1. Difference Between Nanocomposites and Nanohybrids

There is no much difference in the terms "hybrids" and "nanocomposites" and therefore it is rather confusing to recognize whether materials are "nanocomposites" or not. The more extensive description of a hybrid is a matter that contains two substances mix together on molecular level. The term hybrid is generally utilized for inorganic materials which usually prepare *in situ* by sol-gel method [22] while the term "nanocomposites" stands for the materials which contain different phases with various compositions and among different phases at least one phase is in nanometer. For example, in polymer/silica nanocomposites, the phase in nanoscale is usually silica). Thus in fact there is no apparent borderline connecting "hybrids" and "nanocomposites" [22]. In spite of the above definition for "nanocomposites", there are several reports that claim nanocomposites instead of containing all phases above 100 nm. For example, in polymer/silica composites, all dimensions of silica particles are higher than 100 nm but still the composites are frequently called "nanocomposites" in numerous literatures.

Organic/inorganic composites that contain inorganic segments in nano size are named as nanocomposites. Organic/inorganic nanocomposites unite the particular properties of the inorganic material (*e.g.*, rigidity, thermal stability) and organic polymer (*e.g.*, flexibility, dielectric, ductility, and processibility) leading to materials with better properties. Considering polymer nanocomposites, a small size of the fillers direct to a remarkable increase in interfacial area as compared to the traditional composites [23, 24].

## 5. NANOFILLERS

Various inorganic nanomaterials were used as nanofillers such as zinc oxide, graphene oxide, graphite, carbon nanotubes, aluminum oxide, cuppor oxide, cerium oxide, tin oxide, silica nanoparticles, titanium oxide, layered double hydroxide, carbon nanofibers, naturally occuring fibers (hemp or flax), and POSS (polyhedral oligomeric silsesquioxane) and nanoclays [25] but the most common nanomaterials used so far is nanoclay. Among nanoclays, montmorillonite (a layered smectite) is mostly used in the preparation of nanocomposites.

Clays are inorganic materials with layered structures having negatively charged silica areas bind with each other by ions such as $Na^+$ or $K^+$ *etc* (Fig. **4**). These counter ions in the interlayer of mica and montmorillonite can be switched simply by organic cations *e.g.* alkyl ammonium ion or aryl ammonium ions by a method called intercalation reaction. The introduction of alkyl ammonium ions enlarges the space between the silicates layer of the clay and concurrently alter the hydrophilic behavior to hydrophobic which enhance the delamination of clay in polymer matrix [26, 27].

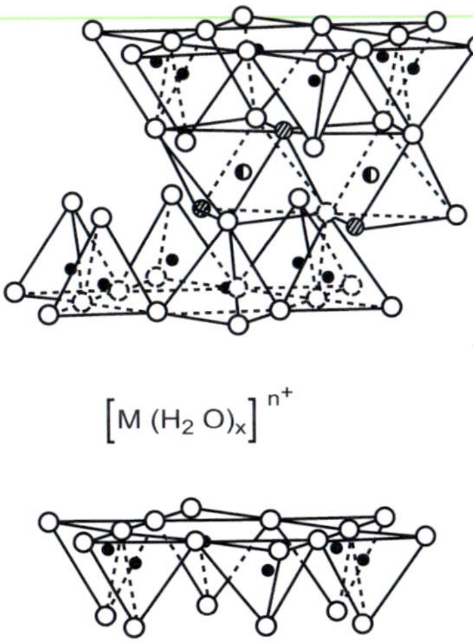


Schematic drawing of smectite clay: O, octahedrally coordinated $Mg^{2+}$ or tetrahedrally coordinated ion (Si); O, hydroxyl ion; and O, oxygen.

**Figure 4:** Clay structure.

## 6. ORGANO-CLAY

Organo-clays are mainly prepared by intercalation of quaternary amines into the silicate layers of the clay. The nitrogen part of the quaternary amine is the hydrophilic end and is positively charged. These positively charged quaternary amine exchange the sodium or calcium ions. Usually long chain amines are utilized to make the clay organophilic [21-25].

Organophilic clay can be used for different purposes [28] and one of the major frequently utilized organoclay is gained from montmorillonite (MMT). Organo-montmorillonite is composed of a number of silicate sheets with the 1.2-1.5 nm thickness and 100– 200 nm of lateral dimension [29, 30]. These sheets stake together with a gap among them, called interlayer. Thc thickness of single layer of the crystal and the spacing between the layers known as d-spacing ($d_{001}$) are measured with the help of X-ray diffraction. In polymer matrix, exfoliation is harder, because the hydrophilic nature of clay. In order to facilitate the process of exfoliation, the surface silicate layer is treated to make the surface more hydrophobic.

In general the exfoliation may be happen by reacting the cationic surfactant with alkylammonium cation (primary, secondary, tertiary and quaternary) [31, 32]. Such adjustment directs to enlarge d-spacing, connecting various layers of silicate caused by alkyl chain inserted between the layer and attain organo clay (OMMT).

## 7. TYPES OF NANOCOMPOSITES

There are two major types of nanocomposites (Fig. **5**).

### 7.1. Exfoliated Nanocomposites

The type of nanocomposites in which the clay is dispersed at the point of individual disks [29-33].

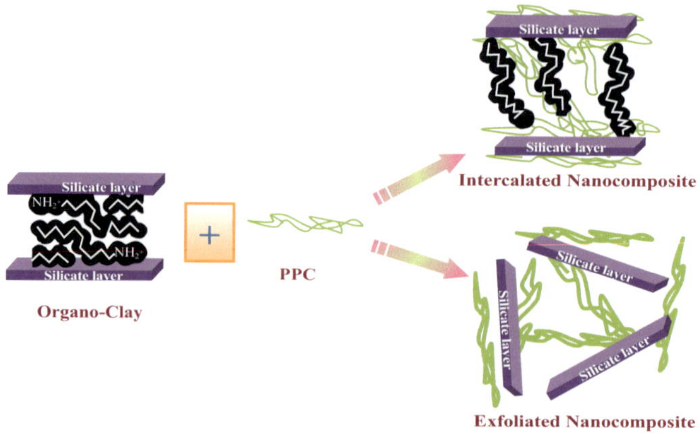


**Figure 5:** Types of nanocomposites.

### 7.2. Intercalated Nanocomposites

The nanocomposites in which the clay holds interlayer arrangement in a specific tactoid; but, intergallery spacing is enlarged compared to normal clay for the reason that stabilizing surfactants, compatibilizers, and/ or matrix polymers are penetrated inside the clay layers [33].

## 8. PREPARATION OF NANOCOMPOSITES

Preparation methods of polymer/silica nanocomposites can be divided into three general processes based on starting materials and preparation methods. These processes are:

1. Blending: The process in which polymer just mix with silica nanoparticles.

2. Sol-gel: This process is employing *in situ* either with polymer or at the time of monomer polymerization.

3. *In situ* polymerization: In this process, first disperse nanosilica inside the monomer then followed by the process of polymerization.

In case of polymer/clay nanocomposites, the blending method can further be divided into solution and melting method. Clay/polypropylene nanocomposites are nanocomposites prepared by melt intercalation process [21-33]. In both solution and melting, the clay (natural and synthetic) incorporation takes place at molecular level [32-34].

The incorporation of polymer into clay layered structure results in synthesis of either intercalated or exfoliated or partial exfoliated nanocomposites.

*Intercalated nanocomsites: In this type of nanocomposites, the polymer chain is inserted inside the silicate layers where the polymer expands the interlayer space of the silicate layers by several nanometers. This type of nanocomposites have multilayered structure containing polymer chains inside silicate layers.*

*Exfoliated nanocomposites: In this type of nanocomposites, silicate layers remain scattered and far from each other in such a way that can not interact with each other* [21-34].

> *Partially exfoliated nanocomposites: In this type of structures, nanocomposites contain both intercalated and exfoliated structures. Both intercalated and exfoliated structure exists at the same time in polymer matrix.*

In most of the cases, clay based nanocomposites have partially exfoliated structures [28-31].

Since long time it has remained acknowledged, that changes could be made in polymer properties by mixing with synthetic and minerals clay but the subject of polymer/silicate nanocomposites has achieved great impetus lately [25-34]. One of the examples is polypropylene based nanocomposites. The big marketable significance of PP and the recent advancement of nanocomposites have motivated the scientists toward the preparation and deep analysis of PP composites [25-34].

The preparation of nylon-6/montmorillonite showing amazing improvement in thermal and mechanical properties with very small content of montmorillonite and the development of melting method for the preparation of polymer/clay nanocomposites were the key findings in the field of nanocomposite. These were the breakthrough which inspired the scientists for dynamic research and brought vigorous motivation for the industrial applications of nanocomposites. These motivations brought dramatic improvements in the field of nanocomposite and resulted in the preparation of many composites with different properties using clays as inorganic silicate layers. Some of the achievements were the dramatic enhancement in flame retardant and barrier properties by using clay as nanofillers [27-35].

The nature of the clay is also crucial for the synthesis of nanocomposites because some clays are miscible with hydrophilic polymers while the other type of clays are miscible for hydrophobic polymers. For the synthesis of poly (ethylene oxide) and poly(vinyl alcohol) nanocomposite, natural montmorillonite can be used because of its miscible nature with hydrophilic polymers while for other type of polymers, organophilic clays can be used which can prepare by replacing alkali counterions of pristine clay with cationic-organic surfactants, such as alkylammoniums.

One of the examples of nanocomposite using organo-clay is polyprophylene carbonate (PPC)/ organophilic clay (C-20B) nanocomposite which prepared by solution blending method [21-37]. The PPC/C-20B nanocomposite was prepared by simple dissolution of PPC in acetone and mixed with dispersed solution of C-20B. The detail preparation process of nanocomposite is shown in Fig. (**6**).

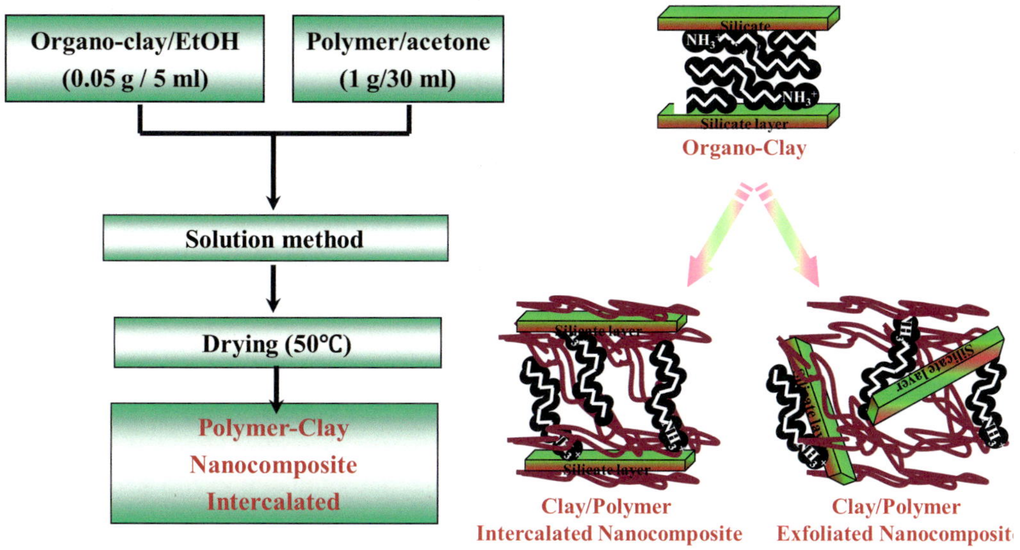


**Figure 6:** Preparation process of polymer-clay nanocomposite.

## 9. CHARACTERIZATION OF NANOCOMPOSITES

### 9.1. X-Ray Diffraction (XRD)

X-ray powder diffraction (XRD), is one of the fast and consistent technique for the determination of phase, unit cell, crystal arrangements, and spacing of atomic layers in various substances [21-42]. It is one of the best techniques for the morphological and structural characterization of nanocomposites and a number of other crystalline materials, providing a fast and robust way for nanocomposites characterization and classification. Amorphous and crystalline solid material is primarily classifying by this method as well as it discriminate specifically the intercalated and exfoliated nanocomposites.

Generally nanocomposites disrupt the stream of X-rays from a source and causing scattering of X-rays. This type of scattering is called diffraction. These scattered X-rays are detected by a detector. Different crystalline solids scatter the X-rays in different way and thus every crystalline solid generates a unique diffraction pattern. XRD differentiate crystalline and non-crystalline materials which lies at atomic level within the molecular make up of each type of solid material.

In this method a beam is directed to the plane of crystalline materials, some of the portion of the incident beam is transmitted, some is absorbed by the material, while the remaining part is refracted as well as diffracted in various directions.

The phenomenon of the diffraction caused by incident beam on the crystalline material is closely related to rainbow, in which the light is diffracted by water droplets. X-rays are diffracted by each one mineral in a different way, depending on crystal lattice composition and arrangement [26-29].

In X-ray powder diffractometry, filament (generally made of copper) inside X-ray tube emits X-rays by applying current and approximately a voltage of 15-60 kilovolts. The applied current warm up a filament and the emitted electrons number depends on the current. Higher current results in higher number of electrons emission from the filament. The electrons were accelerated with high speed by the application of high voltage and directly hit the target. When these electrons hit the sample, it disrupts the stream of X-rays from a source and causing scattering of X-rays. This type of scattering is called diffraction. These scattered X-rays are detected by a detector. The detector sense the scattered X-rays and convert to a count rate either by microprocessor or electronically. By changing the angle between the X-ray source, sample and the detector at a specific rate between limits is X-ray scan. These X-rays are focused at the sample, and the diffracted rays are collected. A critical and important component of all diffraction is the angle among the incident and diffracted rays.

Bragg's Law is $n\lambda = 2d\sin\theta$, can successfully be applied to determine the distance between the plane of the crystalline sample (d spacing) where the integer $n$ is the order of the diffracted beam, 1 is the wavelength of the incident X-ray beam, $d$ is the distance between adjacent planes of atoms ($d$-spacing), and $\theta$ is the angle of incidence of the X-ray beam. Since we know $\lambda$ and we can measure $\theta$, we can calculate the $d$-spacing.

In crystalline phase the atom are arranged in regular pattern. The X-rays are diffracted through the inter planer spacing called miller indices. Simple Bragg equation is needed to generalize the principle of XRD. Two parallel rays are incident at an angle of theta $\theta$ at these planes. If the path length difference of the two waves are integral multiple of the wavelength than the reflected beam of maximum intensity will result (Fig. **7**).

In powder XRD the crystals are present in random arrangement and therefore lattice planes are present in every direction. For each set of planes, therefore, at least some crystals must be oriented at Bragg's angle to the incident beam thus diffraction occurs for these crystals and planes. Geiger counter or scintillation counter can detect these diffracted beams.

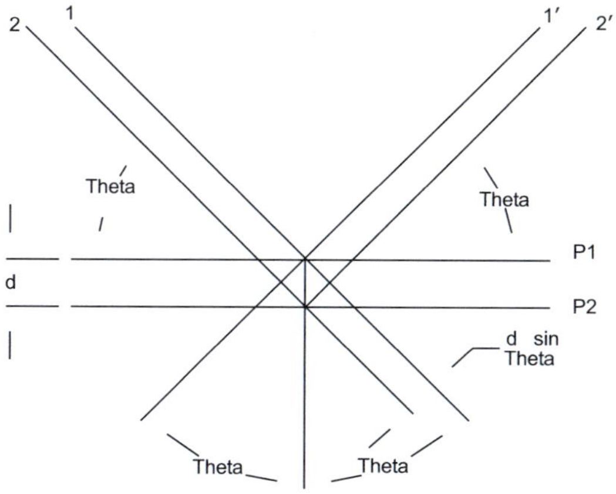


**Figure 7:** Principle of XRD.

By using small angle X ray scattering technique d-spacing in range of nanometers can be detected. According to Bragg's equation smaller angle of incident can detect large d-spacing by keeping the wavelength of rays constant. Due to this considerable ability SAXS is widely used in nanoscience.

Some of the examples in which XRD is applied to study the morphology of nanocomposites are shown in Fig. (**8**).

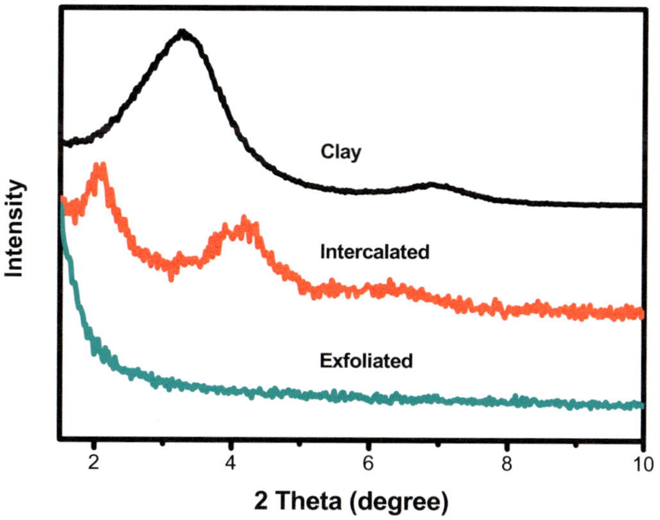


**Figure 8:** Typical XRD pattern of polymer/clay nanocomposites.

XRD pattern of clay display an intense peak at above 3° which shift in the intercalated nanocomposite after treating with polymer which indicates that interlayer basal spacing of the clay sheets has increased by intercalation of polymer. In intercalated nanocomposite, all the crystalline peaks of clay exist while in partially exfoliated nanocomposite, a weak crystalline peak appeared at above 6° which display that the sheets of the clay are disturbed containing both single sheet along with layered structure. In the exfoliated nanocomposite, the crystalline peaks disappeared after treatment with polymer which indicates that the sheets of the clay are well dispersed among the polymer matrix [38].

The XRD pattern which was observed for metal oxide/polymer nanocomposite is shown in Fig. (**9**).

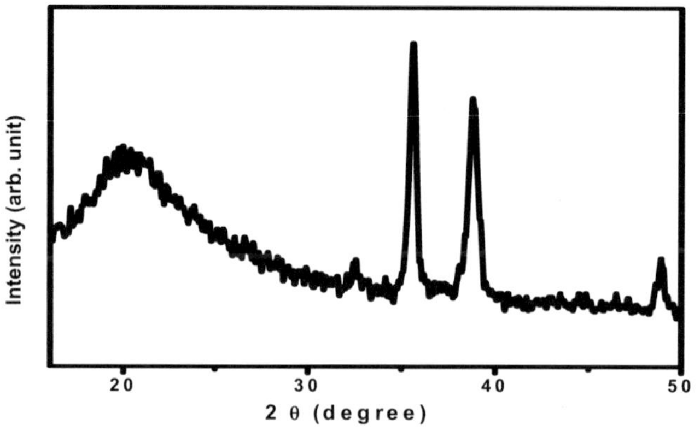


**Figure 9:** Typical XRD pattern of metal oxide/polymer nanocomposite.

The PPC/CuO diffraction pattern clearly indicates a hump at 20° which confirm existence of PPC in amorphous phase. Several other signals in spectrum were appeared for CuO at 32.5, 35.6, 38.7, 49.2, 58.5, 61.6, 66.4 and 68.2° [39]. The absence of any other peak in the spectrum except PPC and CuO confirmed PPC and CuO nanocomposite [39].

The XRD pattern which was observed for PPC/$SiO_2$ nanocomposite is shown in Fig. (**10**). XRD of PPC and PPC/silica nanocomposite did not show any typical crystalline peak but exhibit broad peak at 20° which indicate that PPC and PPC/silica nanocomposite are amorphous in their morphology. The XRD peak of PPC/silica nanocomposite demonstrated lower intensity as compared to pristine PPC which specify that silica may reduce the crystallinity of the PPC [40].

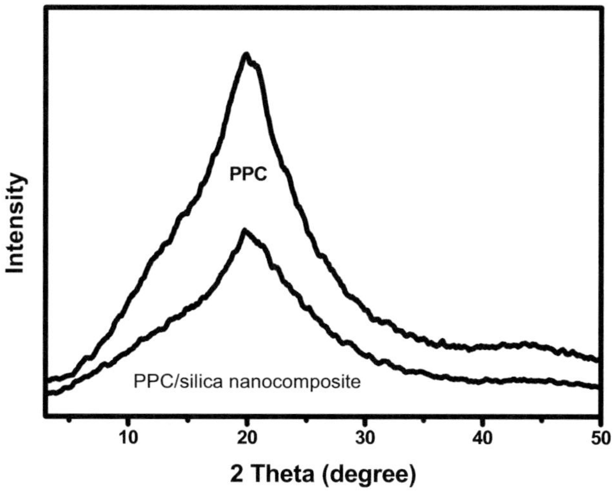


**Figure 10:** Typical XRD pattern of PPC/$SiO_2$ nanocomposite.

## CONCLUSION

Nanocomposite showed superior properties as compared to polymer and clay has a vital role in improving various properties of polymer and accordingly widely increases its application in various fields. Using simple intercalation reaction, polymer/clay nanocomposite can easily synthesize. The structure of nanocomposite can evaluate by XRD.

## ACKNOWLEDGEMENTS

The authors are grateful to the Center of Excellence for Advanced Materials Research (CEAMR) and Chemistry Department at King Abdulaziz University for providing research facilities.

## CONFLICT OF INTEREST

The author confirms that he has no conflict of interest to declare for this publication.

## REFERENCES

[1] P.G. Luo, and F.J. Stutzenberger, Chaptar 4 - Nanotechnology in the Detection and Control of Microorganisms, in Nanomaterials in Microbial Detection and Control. 2008, Advances in Applied Microbiology. pp. 145-181.

[2] Bhushan, B., Part 1 - Introduction to Nanotechnology, in Springer Handbook of Nanotechnology, B. Bhushan, Editor. 2010, Springer Berlin Heidelberg.

[3] W. Ahmed, M.J. Jackson, and I.U. Hassan, Chapter 1 - Nanotechnology to Nanomanufacturing, in Emerging Nanotechnologies for Manufacturing, W. Ahmed and M.J. Jackson, Editors. 2010, William Andrew Publishing: Boston. pp. 1-15.

[4] F. Hussain, *et al.*, "Review article: Polymer-matrix Nanocomposites, Processing, Manufacturing, and Application: An Overview," *J. Composite. Mater.*, vol. 40: pp. 1512-1559, 2006.

[5] T. Athar, Chapter 14 - Metal Oxide Nanopowder, in Emerging Nanotechnologies for Manufacturing, W. Ahmed and M.J. Jackson, Editors. 2010, William Andrew Publishing: Boston. pp. 325-385.

[6] A.B.R. Mayer, Polymer Nanocomposites with Metal Dispersions. Encyclopedia of Materials: Science and Technology, 2001.

[7] D.L. Ciprari, Mechanical Characterization of Polymer Nanocomposites and the Role of Interphase, in Materials Science and Engineering. 2004, Georgia Institute of Technology: Atlanta, USA. p. 100.

[8] D. William, J. Callister, Composites, in Fundamentals of Materials Science and Engineering, W. Anderson, Editor. 2001, John Wiely & Sons, Inc.: New York. p. S-162-S-203.

[9] D. Bondeson, Biopolymer-based Nanocomposites: Processing and Properties, in Daniel Bondeson. 2007, Norwegian University of Science and Technology: Trondheim. p. 114.

[10] T.P. Selvin, *et al.,* Polymer nanocomposites: Preparation properties and applications, in International Conference Polymeric Materials 2004. 2007: Halle/Saale, Germany. p. 49-56.

[11] E.T. Thostenson, C. Li, and T.-W. Chou, "Nanocomposites in context," *Composites.Sci.Technol.,* vol.65: pp. 491–516, 2005.

[12] M.S.Niasari, and D. Ghanbari, Chaptar 21 - Polymeric Nanocomposite Materials, in Advances in Diverse Industrial Applications of Nanocomposites, B. Reddy, Editor. 2011, InTech. pp. 502-520.

[13] G. Kickelbick, Introduction to Hybrid Materials, in Chapter 1 - In Hybrid Materials. Synthesis, Characterization, and Applications, G. Kickelbick, Editor. 2007, Wiley-VCH Verlag GmbH & Co. KGaA Weinheim, Germany. p. 1-48.

[14] A.M. Mazrouaa, Chaptar 14 - Polypropylene Nanocomposites, in Polypropylene, F. Dogan, Editor. 2012, InTech. p. 266-286.

[15] G.E. William, B.A. Aldo, and Z. Jinwen, "Polymer Nanocomposites: Synthetic And Natural Fillers: A Review," *Science and technology*, vol.7(3), pp. 159-178, 2005.

[16] C.D. Delhom, Development and Thermal Characterization of Cellulose/Clay Nanocomposites, in Engineering Science. 2009, University of New Orleans. pp. 62.

[17] D.R. Paul, and L.M. Robeson, "Polymer nanotechnology: Nanocomposites," *Polymer.*, vol. 49, pp. 3187–3204, 2008

[18] A. Kutvonen, *et al.*, "Influence of nanoparticle size, loading, and shape on the mechanical properties of polymer nanocomposites," *J. Chem. Phys.,* vol. 137(214901), pp. 1-8, 2012.

[19] V. Rangari, Chaptar 4 - Polymer Nanocomposite Materials for Structural Applications, in Advances in Nanocomposites - Synthesis, Characterization and Industrial Applications, B. Reddy, Editor. 2011, InTech. p. 61-84.

[20] D. Brabazon, and A. Raffer, Chapter 3 - Advanced Characterization Techniques for Nanostructures, in Emerging Nanotechnologies for Manufacturing, W. Ahmed and M.J. Jackson, Editors. 2010, William Andrew Publishing: Boston. p. 59-91.

[21] A. M. Asiri, S. B. Khan, K. A. Alamry, H. M. Marwani, M. M. Rahman."Growth of $Mn_3O_4$ on cellulose matrix: Nanohybrid as a solid phase adsorbent for trivalent chromium," *App.Sur. Sci.,* vol.270, pp. 539– 544, 2013.

[22] J. S. Choi, S. B. Khan, E. S. Jang, J. Seo, H. Han. "Synthesis and Characteristics of UV-Curable Polyurethane-Co-Acrylic Coatings for Stainless Steel," J.Chem.Soc.Pak. vol. 33, pp. 726– 732, 2011,

[23] S. B. Khan, J.W Lee, H. M. Marwani, K. Akhtar, A. M. Asiri, J. Seo, A. A. P. Khan, H. Han, "Polybenzimidazole hybrid membranes as a selective adsorbent of mercury," *Composites: Part B.* vol.56, pp.392–396, 2014.

[24] D. Kim, M. Jang, J. Seo, K. H. Namb, H. Han, S. B. Khan. "UV-cured poly(urethane acrylate) composite films containing surface-modified tetrapod ZnO whiskers," *Comp. Sci. Tech.,* vol.75, pp.84–92, 2013.

[25] J.W. Lee, S. B. Khan, K. Akhtar, K. I. Kim, T. W.Yoo, K. W. Seo, H. Han, A. M. Asiri. "Fabrication of Composite Membrane Based on Silicotungstic Heteropolyacid Doped Polybenzimidazole for High Temperature PEMFC," *Int. J. Electrochem. Sci.,* vol.7, pp.6276 – 6288, 2012.

[26] E. S. Jang, S. B. Khan, J. Choi, J.Seo, H. Han. "Synthesis and Characterization of PU-Si Nanohybrid: Influence of Silica on Properties of UV-curable Polyurethane Acrylate", *J.Chem.Soc.Pak.,* vol.33, pp.549– 554, 2011.

[27] A. Khan, A. M. Asiri, A. A. P. Khan, M. A. Rub, N. Azum, M. M. Rahman, S. B. Khan, K. A. Alamry, S. A. Ghani. "Sol–gel synthesis and characterization of conducting polythiophene/tin phosphate nano tetrapod composite cation-exchanger and its application as Hg(II) selective membrane electrode," *J. Sol-Gel. Sci. Technol.,* vol.65, pp.160–169, 2013.

[28] A. Khan, A. M. Asiri, M. A. Rub, N. Azum, A. A. P. Khan, S. B. Khan, M. M. Rahman, I. Khan. "Synthesis, characterization of silver nanoparticle embedded polyaniline tungstophosphate-nanocomposite cation exchanger and its application for heavy metal selective membrane," *Composites: Part B.*, vol.45, pp.1486–1492, 2013.

[29] S. B. Khan, K. A. Alamry, H. M. Marwani, A. M. Asiri, M. M. Rahman. "Synthesis and environmental applications of cellulose/ZrO2 nanohybrid as a selective adsorbent for nickel ion," *Composites: Part B.,* vol.50, pp.253–258, 2013.

[30] S. B. Khan, J. Seo, E. S. Jang, K. Akhtar, K. I. Kim, H. Han. "Synthesis and Characterization of Novel PPC-Silica Hybrid with Improved Thermal, Mechanical, and Water Sorption Properties," *Macromol. Res.,* vol. 19, pp.876-882, 2011.

[31] E. S. Jang, S. B. Khan, J. Seo, , K. Akhtar, J. Seo, K. I. Kim, H. Han. "Synthesis and Characterization of Novel UV-Curable PU-Si Hybrids: Influence of Silica on Thermal, Mechanical, and Water Sorption Properties of Polyurethane Acrylates," *Macromol. Res.,* vol. 19, pp.1006-1013, 2011.

[32] E. S. Jang, S. B. Khan, J. Seo, K. Akhtar, Y. H. Nam, K. W. Seo, H. Han. "Preparation of Cationic Latent Initiators Containing Imidazole Group and Their Effects on the Properties of DGEBA Epoxy Resin," *Macromol. Res.,* vol.19, pp. 989-997, 2011.

[33] E. S. Jang, S. B. Khan, J. Seo, Y. H. Nam, W. J. Choi, K. Akhtar, H. Han. "Synthesis and characterization of novel UV-curable polyurethane–clay nanohybrid: Influence of organically modified layered silicates on the properties of polyurethane," *Prog. Org. Coat.,* vol.71, pp. 36–42, 2011.

[34] J. S. Choi, J. Seo, S. B. Khan, E. S. Jang, H. Han. "Synthesis and characterization of novel UV-curable polyurethane–clay nanohybrid: Influence of organically modified layered silicates on the properties of polyurethane," *Prog. Org. Coat.,* vol.71, pp. 36–42, 2011.

[35] D. Kim, Y. Lee, J. Seo, H. Han, S. B. Khan. "Preparation and properties of poly(urethane acrylate) (PUA) and tetrapod ZnO whisker (TZnO-W) composite films," *Polym. Int.,* vol.62, pp. 257–265, 2013.

[36] D. Kim, K. Jeon, Y. Lee, J. Seo, K. Seo, H. Han, S. B. Khan. "Preparation and characterization of UV-cured polyurethane acrylate/ZnO nanocomposite films based on surface modified ZnO," *Prog. Org. Coat.,* vol.74, pp. 435– 442, 2012.

[37] S. B. Khan, M. M. Rahman, E. S. Jang, K. Akhtar, H. Han. "Special susceptive aqueous ammonia chemi-sensor: Extended applications of novel UV-curable polyurethane-clay nanohybrid," *Talanta.,* vol.84, pp.1005–1010, 2011.

[38] S. B. Khan, K. Akhtar, M. M. Rahman, A. M. Asiri, J. Seo, K. A. Alamry, H. Han. "A thermally and mechanically stable eco-friendly nanocomposite for chemical sensor applications," *New. J. Chem.,* vol.36, pp. 2368–2375, 2012.

[39] S. B. Khan, M. M. Rahman, K. Akhtar, A. M. Asiri, K. A. Alamry, J. Seo, H. Han. "Copper Oxide Based Polymer Nanohybrid for Chemical Sensor Applications," *Int. J. Electrochem. Sci.*, vol.7, pp. 10965 – 10975, 2012.

[40] S. B. Khan, M. M. Rahman, K. Akhtar, A. M. Asiri, J. Seo, H. Han, K. A. Alamry. "Novel and Sensitive Ethanol Chemi-Sensor Based on Nanohybrid Materials," *Int. J. Electrochem. Sci.,* vol.7, pp. 4030 – 4038, 2012.

[41] Y. Lee, D. Kim, J. Seo, H. Han and S. B. Khan. "Preparation and characterization of poly(propylene carbonate)/exfoliated graphite nanocomposite films with improved thermal stability, mechanical properties and barrier properties," *Polym. Int.,* vol. 62, pp. 1386–1394, 2013.

[42] J. Seo, G. Jeon, E. S. Jang, S. B. Khan, H. Han. "Preparation and Properties of Poly(propylene carbonate) and Nanosized ZnO Composite Films for Packaging Applications," *J. App. Poly. Sci.,* vol.122, pp. 1101–1108, 2011.

# Development and Prospective Applications of Nanoscience and Nanotechnology

*Volume 1: Nanomaterials and their Fascinating Attributes*

CHAPTER 5

# Solvothermal/Hydrothermal Synthetic Methods for Nanomaterials

**Shahida B. Niazi***

*Institute of Chemical Sciences, Bahauddin Zakariya University, Multan, Pakistan*

**Abstract**: Synthesis of nanoparticles of various compositions and exciting properties, sizes and a variety of morphologies is a vast expanding subject of nanoscience because of their very interesting and a variety of applications. Preparation of nanoparticles with remarkable characteristics along with desired size and morphology in a precisely controlled and reproducible manner is an art by itself. Solvothermal/hydrothermal based synthetic methods are economical and environmental friendly because they are carried out in a close system avoiding toxic solvents and reagents. In this chapter, solvothermal/hydrothermal syntheses of magnetic nanomaterials especially iron oxide nanoparticles and other nanostructures, metal ferrites and mixed metal ferrites and composites have been reviewed.

**Keywords:** Hydrothermal, iron oxide, metal ferrites, metal oxides, mixed metal ferrites, nano-composites, nanomaterials, nanoparticles, solvothermal

## 1. INTRODUCTION

Tremendous advancement in the growth of nanoscience and technology is due to a wide variety of synthetic strategies discovered for nanomaterials and established techniques for their characterization and manipulation. Nanoparticles, nanoporous solids, nanocomposites, nanocrystals, nanocapsules, nanostructured polymers, nanotubes, nanorods and nanowires and their assemblies have been prepared for metals, alloys, non-metals and other inorganic materials. Various aspects of the chemistry of nanomaterials and nanochemistry have been reviewed [1, 2].

Chemists use bottom up approach for the synthesis of nanomatreials. A number of specific methods based on solid state or solution or vapour phase reactions have been reported for the synthesis of stable isolated particle or composite nanomaterials. A sound synthetic method produces high purity crystalline nanomaterial of desired characteristics. Wet chemical methods include conventional hydrothermal,

***Corresponding author Shahida B. Niazi:** Institute of Chemical Sciences, Bahauddin Zakariya University, Multan, Pakistan; E-mail: sbniazi@bzu.edu.pk

**Sher Bahadar Khan, Abdullah M. Asiri & Kalsoom Akhtar (Eds)**

microwave-hydrothermal, co-precipitation, sonochemical reactions, sol-gel *etc.* Some of the reported methods are summarized as under:

- Classical co-precipitation in aqueous media
- Reduction of metal salt or solution
- Solvothermal/hydrothermal
- Hot thermolysis/colloidal synthesis
- Flame synthesis
- Photochemical synthesis
- Liquid-liquid interface synthesis
- Sol-gel synthesis
- Synthesis in structural mediums
- Vapour phase/gas phase reactions *etc.*

Use of high boiling point solvents for the synthesis of nanomaterials has some inherent problems. They are generally toxic, expensive and sometimes unable to dissolve simple salts. To overcome such problems, solvothermal methods are used in the preparation of wide ranging nanomaterials of different shapes and sizes. Use of solvothermal technique is one of the solution based methods to develop single crystal particles and grains without any defect and with improved crystallinty. It is a beneficial synthetic route providing increase in the solubility and crystallization, and control over morphology, size and phase change. Synthesis of single crystal/particle by this method depends upon the solubility of salt/material in hot solvent/water under high pressure. Hydrothermal procedures are dependent on the property of hydrolysis and dehydration of metal salts by water at higher temperatures and pressures. Under hydrothermal conditions, significant decrease in the solubility of metal oxides and hydroxides results in higher super-saturation leading to very fine crystals. Depending upon the starting material, precursor complex and final product, chemical reactions such as oxidation, reduction, metathesis, hydrolysis and thermolysis are generally used in the synthesis of nanoparticles. Parameters such as temperatures and pressures used in the synthesis procedure can easily differentiate a solvothermal process from other processes, *e.g.*, sol–gel and coprecipitation/calcination processes.

In Solvothermal (in case of water, hydrothermal) methods, low boiling point solvents are used. A sealed reaction vessel (autoclave or bomb type) can make solvents work well above their boiling point at atmospheric pressure. This increases the autogenous pressure over the ambient pressure and raises the boiling point of the solvent. The properties of nanomaterials are dependent on the size and shape of nanoparticles. Desired control of size, shape and crystallinity can be achieved for nanoparticles/nanostructures. For this purpose, experimental conditions, *e.g.*, pressure, reaction temperature and time, type of starting material, solvent type and surfactant type, need to be adjusted for maintaining elevated nucleation rates for controlling the growth. Good quality monodispersed nanoparticles can be prepared by use of suitable reagents and solvents in solvothermal route and this makes it relatively safe and economical. Using this technique various nanostructures can be prepared at temperatures considerably lower than those needed in traditional solid- or vapour-phase reactions. The nanoparticles synthesized by hydrothermal method are pure, ultrafine, relatively strain free with narrow size distribution and characterized by high reactivity and sinter-ability. Furthermore an important point in favor of hydrothermal process is that it can promote formation of crystalline products without any post-treatment. Hydrothermal process is eco-friendly as reactions are performed in a closed system and the products can be isolated for further use on cooling to room temperature.

In hydrothermal process, basic chemical reactions such as: i) alkali neutralization of trivalent iron and bivalent metal cations, ii) neutralization and oxidation of bivalent iron and other metal cations, iii) neutralization of $Fe^{3+}$ by cationic (*e.g.*, $Li^{+}$) hydroxides.

## 2. INNOVATIONS IN SOLVOTHERMAL/HYDROTHERMAL METHODS

Solvothermal method is an efficient and versatile low temperature technique. At a given temperature, use of conventional solvothermal method shows slow reaction kinetics. Innovations in the conventional method (such as oven heated autoclave) are also a focus of research. For conventional hydrothermal method, the outer shell of autoclave is made up of stainless steel and inner liner is composed of Teflon. In many cases, distinctive features of nanomaterials can be changed by use of added external aids, like a) microwave heating and b) ultrasonically assisted hydrothermal treatment.

Use of microwave heating in solvothermal method has advantages over conventional hydrothermal in providing local super-heating (rapid heating to hydrothermal temperature), and thus can enhance the reaction kinetics up to two folds and save time and energy. It also results in formation of novel phases and selective crystallization. For microwave hydrothermal method, the outer shell of the autoclave is made up of polyetherimide (a highly strong polymer) with internal Teflon lining. The reaction vessel is transparent to microwaves and can be heated hydrothermally. It is attached to a pressure transducer to monitor and control the pressure in hydrothermal treatment. Microwave heating coupled with conventional hydrothermal method is eco-friendly and energy efficient in that it requires very low reaction temperature along with a short reaction time (in minutes) having higher selectivity and yield.

In addition to microwave heating, the autoclave systems are available which are equipped with magnetic stirrer and/or in-vessel temperature sensors (thermocouple) to control heating rate and maintain the desired temperature of reaction mixture in the reaction vessel. In some cases temperature and pressure in autoclave is controlled and monitored by modular controller attached to it. A microwave digestion system with 2.45 GHz microwaves is designed to work at 0–100% full power (630 ± 50W). The pressure controls the system and a maximum pressure of 200 psi (equivalent to 194 °C, based on steam tables) could be attained. During the preparation, the control and monitoring of pressure is done by connecting a pressure transducer to reaction vessel. The parameters such as time, pressure and power are regulated by computer attached to system.

Sometime inert atmosphere (argon or nitrogen gas) is required to avoid undesirable oxidation or reduction during mixing of reagents prior to being autoclaved for hydrothermal reaction. Occasionally isolation/separation/recovery of magnetic nanomaterials are facilitated by use of external magnetic field rather than conventional separation techniques.

Magnetic field-supported hydrothermal growth of one-dimensional magnetic materials with nanostructures is an innovation. In fact it has been observed that magnetic field has a noteworthy effect on the mobility of magnetic particles. In this innovation, external magnetic field is generally applied using two permanent rare earth NdFeB magnets beneath and above the autoclave while undergoing hydrothermal treatment. It has find use in the preparation of Co and Ni microwires by the self-assembly of Co and Ni nanocrystallites in an applied independent 0.25T magnetic field. Magnetite ($Fe_3O_4$) single-crystal nanowires and chain-like nanostructure are also prepared this way.

Hydrothermal sonochemical treatment is another innovation which has enhanced the usefulness and efficiency of conventional hydrothermal synthetic method. By sonication of hydrothermal solutions, degree of decay of organic waste considerably rises. It is proposed that structure and growth of nanophase (for example, oxide phase) during hydrothermal treatment will be influenced by all the stages related to ultrasonic action on liquid in typical conditions, such as:

i) Formation of additional nucleation points on cavitation bubbles

ii) Enhancement in the rate of crystal growth owing to increase of effective diffusion coefficients and uninterrupted regeneration of the surface of growing particles

iii) Breaking up of particle combined masses owing to shock waves and micro-jets of liquid

For ultrasonically supported hydrothermal treatment, a specific set up is required. A high-pressure titanium reactor is attached to ultrasonic generator of specific power through magnetostriction transducer. The ultrasonic vibration frequency is maintained at a certain magnitude with the help of a master generator and a frequency measuring device. To maintain the temperature in the reaction cell, a resistance furnace fitted with standardized chromel–alumel thermocouple is used. *In situ* recording of acoustic noise level shows that the cavitation intensity in hydrothermal process at 523 K is nearly 70–80% of its magnitude at 298 K.

The method based on uninterrupted hydrothermal process at supercritical water conditions ($Tc$ = 647 K, $Pc$ = 22.1 MPa) using a flow reactor is environment friendly, fast and simple and does not require addition of solvents or chemicals. Precursor solution is introduced using a high-pressure pump at a known flow rate; at a mixing point, solution is combined with preheated water allowed in from a second pump at a known rate. A micro-mixing unit is used for quick heating of the mixture to reaction temperature, which is basically estimated using an enthalpy balance with the enthalpy of the water at the given condition and the flow rate of each solution. Residence time, $t$ (in seconds) is estimated from total flow rate, reaction volume and densities of water at 298 K and 673 K at system pressure of 30 MPa. At the leaving point of the reactor, the fluid is quickly cooled by i) allowing cooling water in at a known rate and ii) using an outer water jacket. The variation in the system pressure is kept in the limit not exceeding than 0.1 MPa with use of back-pressure regulator located behind the cooler. Nanoparticles obtained in slurry solution form are subjected to membrane filter for separation, and after washing with pure water, the drying was carried out at 333 K in electrically heated oven for 24 hours.

A novel development in hydrothermal process is use of central-collision type micro-mixer (CCM). It is aimed at swift heating of precursor solution to the reaction temperature and homogeneous nucleation for uninterrupted hydrothermal preparation of metal oxide (MO) and mixed metal oxide (MMO) nanoparticles. This innovation is a production technology for producing sufficient amounts of nanoparticles having desirable properties in very short residence times not exceeding from several milliseconds to several tens of seconds on the concept of green chemistry.

## 3. SOME STRATEGIES IN THE SOLVOTHERMAL/HYDROTHERMAL ROUTE

Due to improvements in experimental skills and tools, stabilization of isolated nanoparticles has been achieved by capping or passivation agents. The role of stabilizing reagents is to control the growth rate and particle size, inhibit particle aggregation and to improve the solubility in various solvents. Mode of action is encapsulation through electrostatic stabilization or steric stabilization depending on the nature of stabilizing agent used. Surfactants, polymers, ligands such as thiols and amines, solvents such as ethers and thioethers and some polyoxoanions of W, Nb and Mo are normally used as stabilizing agents. Nanoparticles thus obtained are easily separated by precipitation or evaporation of solvent. They can be re-dissolved in contrast to colloidal sols, which after precipitation cannot be easily dispersed.

Hydrothermal treatment of the precursors prepared by co-precipitation, results in controlled stoichiometry of nanoparticles, improved chemical stability and magnetic properties.

Biomolecule-assisted route is a new, environmental-friendly, and future strategy for the synthesis of a variety of materials in which biomolecules, such as aspartic, commonly acts as structure-directing reagents/templates. Aspartic acid as reducing agent is distinctively non-toxic, cost effective, and environmental-friendly. Hydroxyl rich polysaccharide, such as starch, has been used for the synthesis of complex nanostructures in the hydrothermal conditions, where -OH groups act as the necessary reducing agent. DNA with peculiar cylindrical and helical structure is also established to be efficient in assembling nanoparticles into two- or three-dimensional arrangements and in the alignment of discontinuous one-dimensional nanomaterials.

Interactive growth of spinel ferrite and M-hexaferrite has been achieved using a new strategy in hydrothermal approach, hydrothermal decomposition, which has also been employed to prepare plate-like spinel ferrite ($Fe_3O_4$) particles from pre-synthesized barium hexaferrite powder using high concentration of sodium bi-carbonate and a suitable reducing agent such as hydrazine.

Seed-hydrothermal route strategy is basically hydrothermal method combined with seed growth method. This is a facile approach which is applied to control the morphology of nanoparticles of metals (Ag, Au), nanometal oxides of Zr, Ce, and Fe, and ferrites *etc.*

For solvothermal synthesis of magnetic nanoparticles, some reagents are so chosen that act bi-functionally. For the synthesis of $MnFe_2O_4$ ferrite nanoparticles, iron (II) salt is taken as an iron source and $KMnO_4$ as Mn source and an oxidant for iron (II) to convert it to Fe (III) and itself to Mn (II). Hydrazine is added to control the oxidation of iron (II). This solvothermal redox reaction is carried out in ethylene glycol (EG) solvent.

For the formation of magnetic spinel nanoparticles, in solvothermal reduction method, some reducing agent (such as trisodium citrate dihydrate, hydrazine hydrate, formamide *etc.*) is added along with other required reactants before proceeding for the solvothermal step. Ethylene glycol (EG) with a comparatively high boiling point is the only solvent used in this approach. In the synthesis of ferrites, role of ethylene glycol is important, and it acts as a strong reducing agent. It broadly finds a use in the polyol process to produce monodisperse fine metal or metal oxide nanoparticles. To control morphology, specifically nanosphere, some surfactant such as polyethylene glycol (PEG) or polyvinylpyrolidone (PVP) is also required to prevent agglomeration of magnetic particle during solvothermal reaction.

For bulk synthesis of monodisperse $M^{II}Fe_2O_4$ (M = Ni, Co, Mn) and $\gamma$-$Fe_2O_3$ nanoparticles, water-non polar organic solvent such as toluene interface is used. Reaction is carried out with conventional and microwave hydrothermal conditions using metal nitrate or chloride precursors and oleic acid as dispersing agent. Advantage of this strategy is that a simple separating funnel is required to collect as- synthesized nanoparticles in a non-polar organic solvent from reaction mixture. The organic solvent can be reused after concentrating it under reduced pressure. This method is suitable for large scale and economical production of monodisperse ferrites under conventional hydrothermal and microwave hydrothermal conditions.

Dielectric constant of a solvent is an important property which strongly influences the particle formation under hydrothermal conditions. Addition of low dielectric medium (such as alcohols: isopropyl alcohol) to the aqueous solutions (particle forming medium) of inorganic salts to speed up particle formation kinetics is a strategy that has been used to prepare good quality $NiFe_2O_4$ nanopowder having perfect morphology at low temperature hydrothermal.

An important focus of research in hydrothermal synthetic methods is on the development of facile strategies for shape/morphology-controlled synthesis of magnetic spinel ferrites to improve their intrinsic properties. One of the strategies is template assisted-route synthesis. Polyethylene glycol 400 (PEG 400) is used as a template reagent to prepare $NiFe_2O_4$ magnetic nanorods and $ZnFe_2O_4$ spherical nanoparticles under hydrothermal conditions. Amines are also used as templates for the synthesis of microporous zinc ferrite.

Another synthetic strategy for shape-controlled synthesis is to use metal dodecyl sulphate (M = Ni, Cu, Pb, Co) as a precursor and a stabilizing surface active reagent in hydrothermal synthetic route of nanomaterials of a variety of morphologies.

Use of reverse emulsion-supported hydrothermal treatment to prepare nanomaterials of different kinds is another strategy. The size and its distribution in reverse emulsion are restricted by varying water-surfactant ratio which in turn controls the size and morphology of product.

Interface reaction between a solid precursor and alkali solution followed by hydrothermal crystallization is an important strategy for the controlled synthesis of magnetic ferrites. The solid precursor is synthesized by co-precipitation and subjected to hydrothermal crystallization in the presence of an alkali. The chemical reaction occurs at the junction of solid-liquid phase and released particles are scattered in aqueous medium by Brownian movement. Sometimes, solid is a soft metal, like Zn metal sheet, is brought in contact with solution of $Fe^{2+}$ in ammonia solution during hydrothermal treatment and it has been used for the production of zinc ferrite.

In a typical solvothermal/hydrothermal procedure, reagents are taken in appropriate stoichiometric ratios, dissolved in solvent ($H_2O$ in hydrothermal), mixed with constant stirring and desired pH is adjusted with suitable alkali. If the reaction demands, inert atmosphere and addition of stabilizing agent is used. Reaction contents are then sealed in an autoclave and subjected to heating to the required temperature and time. After naturally cooling the contents, the product is separated magnetically or by centrifugation and thoroughly washed to remove unwanted chemicals/impurities present in the final formulation. For this purpose water and alcohol, ether or acetone are generally used. Finally product is dried (air drying or vacuum drying or in oven at about (60-100 °C) and stored for further study and characterization.

## 4. OXIDE NANOPARTICLES AND OTHER NANOSTRUCTURES

Oxides are a unique class of nanomaterials which exhibit a variety of structures and assemblies and show very interesting and novel properties. Magnetic properties of nanoparticles are very much influenced by synthetic conditions, particle size, shape and composition. Oxides, especially of transition metals, find application in electronics, magnetic and superconducting devices, integrated semiconducting devices, fuel cells, battery materials, decoupling capacitors, devices for dynamic access memories and in biology and biomedical science. During the synthesis of oxide nanoparticles, chemical reactions such as oxidation, hydrolysis and thermolysis may occur in sovothemal methods depending upon the starting material, precursor complex and final product. It should be pointed out that hydrothermal treatment permits synthesizing simple and mixed oxides with restrained morphology at comparatively low temperatures (373–623 K).

Important factors that result in nanomaterials to exhibit unlike behavior as compared to those of bulk materials are: (1) surface effects, which cause smooth characteristics scaling owing to nano size of particles; (2) quantum effects, which cause discontinuous character owing to quantum restricted effects in materials having delocalized electrons [3-4(a)]. Effect of these factors is observed on the chemical and physical properties (mechanical, optical, electrical and magnetic) of nanomaterials.

### 4.1. Iron Oxides

Iron oxide nanoparticles ($\gamma$-$Fe_2O_3$, $Fe_3O_4$) show variety of magnetic properties and are extensively studied for many technological uses such as biomedical applications, ferrofluids, magnetic refrigeration systems, catalysts and high density storage devices. They are termed as SPION (super-paramagnetic iron oxide nanoparticles).

Superparamagnetism is associated with random variations in the magnetization of the single magnetic domain particles, due to thermal energy overpowering the anisotropy energy. At a temperature above a specific value (blocking temperature, $T_B$) the thermal energy overcomes the anisotropy energy, $E_A$, and magnetization of each particle starts fluctuations within the two directions of a relaxed axis and these conditions lead to observation of the classical paramagnetic behavior of magnetic system. The magnetic character of the nanoparticles is an outcome of interactions between bulk magnetocrystalline features (finite size effect and collective magnetostatic inter-particle reciprocal actions).

A modified solvothermal reduction treatment for the synthesis of monodisperse magnetic single-crystal $Fe_3O_4$ microspheres has been reported [5(a)]. It has three added critical features: i) addition of sodium acetate for electrostatic stabilization to inhibit disordered particle accumulation. It has been used in similar preparations of Ru, Pt, Rh metals. For monodisperse $Fe_3O_4$, its presence appears to facilitate in the ethylene glycol assisted reduction of $Fe^{3+}$ to $Fe_3O_4$ as has been observed in the control experiment that without sodium acetate, under same reaction conditions, ethylene glycol alone could not reduce the $Fe^{3+}$ to $Fe_3O_4$; ii) addition of polyethylene glycol as surfactant and as further preventive measure against particle agglomeration; iii) increase in reaction temperature of 200 °C, essential for the preparation of $Fe_3O_4$. In addition to $Fe_3O_4$, this generalized simple, one step and inexpensive method has been successfully used for the production of monodisperse microspheres ferrite series *i.e.*, $M^{II}Fe_2O_4$ ($M^{II}$ = Fe, Mn, Co, Zn) with a controllable diameter ranged in ~200-800 nm. For microspheres with diameter of ~200nm, the saturation magnetization ($M_s$) values are reported as; $Fe_3O_4$, 81.9 emu/g; $MnFe_2O_4$, 53.2 emu/g; $CoFe_2O_4$, 61.6 emu/g; $ZnFe_2O_4$, 60.0 emu/g at 300 K. The prepared magnetic ferrite microspheres exhibit good colloidal stability. It is believed that these hydrophilic and biocompatible microspheres may be significantly useful in advanced magnetic materials and ferrofluid technology, biomedical areas like biomolecular separations, targeted drug delivery, cancer diagnosis and magnetic resonance imaging (MRI).

Magnetite ($Fe_3O_4$) hollow spheres are synthesized using a facile solution-phase solvothermal route using ethylene glycol as solvent and iron (III) chloride and urea as precursors without addition of any surfactant or template [5(b)]. Iron (III) and urea (4:12 mmole ratio) precursors and a known volume of absolute ethylene glycol were autoclaved and treated at 200 °C for 12 hours. The resulting material cooled to room temperature, thoroughly washed with water and absolute ethanol and left to vacuum drying at 50 °C for 6 hours. Magnetite hollow spheres with diameter in the range 200-500 nm, showed ferromagnetic behavior at room temperature, and have saturation magnetization ($M_s$) 60.7 emu/g, remnant magnetization ($M_r$) 14.9 emu/g and coercivity ($H_c$) 148.8 Oe. When the above reaction was conducted in the presence of PVP keeping other reagents and reaction conditions the same, it resulted in hollow monodisperse spheres similar to the results reported earlier [5(a)]. Hydrothermal reaction between aqueous solution of iron (III) chloride and urea at 100 °C for 12 hours resulted in pure phase of β-FeOOH consisting of flower like superstructures constructed with single-crystalline nanorods. Above hydrothermal reaction at 180 °C produced α-$Fe_2O_3$ nanoparticles consisting of polyhedra having an average diameter 120 nm. Similarly solvothermal reaction of iron (III) chloride and urea in absolute ethanol in the absence of any surfactant or template at 120 °C for 12 hours also resulted in α-$Fe_2O_3$ nanoparticles (10 nm size) assembling into 3D quasi-spheres with a mean diameter of 120 nm. Mono-disperse 3-D spheres with mean diameter of 200 nm were the product of solvothermal reaction of iron (III) chloride and urea in absolute ethanol in the presence of PVP at 120 °C for 12 hours. It can be inferred from the above observations that by changing the solvent in solvothermal route, morphologies can be selectively designed.

Nanotube, nanorods and nanowire are one-dimensional nanostructured materials that are the focus of materials research owing to their peculiar characteristics and possible use in molecular-based electronic devices. Magnetic nanomaterials with chain-like structures, as well, belong to the group of one-dimensional nanostructured materials. In a number of uses, especially those concerned with electrical transport and magnetic switching, the specified sequencing of nanoparticles has a crucial function. Polycrystalline Co and Ni microwires are prepared using the self-assembling of Co and Ni nanocrystallites in the influence of an applied magnetic field [6(a)] and single crystalline $Fe_3O_4$ nanowires may be fabricated in an applied magnetic field [6(b)]. It is observed that magnetic field can exert a noteworthy effect on the mobility of magnetic particles.

Chain-like nanostructures with exceeding than 10 magnetite nanoparticles having a mean size of 80 nm are hydrothermally produced at 200 °C in the influence of a 0.18 T applied magnetic field [6(c)]. These structure exhibited fairly high saturation magnetization ($M_s$), remnant magnetization ($M_r$), coercivity ($H_c$) and lower magnetization ($M_r$)/($M_s$), in comparison with those of disorderly scattered magnetite nanoparticles produced in the absence of applied magnetic field. These properties were experimentally associated to the aligned assembly character, lowered demagnetizing factor, improved crystallinity in the presence of an applied magnetic field and the inter-particles dipole reciprocating-action.

Purpose-defined biological uses require magnetic particles to be spherical with even surface, large surface area and narrow size distribution for maximum binding of proteins and enzymes. Furthermore high saturation magnetization ($M_S$) is also desired for the highest attainable signal strength and better dispersion in liquid medium. Focusing on these characteristics, preparation of maghemite sub-microspheres by simple solvothermal reduction treatment at low temperature is reported [7(a)]. Iron (III) is reduced with formamide dissolved in ethylene glycol solvent and polyvinylpyrolidone as surfactant to prevent agglomeration of very active magnetic particles during chemical reaction in liquid phase process and control spherical morphology. In a typical procedure the required amount of ferric chloride and polyvinylpyrolidone (PVP, MW 58 000) are added in ethylene glycol and formamide and vigorously stirred for 10 minutes to get a clear solution. The resulting solution was autoclaved (Teflon coated stainless steel autoclave) for 8 hours at 160 °C temperature. The product was naturally cooled to room temperature, centrifuged for isolation, thoroughly washed with absolute ethanol and then deionized water and vacuum dried in at 60 °C to obtain brown product. The prepared $\gamma$-$Fe_2O_3$ (cubic maghemite) particles were spherical with 200-400 nm diameter and having large saturation magnetization of 81 emu/g at 300 K. The product showed ferromagnetic behavior and Curie temperature above 400K.

Interest in the study of nanosized $\gamma$-$Fe_2O_3$ (maghemite) particles is due to its important use in technological and biomedical applications. Conventional hydrothermal method for its synthesis has been improved. It has been prepared by microwave-hydrothermal method [7(b)] using aqueous solutions of iron sulphate and ferric chloride salts in 1:2 molar ratio in a Teflon lined reaction vessel at pH 12 adjusted with 20% aq. NaOH solution. Using in-vessel temperature sensor, the reaction was conducted at 150 °C and 150 psi for 25 minutes and resultant brown slurry was centrifuged to get solid product, which after thorough washing with distilled water to eliminate excess salts was finally air dried. Synthesized $\gamma$-$Fe_2O_3$ nanoparticles with average size of 10 nm are super-paramagnetic at room temperature with super-paramagnetic blocking temperature of 200 K. Large maghemite nanoparticles are prepared under hydrothermal conditions and effect of pH on the particle size is studied [8].

Size-restrained preparation of metal oxide nanoparticles has also been achieved with a flow through supercritical water procedure [9]. Metal oxides; $AlOOH/Al_2O_3$, CuO, $Fe_2O_3$, NiO, $ZrO_2$ nanoparticles has been synthesized through hydrothermal route using aqueous solutions of metal nitrates at 673 K in a pressure range 25-37.5 MPa. Effect of different metals in the starting material, pressure and concentration on the particle size crystallinity was studied on the basis of super- saturation evaluated from the solubility of metal oxides. Generally the order of decrease in the solubility is as ZnO > CuO > NiO > $Fe_2O_3$. The results suggest that supersaturation should be maintained at higher than $1x10^4$ in flow through method to get particles under 10 nm of diameter. Furthermore, higher supersaturation resulted in decrease in crystallinity.

On the basis of magnetite ($Fe_3O_4$) and maghemite ($\gamma$-$Fe_2O_3$) phases, uniform iron oxide nanocrystals are prepared by using fatty acid modified iron oxide precursor subjected to hydrothermal growth at 230 °C for 3 hours. The precursor was synthesized by oleate stabilized precipitation treatment taking aqueous solutions of $FeCl_2$ and $FeCl_3$ salts in 1:2 molar ratio and ammonia [10]. Elevated hydrothermal temperature and reduced [oleate]/[Fe] ratio 0.05 promoted nanocrystals of uniform size and shape with an average particle size 14.5 nm. Chemical composition of resultant product is approximately $Fe_{3-\delta}O_4$ ($\delta$ = 0.15) indicating incomplete oxidation of $Fe^{2+}$ ions during the preparation. For samples obtained with [oleate]/[Fe] ratios of 0-0.5 at 230 °C for 3 hours, room temperature saturation magnetization $M_s$ values of 71.7, 65.7, 59.3 and 58.9 emu/g are observed for the 21.1, 16.4, 10.8 and 10.5 nm sized nanocrystals respectively; which are lower than the respective bulk values, *i.e.*, 73.5 emu/g and 92 emu/g for maghemite and magnetite respectively.

Surfactants can effectively stabilize and modify the shape and size of nanoparticles. Four different surfactants, sodium dodecyl sulfonate (SDS), sodium dodecylbenzene sulfonate (DBS), cetyltrimethyl ammonium bromide (CTAB) and hexadecylpyridinium chloride (HPC) are reported for use in surfactant modified hydrothermal approach for the synthesis of hematite; $\alpha$-$Fe_2O_3$ monodisperse nanoparticles using $FeC_2O_4$ and NaOH precursors [11]. It has been observed that surfactant modified/stabilized products has rhombohedral structure in contrast to unmodified $\alpha$- $Fe_2O_3$ (synthesized without surfactant under same reaction conditions), which has monodisperse rod like shape. Both the products show weak ferromagnetic properties with some variations in coercivity values due to shape anisotropy. It is observed that surfactants can affect size and shape of particles but not their magnetic properties.

In structures grown *via* hydrothermal route, control of morphology can be achieved by monitoring aqueous hydrolysis-condensation growth onto substrates [12]. Solution pH greatly affects the morphology. A highly oriented faceted nanostructure with an array of nanorods with superimposed flower like structures (with a size of around 40 nm x 200 nm) was obtained over the intact dip-coated pristine hematite thin film using amino acid (L-arginine) assisted hydrothermal one-pot treatment [13(a)]. Hematite solid and hollow spindles have been studied for gas sensing and photocatalysis [13(b)]. Such nanostructures are used in gas sensors, batteries and photo-catalysts. Hydrothermal treatment in the presence of L-arginine has modified the structure, size and shape of nanostructure of the hematite, $\alpha$-$Fe_2O_3$, film and thus has enhanced its optical and surface properties. Other morphologies are also reported [14-16].

Magnetite ($Fe_3O_4$), due to presence of 2+ and 3+ oxidation states of iron in inverse spinel structure, shows different properties among all iron oxides. $Fe_3O_4$ nanoparticles with grain size greater than 20 nm behave ferrimagnetic at room temperature. Monodisperse magnetite nanoparticles with stoichiometry $Fe_{2.95}O_4$ have been synthesized by co-precipitation coupled with hydrothermal treatment at 250 °C for 24 hours in argon atmosphere [17]. For co-precipitation, solutions of chlorides of $Fe^{2+}$ and $Fe^{3+}$ were made in 2M HCl prepared in distilled water degassed with argon to avoid oxidation of $Fe^{2+}$ ions and precipitation was carried with addition of aqueous tetramethylammonium hydroxide under constant stirring. The mixture was autoclaved and subjected to hydrothermal treatment at 250 °C for 24 hours in argon atmosphere. Maghemite, $\gamma$-$Fe_2O_3$, was prepared by heating $Fe_3O_4$ at 300 °C for 12 hours. Average crystallite sizes in pre- and post-hydrothermal process are 12±2 and 39±5 nm respectively. Room temperature saturation magnetization ($M_s$) of nanoparticles in pre- and post- hydrothermal process is 59.8 and 82.5 emu/g respectively. It suggests that $M_s$ value is lower than that of bulk magnetite (92 emu/g) before and near that of bulk after hydrothermal treatment. Co-precipitation coupled hydrothermal modification in synthetic strategy has produced magnetite nanoparticles with good monodispersity and ferrimagnetic properties at room temperature.

Applications of magnetite, $Fe_3O_4$, nanoparticles are dependent on particle size and surface characteristics. The particle size and its spread is an important decisive factor for predicting magnetic behavior of particles, *i.e.*, ferromagnetic or SPION with distinct saturation magnetization. Hydrothermal-reduction route has been developed for the preparation of size-controlled $Fe_3O_4$ SPION nanoparticles (size range 4–16 nm) stabilized with glucose and gluconic acid using ferric chloride precursor only [18(a)]. Sucrose and $FeCl_3$ reactants are taken in 7:4 molar ratio and pH of their solution is maintained at 10 with drop wise addition of 25% ammonia solution with strong stirring. The mixture thus obtained was autoclaved for hydrothermal treatment at 180 °C for 48 hours. After natural cooling of reaction system to normal temperature, the product was centrifuged to separate, thoroughly washed with deionized water to remove water soluble byproducts and dried at 60 °C over nightly. Hydrothermal reaction in the absence of sucrose resulted in hematite ($\alpha$-$Fe_2O_3$) as product instead of magnetite ($Fe_3O_4$). Sucrose (in the form of its hydrolyzed product glucose) not only acts as a reducing agent to promote magnetite phase but also a capping agent to stabilize nanoparticles and improve their surface properties. The resultant nanoparticles exhibit immeasurable coercivity suggesting that the product is superparamagnetic at room temperature. Particle size is dependent on the concentration of sucrose and can be limited in 4-16 nm range by changing its concentration in the reaction. Room temperature saturation magnetization, $M_s$, relates to particle size and it goes up from 14.82 to 21.95 to 29.55 emu/g with decrease in sucrose concentration. These oxides are likely to be used in biotechnology, biomedicine, magnetic resonance imaging and catalysis.

Magnetic nanoparticles are made up of magnetic cores and polymeric shell, the former provides strong magnetic action and later suitable functional group and properties for different applications. Due to Strong dipole-dipole interactions between particles, SPION; Magnetite ($Fe_3O_4$) nanoparticles show aggregate property. Their stabilization has been achieved by using surfactant and polymeric compound stabilizers. Chitosan, a polyaminosaccharide, has been used as a polymeric shell over magnetic core made up of $Fe_3O_4$ nanoparticle synthesized from ferrous complex and $H_2O_2$ as oxidant under hydrothermal conditions [18(b)]. Chitosan-modified and untreated $Fe_3O_4$ nanoparticles have regular spherical shape with spinel structure having mean particle diameter 25 nm and 23 nm respectively. Chitosan-modified $Fe_3O_4$ nanoparticles are monodisperse and superparamagnetic in nature and show saturation magnetization, $M_s$, 21.5 emu/g at room temperature. They may find application in biomedical science, *e.g.*, magnetic assisted drug delivery systems and cell/enzyme immobilization in addition to industrial operations.

Biomolecule-assisted route is a new, environmental-friendly, and future strategy for the synthesis of a variety of materials and biomolecules commonly act as structure-directing agents/templates. Hydroxyl rich polysaccharide, such as starch, has been used for the synthesis of complex nanostructures in the hydrothermal conditions and -OH group acts as the indispensible reducing agent [18(c)]. DNA with specific helical structure has also been established to be efficient in assembling nanoparticles into two- or three-dimensional structures and in the alignment of discontinuous one-dimensional nanomaterials [19(a)].

A facile synthesis with aspartic acid as moderate reducing agent for the preparation of uniform magnetite ($Fe_3O_4$) octahedra has been reported [19 (b)]. Aspartic acid (Asp) is an $\alpha$- amino acid and its -COOH group is electron donating particularly in anionic form. In comparison with other reducing agents (such as hydrazine hydrate and sodium borohydride) in majority of hydrothermal preparations for magnetite, aspartic acid (one of amine containing compounds) is used as potential reducing agent in the preparation of nanomaterials [19(c)]. It is specifically non-toxic, low cost, and environmental-friendly, thus, this biomolecule supported treatment may give an eco-friendly suitable circumstances for the preparation of other transition metal oxides with unvarying size and restrained morphology.

The synthesis, protection, functionalization and application of magnetic nanoparticles have been reviewed [4 (a) and (b)]. Another review has been reported on the preparation, stabilization, vectorization, physicochemical characterization, and biological uses of magnetic iron oxide nanoparticles [20].

## 4.2. Ferrites and Related Compounds

Among technologically important magnetic materials, ferrites, because of their unusual electronic and physical structures and ferromagnetic properties, are very important class of compounds and find applications in variety of magnetic and electronic devices. Ferrite nanoparticles are basically functional nanoparticles of mixed metal oxides (MMOs). Since hydrothermal method provides improved control for uniform particle size and powder homogeneity, it may take over as a preferred method for the synthesis of ferrites in future. Hydrothermal preparation has been reported for ultrafine ferrites with subsequent sintering or calcination at high temperature [21].

Ferrites show two distinctive structural symmetries, cubic and hexagonal. Cubic ferrites, with a basic formula $M^{II}Fe_2O_4$ (M = Mn, Fe, Co, Ni, Zn and alkaline earths *etc*.), are magnetically soft and exhibit spinel crystalline structure that can easily be magnetized and demagnetized. To molecularly design a wide range of magnetic properties, chemical identity of M (II) cation can be adjusted. Pure nanoferrites including $Fe_3O_4$ and $M^{II}Fe_2O_4$ (M = Mn, Co, Ni, Zn) have shown their importance and supremacy of nanosize in terms of properties and uses. Hexagonal ferrites, with a general formula $M^{II}Fe_{12}O_{19}$, for example barium hexaferrite, behave like permanent magnets and have high Curie temperature, very high magneto-crystalline anisotropy, high coercivity, high saturation magnetization, superior chemical stability and corrosion resistance. It is a frequently used material in magnetic recording devices. Control of magnetic properties of ferrite particles can be achieved by monitoring particle size and its distribution; morphology and micro-structure which further depend upon reaction time and temperature, pH and concentration of starting materials.

### *4.2.1. Barium Ferrites*

Well crystalline barium ferrite ($BaFe_{12}O_{19}$) has been prepared *via* direct one step hydrothermal process using chlorides of barium and ferrous, potassium nitrate and sodium hydroxide with $OH^-/Cl^-$ ratio 2:1 [22]. Particle size and saturation magnetization and intrinsic coercivity increase with increasing hydrothermal temperature (180 to 260°C) and later is maximum (685 Oe) at 230°C. Morphology changes from barium ferrite particles to hexaferrite particles with increasing hydrothermal reaction time up to 16 hours at 230°C [23].

Synthesis of amorphous barium ferrite nanoparticles was carried out using aqueous solutions of ferric nitrate, barium chloride, sodium citrate and sodium hydroxide followed by complete hydrolysis under refluxing conditions at 100 ° C in argon atmosphere for 3 hours [24]. Sodium citrate acted in two ways, allowing homogenous mixing of $Ba^{2+}$ and $Fe^{3+}$ ions in as synthesized amorphous nanoparticles and slowing down particle growth by stabilization through the formation of surface citrate complex preventing condensation of nanoparticles. After washing and drying, the resultant product was calcinated in air in temperature range 500-750 °C for100 minutes using heating rate of 100 °C/min. As synthesized and calcinated barium ferrite nanoparticles showed complete transformation in morphologies at 600 °C from amorphous to hexagonal phase with no intermediate phase. Mean particle diameter of as synthesized (amorphous form) barium ferrite is 4.7 nm with specific surface area 137.8 $m^2/g$. Calcinated hexagonal barium ferrite monodisperse nanoparticles show coercivity and saturation magnetization values 3580 Oe and 43 emu/g respectively.

Barium hexaferrite, $BaFe_{12}O_{19}$, is a prospective candidate magnetic nanocomposite material that can absorb microwave radiations [25]. The idea of low temperature chemical process has been used for preparation of superpararmagnetic barium hexaferrite in which strength of [$OH^-$] ions is one of the indispensible variables to control the nucleation and growth of this material at low temperature [26]. A modified hydrothermal approach has been used to obtain nanoparticles with superpararmagnetic behavior using nitrates of ferric and barium, and sodium hydroxide with optimized $OH^-/NO_3^-$ ratio at 16:1 [27]. Temperature of autoclave was adjusted to synthesis temperatures, $T_s$, (160 and

150 °C) at heating rate of 200 °C/hr during the reaction. After attaining the target temperatures, heaters were switched off so that $T_s$ = 0h, meaning no soaking time was given to synthesis. At $T_s$ of 160 °C, morphology of barium hexaferrite particles shows bimodal grain size distribution with larger platelets mainly hexagonal in shape having several hundred nm width and less than 10 nm thickness. Whereas barium hexaferrite particles prepared at $T_s$ of 150 °C, shows monomodal grain size distribution and plate like shape with an average 11.2 nm width and approximate 3 nm thickness. This Variation in morphology at low temperatures is attributed to slow crystallization under these conditions, where supersaturation and nucleation makes it possible to reduce post nucleation growth eliminating coarsening of crystallites as in this case monomodal grain size spread occurs at 150 °C.

Functional and intrinsic properties of particle materials are inter-related. Modification in functional properties of these materials is achievable by controlling the intrinsic properties of constituent components (by changing composition or by doping) and their morphologies. Thus by coupling inherent characteristics of unlike materials in composite particles, their functional properties may be enhanced or modified. Particle sizes at nano-level and particle shape also considerably affect the properties of nanomaterial. Nanoparticles with highly anisotropic shapes, such as, nanorods, nanotubes, nanoneedles, nanowires, nanobelts and nanoplatelets are particularly of technological and scientific interest. In many instances particle size can be easily controlled by wet chemical methods. However, particle shape is not easy to control and it requires specific strategies, such as hydrothermal disintegration of hexaferrite particles, to synthesize highly anisotropic particles.

Nanocomposite particles with intergrowth of two different ferrites (cubic spinel ferrite and M-hexaferrite) have been produced using a new strategy in hydrothermal approach, hydrothermal decomposition, which has also been employed to prepare plate-like spinel ferrite ($Fe_3O_4$) particles from pre-synthesized barium hexaferrite powder using high concentration of sodium bi-carbonate and a suitable reducing agent such as hydrazine [28]. The fact, that barium and iron ions have large reactivity differences with carbonate to preferentially form $BaCO_3$, was used that led to break up of hexaferrite into plate-like crystallites, which partially or wholly converted into magnetite ($Fe_3O_4$). Nanoparticles resulting after Partial decomposition demonstrated an intergrowth of spinel and hexaferrite structures. For partial treatment, reaction variables that inhibit to disunite two phases are desired. In this case, it is the concentration of hydrazine that controls the partial or complete conversion of barium hexaferrite to spinel form. Intergrowth proves that hard magnetic properties of hexaferrite can be coupled with soft magnetic characteristics of spinel cubic ferrite in varying ratios depending on the requirement of specific application. Such composites can possibly be used in devices developed for the absorption of electromagnetic waves with a wide absorption frequency bandwidth and magnetic mediators for cancer treatment [29, 30].

### *4.2.2. Manganese, Cobalt, Nickel, Copper and Zinc Ferrites*

Due to excellent magnetic and electrical characteristics and chemical and thermal stabilities, spinel ferrites, $M^{II}Fe_2O_4$ (M = Mg, Mn, Fe, Co, Ni, Cu, Zn), are technologically important magnetic materials. They are promising contenders for various practical applications in important fields such as catalysis, high-density data storage, magnetic resonance imaging (MRI) and magnetic carriers for site specific drug delivery. Some specific examples are microwave adsorption [31], electrode material [32], environmental studies [33, 34]. In the removal of dyes and pollutants from waste waters, processes such as ozonation, adsorption, flocculation, membrane filtration, photooxidation *etc.*, are used. Separation problems make their reutilization difficult in the industry. Combinations of magnetic nanoparticles with photocatalytic activity are focus of attention [35, 36]. It is observed that magnetic behavior of nanoparticles is very much dependent on its size, shape and synthetic method.

$MnFe_2O_4$ is one of the most common soft spinel magnetic ferrites with good magnetic characteristics. It is partially inverse spinel ferrite [37] characterized by high magnetic permeability, high electrical resistance and low losses. It finds application in microwave devices, ferrofluid technology, radiofrequency coil fabrication, rod antennas and transformer cores, devices used in electronic engineering and telecommunication and in drug delivery systems. $MnFe_2O_4$ magnetic nanoparticles have been reported as an excellent adsorbent in water treatment for removal of the azo dye Acid Red B [34], water treatment [38, 39], and for $CO_2$ emission reduction [40]. The microwave-hydrothermal synthesis of nanoferrites ($MnFe_2O_4$, $CoFe_2O_4$, $NiFe_2O_4$ and $ZnFe_2O_4$) has been reported [41(a)].

A spinel structure of ferrites is of high symmetry and it causes unfavorably one dimensional growth very freely and this makes synthesis of one dimensional nanostructured ferrites problematic. This problem is solved by developing template-based synthetic methods and $MFe_2O_4$ (M= Fe, Co, Ni, *etc.*) nanowires are prepared. This strategy has an apparent deficiency of producing polycrystalline structures, which affect magnetic domains while fabricating such nanomaterial. One dimensional single crystalline spinel nanostructure using hydrothermal approach for single crystalline $MnFe_2O_4$ nanorods having 20 nm average diameter and 250 nm average lengths has been reported [41(b)]. Synthesis of single crystalline $MnFe_2O_4$ nanorods has been achieved using a surfactant free simple hydrothermal route [42]. In a typical experiment, chlorides of $Fe^{3+}$ (3.5 mmole) and $Mn^{2+}$ (1.75 mmole) are dissolved in 35 ml distilled water with stirring and 1.20 g NaOH in 5 ml water added drop wise. The resultant mixture is autoclaved and heated in an oven kept at 180 °C for 12 hours. After isolating by centrifugation and washing with water and ethanol, vacuum drying of prepared material is carried out at 60 °C. As-prepared $MnFe_2O_4$ is mainly nanorods with uniform diameter of around 25 nm and up to 500 nm in length with spinel cubic structure. It is observed that it is the $Mn^{2+}$ which is responsible for nanorod morphology because replacing $Mn^{2+}$ with $Fe^{2+}$ or $Co^{2+}$ in $MnFe_2O_4$ did not result in this specific morphology. Room temperature Saturation magnetization of $MnFe_2O_4$ nanords is 74.0 emu/g and is claimed to be the highest so far described.

The seed growth method is a common and easy practice for preparing shape controlled nanoparticles of precious metals [43, 44], and metal oxides of Zr, Ce, Fe [45-47]. Preparation of spinel $MnFe_2O_4$ nanorods is reported using seed-hydrothermal route strategy [48]. This is basically hydrothermal method combined with seed growth method. Rod like seed crystals of $Mn_2O_3$ are grown by precipitation of $Mn^{2+}$ as oxalate and calcination of thoroughly washed and dried precipitates at 800 °C for 6 hours. For the synthesis of $MnFe_2O_4$ nanorods, $Fe^{3+}$ and $Mn^{2+}$ salt solutions are taken in 2:1 mole ratio, pH 12 is maintained with NaOH and seed crystals of $Mn_2O_3$ are introduced. Contents are autoclaved and heated under hydrothermal conditions at 200 °C for 12 hours. On cooling the product is filtered, washed thoroughly with water and anhydrous ethanol. The resultant $MnFe_2O_4$ product has crystal structure with high symmetry and partial spinel inversion, uniform one dimensional nanorod shape with 25-40 nm diameter and 300-400 nm length. Without addition of seed crystals, $MnFe_2O_4$ nanorods are obtained at 250 °C under hydrothermal conditions suggesting that seed-hydrothermal route is a low temperature method. As prepared nanoferrite has high saturation magnetization value of 72.45 emu/g at room temperature. During the study of photo-catalytic activity of synthesized $MnFe_2O_4$ ferrite to methylene blue, an external magnetic field was applied for its separation and it exhibited high photo-catalytic activity (67.18 %).

Water pollution is one of the serious environmental problems. Since magnetic nanoparticles can be magnetically separated with efficiency, they find use in water treatment technology to clean waste waters. Heavy metals such as mercury, chromium, cadmium and lead *etc*., non-metals such as arsenic, cyanide, ammonia, volatile organic compounds like phenols are some common and important water pollutants. Due to efficiency of separation, the use of magnetic nanoparticles can be more effective in detecting and separating toxics of waste waters and shortening the water treatment time. Requirements for most applications are strongly dependent on the shape, size and magnetization of magnetic nanoparticles. Since $MnFe_2O_4$ ferrite nanoparticles find use in waste water treatment, different strategies have been worked out for synthesis of its nanocrystals. A solvothermal redox reaction has been applied to synthesize single crystalline $MnFe_2O_4$ ferrite nanocrystals using ferrous ions and $KMnO_4$ (acting as a reactant and an oxidant for $Fe^{2+}$) taken in 2:1 Fe:Mn mole ratio [49]. The two salts were magnetically stirred in ethylene glycol to dissolve and make a transparent solution and then reacted with NaOH. During vigorous stirring for 15 minutes, hydrazine hydrate was added to above solution to control the oxidation of $Fe^{2+}$. The resultant mixture was autoclaved for solvothermal treatment and placed in oven maintained at 260 °C for 24 hours. For the formation of nano-octahedral $MnFe_2O_4$ ferrite particles, sufficient amount of NaOH and ethylene glycol were necessary, whereas, under low concentration of NaOH and $H_2O$, $MnFe_2O_4$ ferrite nanosphere was the product. Both morphologies are ferromagnetic at room temperature. The $MnFe_2O_4$ nanooctahedra have saturation magnetization ($M_s$) 58.7 emu/g and coercivity ($H_c$) 86 Oe values. The $MnFe_2O_4$ nanpspheres with 59 nm diameter exhibit high value of $M_s$, 79 emu/g, which is better than previously reported 74 emu/g for nanorods [42] and $H_c$, 77 Oe. Due to small size and high $M_s$ the nanospheres showed better adsorption preference for the removal of some pollutant such as congo red, Cr (VI), and Pb (II) even at neutral pH suggesting their future application in water treatment.

Magnetic properties of ferrites are related to their morphologies. Similarly adsorption properties of spinel ferrites are also influenced by morphologies. Adsorption properties of ferrites have immense potential for their use as effective adsorbents for magnetic separation and reusable adsorbents in the environmental and material science fields such as adsorption of As(III/V) [50] and Cr(VI) [51] on $MnFe_2O_4$ nanoparticles. The coercivity and magnetization of $MnFe_2O_4$ nanoparticles increase with increasing crystallite size [52].

An investigation on adsorption and magnetic behavior of $MnFe_2O_4$ ferrite nanoparticles, porous bulks and nanowires has been carried out [53]. Spinel ferrite

$MnFe_2O_4$ nanowires, with diameter of about 20-25 nm and length about 1 μm, are synthesized by hydrothermal route at pH 12, 250 °C for 12 hours. Spinel nanoparticles and porous $MnFe_2O_4$ are prepared by sol-gel process. The three morphologies of $MnFe_2O_4$ exhibit ferromagnetism, and thus as adsorbents capable of being separated from waste waters under the influence of magnet. Order of magnetization saturation, $M_s$, of different morphologies is as follows: $M_s$ (porous) < $M_s$ (nanoparticles) < $M_s$ (nanowires). Furthermore adsorption properties of porous bulk are much better than the other two forms.

A simple solvothermal method has been reported for $MFe_2O_4$ (M = Mn, Co, Ni) nanoparticles [54(a)]. A Solvothermal reduction method has been used to prepare monodisperse spinel ferrites, $Fe_3O_4$ and $MnFe_2O_4$, nanosperes with average size 300 nm [54(b)]. Required amounts of chlorides of Fe (III) and Mn (II) are dissolved in a known volume of ethylene glycol solvent. Trisodium citrate dihydrate is used as reducing agent and polyethylene glycol as surfactant to control nanosphere morphology. Solvothramal reaction is carried out in an oven kept at 200 °C for 8 hours. After thorough washing with distilled water and ethanol, drying of resultant black powder is done at 100 °C for 2 hours.

$CoFe_2O_4$ is a spinel cubic ferrite and chemically very stable permanent magnetic material with large coercivity ($H_c$), fairly good saturation magnetization ($M_s$), high magnetostrictive coefficient, high magnetic crystal anisotropy, big magneto-optical deflection angle and mechanical hardness [55]. It is generally used in high density magnetic recording media, high performance electromagnetic and spintronic devices, magnetic catalysis, sensors and actuators and biomedical applications [56, 57]. In addition to spinel type single metal ferrites ($CoFe_2O_4$ and $NiFe_2O_4$) their mixed ferrites are significant for application as gas sensors [56(b)]. $CoFe_2O_4$ has been used as ethanol vapour sensitive material [58((a)]. It has been observed that surface area of cobalt ferrite increases when Mn is substituted [58(b)]. Magnetism of ferrite depends upon the method used for its synthesis; furthermore, reaction conditions contribute a crucial role in determining the magnetic characteristics of resultant products. It is, therefore, much care is required in choosing hydrothermal conditions such as metal salts, surfactant and its concentration, type of alkali and its concentration, reaction temperature and time and so on. Inert atmosphere is also desired in many cases. Hydrothermal synthetic method is a preferred synthetic route because it is simple and higher yield producing [59, 60].

Water has specific supercritical properties, *i.e.*, T > 374 °C; P > 22.1 MPa with much lower dielectric constant than that of subcritical water. With decreasing dielectric constant of water, under hydrothermal condition, an increase in the rate of hydration of metal salts is observed whereas solubility of metal oxides is suppressed, *i.e.*, supercritical water conditions accelerate hydrothermal reaction rates and decrease solubility of many metal oxides. A high degree of supersaturation favors nucleation and may lead to formation of fine particles in supercritical water [61]. Hydrothermal synthesis method can be continuous or batch type under subcritical and supercritical conditions.

Batch supercritical hydrothermal method, at 390 °C and pH 12 using $Co^{2+}/Fe^{3+}$ mole ratio 0.5, has been employed for the synthesis of pure and uniform size $CoFe_2O_4$ nanoparticles of about 5nm diameter [62]. Characteristics of resultant product are dependent on reaction pH, temperature, $Co^{2+}/Fe^{3+}$ mole ratio. In subcritical temperature range (25-350 °C) particle size of the product was non uniform whereas supercritical temperature (390 °C) resulted in relatively uniform size distribution of nanoparticles. Products obtained at lower pHs were a mixture of $CoFe_2O_4$ and $Fe_2O_3$ in different proportions. Similar results were obtained with mole ratios < 0.5 for $Co^{2+}/Fe^{3+}$, whereas, the product was $Co_{1-x}\,Fe_{2-x}O_4$ ($0 \geq x \geq 1$) for ratios > 0.5. Reaction time exerts a little influence on the size and morphology of product. Maximum coercivity, $H_c$, of $CoFe_2O_4$ synthesized at supercritical temperature (390 °C) was lower (340.6 Oe) than that of prepared (595.6 Oe) at subcritical temperature (250 °C), however, the magnetization saturation, $M_s$, values were comparable, *i.e.*, 68.9 and 69.2 emu/g respectively. It is believed that for a given magnetic nanomaterial, different synthetic methods may result in different $H_c$ and $M_s$ Values. Continuous hydrothermal synthesis method has also been used for Spinel $CoFe_2O_4$ under subcritical and supercritical conditions and the product characteristics are very much dependent on the relative solubility of cations [63] and pH control by NaOH [64]. This process at supercritical water ($T_c$ = 647 K, $P_c$ = 22.1 MPa) with flow reactor is environment friendly, simple and fast because it does not require addition of solvent or chemicals.

Structural studies of nanoparticle ferrites produced continuously in near critical and supercritical water have been made to identify the type of spinel structure of nanophase [65]. Rapid and continuous preparation of $M^{II}Fe_2O_4$ (M = Ni, Cu, Zn) could be achieved by hydrothermal synthesis using a flow reactor at 673 K and 30 MPa with variation in residence time 0.002-4 seconds [66]. The reaction resulted in solid solution of $MFe_2O_4$ and $\gamma$-$Fe_2O_3$ nanoparticles having non stoichiometric M/Fe molar ratio, with cubic spinel structure and a mean particle size below 10 nm (slightly increased with increase in residence time). With increase in residence time and molality of KOH, variations of M/Fe molar ratio in the resulting ferrite were obtained from 0.0 to 0.5. Increase in M/Fe molar ratio resulted in lattice parameter varying from $\gamma$-$Fe_2O_3$ to $MFe_2O_4$.

Bulk synthesis of monodisperse $M^{II}Fe_2O_4$ (M = Ni, Co, Mn) and $\gamma$-$Fe_2O_3$ nanoparticles at water-toluene interface under conventional and microwave hydrothermal conditions using metal nitrate or chloride precursors and oleic acid as dispersing agent is reported [67]. Advantage of this method is that a simple separating funnel is required to collect as synthesized nanoparticles in a non-polar organic solvent from reaction mixture. The organic solvent can be reused after concentrating it under reduced pressure. This method is suitable for large scale and economical production of monodisperse ferrites under conventional hydrothermal and microwave hydrothermal conditions. The resultant particles are spherical monodisperse having oleic acid surface coating with size distribution in 5-10 nm range and cubic spinel nanostructures with Fe/M mole ratio approximately equal to 2. At room temperature the nanoferrite particles are superparamagnetic having saturation magnetization (20-25 emu/g) and very small coercivity (11-12 Oe). Surface functionalization of the particles with organic moieties (lysine) resulted in increased dispersion in water for their promising use in drug delivery and catalytic applications.

With ferrous sulphate as original iron source, Fe-Co/$CoFe_2O_4$ nanocomposite, bulk composite and $CoFe_2O_4$ nanopowder are prepared using hydrothermal synthesis in the presence of cetyltrimethylammonium bromide (CTAB) [68]. Addition of surfactant assured that cationic distribution in the nanostructure is not affected by other reaction conditions. With ferric salt as original iron source, the resultant products showed low saturation magnetization. This deficiency was overcome by use of ferrous salt because its oxidation could promote the nucleation of ferrite to enhance hydrothermal reaction. Careful control of hydrothermal conditions led to the formation of $CoFe_2O_4$ ferrite with single domain critical size of 70 nm with high coercive force ($H_c$) magnitude of 4.6 kOe. Effective role of inert atmosphere was very much evident and under nitrogen atmosphere only Fe-Co/$CoFe_2O_4$ nanocomposite was produced, without which $CoFe_2O_4$ nanopowder was the resultant product. Nanocomposite showed higher value saturation magnetization, $M_s$, (113emu/g) and moderate coercivity, $H_c$, (1438 Oe). Fe-Co/$CoFe_2O_4$ composites exhibit a variety of physical, chemical and magnetic properties and have gained special attention.

A facile hydrothermal treatment of cobalt dodecyl sulphate, ferric chloride and aqueous sodium hydroxide at 120 °C has been used to prepare nanoplates and nanoparticles of $CoFe_2O_4$ [69]. At 300 K, the $CoFe_2O_4$ nanoferrite crystals show the saturation magnetization ($M_s$) 60.27 emu/g, which is lower than its analogous bulk material (about 80 emu/g) and coercivity ($H_c$) 465 Oe value is also very smaller than its bulk material (5400 Oe). Smaller magnitudes of $H_c$ may be assigned to their nanostructure.

For high density magnetic recording media devices, nanosize magnetic particles with high coercivity are required to prevent the interaction between adjoining grains and to subdue the media noise in material. Generally calcination of as-synthesized $CoFe_2O_4$ particles at higher temperatures is used to accomplish above stated properties [70, 71]. Since physical and magnetic properties are considerably affected by reaction conditions, an easy and rapid hydrothermal approach with no surfactant or template is used to synthesize $CoFe_2O_4$ nanocrystals without calcination [72]. Aqueous solutions of chlorides of cobalt (II) and iron (III) in 1:2 mole ratio prepared in 0.01M HCl were treated with NaOH in an autoclave, fitted with mechanical stirrer and external heating furnace, which was immediately sealed and heated to required temperature (120, 150, 200, 250, 300 °C) for one hour. Mean crystallite size increased from 14.53 to 26.79 nm with increasing temperature from 120 to 300 °C. Single nanocrystals of magnetic $CoFe_2O_4$ were well dispersed with narrow size distribution and well defined in shape. Crystal morphology was spherical for product obtained at hydrothermal temperature of 120 °C, transformed into cube with increase in temperature to 200 °C and finally to rod shape at 300 °C. $CoFe_2O_4$ nanocrystals exhibited maximum $H_c$ (2216 Oe) and $M_s$ (74.8 emu/g).

Applications of magnetic nanoparticles in the areas of imaging and therapy are vast expanding [73]. Hydrothermal synthetic methods are constantly being improved to allow easy control of particle size without using any chemicals that could not be easily removed from the final product. Cobalt ferrite nanoparticles are prepared using hydrothermal method with the addition of trisodium citrate dihydrate [74]. The reaction product is mainly cobalt ferrite nanocrystallite with average size 7.6±0.3 nm along with small amount of FeOOH. Magnetic properties are temperature dependent and cobalt ferrite is superparamagnetic above 380 K due overcoming of energy barrier for flip of spins, arising from magnetocrystalline anisotropy energy and the inter-particle interactions due to aggregation of nanoparticles. Below 20 K, frozen glasslike state is observed which is accompanied with a decrease in coercivity and high field paramagnetic susceptibility, enhanced saturated magnetization and effective magnetic anisotropy constant.

Ionic magnetic fluids based on cobalt nanoferrite particles are synthesized by using aqueous solution of nitrates of cobalt (II) and iron (III) in 1:2 mole ratio and NaOH to maintain pH 13 with constant stirring for 1 hour at room temperature [75]. The reaction contents were autoclaved for hydrothermal treatment and heated in oven at 100 or 200 °C for variable time periods, *i.e.*, 2 to 120 hours incubation times. The nanoparticles after isolation were dispersed for 10 minutes in HCl (0.2 mole $L^{-1}$) and then Citrate ions (5 %) were added as stabilizer and it resulted in decrease in size of nanocrystals. Finally the resultant product was washed with acetone and ether and dispersed in distilled water to prepare a ferrofluid of nanoparticles stable at pH 7. The average diameter of resulting nanoparticles was in the range of 11.9 to 18.7 nm with Spinel structure of $CoFe_2O_4$. Morphology transformed from spherical to cubic with increase in heating time (from 2 to120 hours) for hydrothermal treatment at 200 °C.

Nickel ferrite, $NiFe_2O_4$, is a technologically important ferrite with spin soft-magnetic properties, low conductivity, high electrochemical stability, and inverse spinel structure exhibiting typical ferromagnetism. Nickel ferrite powders are well studied for its potential use in magnetic materials, ferrofluids, catalysts and microwave devices. It has been prepared by hydrothermal processes [66, 76] and microwave hydrothermal method [77]. When $NiFe_2O_4$ is doped with small quantity of Co and Mn ions, its gas sensing property is improved [78(a)]. Magnetic properties of nanoparticles are highly dependent on the size and synthetic method. Synthesis of ultrafine ferrite powders under mild hydrothermal conditions is always a focus of research. Dielectric constant of a solvent is an important property which strongly influences the particle formation under hydrothermal conditions. Addition of low dielectric solvent (such as alcohols) to the aqueous solutions of metal salts to speed up particle formation kinetics is a strategy that has been used to prepare good quality $NiFe_2O_4$ nanopowder having perfect morphology at low temperature hydrothermal conditions [78(b)]. In this method isopropyl alcohol was used to lower the dielectric constant of particle forming medium. Aqueous solutions of nickel nitrate and ferrous ammonium sulphate (in 1:2 mole ratio) were treated with NaOH until pH 9.7 was attained and isopropyl alcohol was added with continuous stirring. The reaction contents were autoclaved and subjected to hydrothermal treatment with temperature range 60 to 120 °C for 1-8 hours. Spinel phase of $NiFe_2O_4$ ferrite, with 10-25 nm particle size, was observed at 60 °C for 3 hours heating time and with 0.05 to 0.10 isopropyl alcohol/water ratio. The crystallization of spinel ferrite was enhanced with increase in alcohol/water ratio 0.10, reaction temperature (up to 120 °C), and heating (holding) time up to 8 hours.

Template strategy is also used to control the morphology of magnetic nanoferrites. Synthesis of $NiFe_2O_4$ nanorods has been achieved by using polyethylene glycol 400(PEG 400) assisted route [79]. In a typical synthesis, $Fe^{3+}$ and $Ni^{2+}$ salt aqueous solutions in 2:1 mole ratio were poured drop wise into Teflon lined stainless steel autoclave, followed by injection of molten PEG 400 and then a known amount of NaOH with vigorous stirring until a homogeneous solution resulted. The autoclave was maintained at a certain temperature (100, 120, 150 or 180 °C) for 12 hours. It was observed that increase in hydrothermal temperature resulted in improving the crystallinity, and pure $NiFe_2O_4$ nanocrystals were obtained in the pH range 8-10. At lower pHs, $Ni^{2+}$ dissolved and excess $Fe^{3+}$ precipitated as FeOOH with incomplete ferrite formation and Higher pHs led to the dissolution of hydroxides of $Ni^{2+}$ and $Fe^{3+}$ ions. Nanorod Morphology was observed for $NiFe_2O_4$ with average diameter of 80 nm and length of 1μm and coercivity ($H_c$) 662 Oe which is greater than that of $NiFe_2O_4$ prepared without PEG 400 (248 Oe). PEG 400 has uniform and a proper chain structure, which with no difficulty is adsorbed on the surface of metal oxide colloids to modify the growth kinetics resulting in anisotropic growth of nanorods.

Ultrasonically facilitated hydrothermal treatment has been used to prepare nanocrystals of $ZrO_2$, $TiO_2$, $NiFe_2O_4$ and mixed ferrite $Ni_{0.5}Zn_{0.5}Fe_2O_4$ powdered materials [80]. Initially amorphous $ZrO_2$. *n* $H_2O$, $TiO_2$. *n* $H_2O$ gels were synthesized by co-precipitation from aqueous solutions of metal salts with $NH_4OH$ under continuous stirring prior to their ultrasonically assisted hydrothermal treatment at respective temperatures 423 and 523 K and times 1 and 3 hours. Similarly binary Ni-Fe-hydroxides and ternary Ni-Zn-Fe-hydroxides were prepared by co-precipitation and subjected to ultrasonically assisted hydrothermal treatment at 523 K for 1 hour. The elevation in temperature during hydrothermal treatment was maintained at 4°C/min in all preparations. Ultrasonic treatment during hydrothermal stage produced well crystalline $ZrO_2$ and $TiO_2$ in contrast to conventional hydrothermal route which resulted in a mixed oxide in amorphous and crystalline phases. Traditional and ultrasonically facilitated hydrothermal process of mixed hydroxides produced strongly aggregated nanoparticles of $NiFe_2O_4$ and mixed ferrite $Ni_{0.5}Zn_{0.5}Fe_2O_4$, perhaps due to smallness of individual crystallites of ~ 10-15 nm size. However, specific magnetic susceptibility curves for ultrasonically prepared materials were sharp and had higher blocking temperature. It is observed from the studies that ultrasonically assisted hydrothermal treatment results in ferrite products of large size crystallites with narrow size spread.

An easy and low cost hydrothermal treatment at a lower temperature is described for nanocrystalline $NiFe_2O_4$ [81]. Aqueous solutions of nickel sulfate and ferric chloride (1:2 mole ratio) were mixed with NaOH solution drop wise with constant stirring. The mixture was autoclaved and subjected to hydrothermal treatment at 180 °C for 10 hours. The resultant solid material was filtered, washed with acetone and deionized water and placed in vacuum at 50 °C for 4 hours for drying. The product was well crystalline with spread of grain size in 50-80 nm range with saturation magnetization ($M_s$) 53.6 emu/g and coercivity ($H_c$) 68 Oe. The magnitude of, $M_s$, compares well with the theoretically calculated value of 50 emu/g using Neel's sublattice theory. It is known that $M_s$ value increases with increase in crystallinity, thus observed higher $M_s$ and lower $H_c$ of $NiFe_2O_4$ may be associated to highly crystalline nature and unvarying morphologies.

Nickel ferrite, $NiFe_2O_4$, nanoparticles are synthesized using sol-gel and hydrothermal treatments [82]. In sol-gel treatment glycolic acid has been used as a chelating compound, it works as an energy source and breaks-up metal complex at low temperature to produce pure $NiFe_2O_4$ nanocrystalline structures with size ~ 27 nm. Hydrothermal synthesis has been carried out using nitrates of $Ni^{2+}$ and $Fe^{3+}$ in molar stoichiometric ratio, at low temperature (~160 °C) for heating time 15 hours and pH ~ 7.5 (maintained with 25% aq. ammonia) to produce nanocrystals with size ~ 9 nm. $NiFe_2O_4$ obtained from sol-gel method showed ferromagnetic properties at normal temperature with saturation magnetization, $M_s$, 31 emu/g and coercivity, $H_c$, 93 Oe, while that of hydrothermal was a superparamagnetic product with saturation magnetization, $M_s$, 46 emu/g and immeasurable coercivity. These observations suggest that magnetic features of these nanoparticles mainly depend on the size and preparation methodology.

Composition, size and shape of a spinel ferrite mainly determine its intrinsic properties. Shape controlled preparation of magnetic spinel ferrites requires development of strategies. Dodecyl sulphate of metals (M = Ni, Cu, Pb, Co) can act as a precursor and surface active reagent to stabilize nanoparticles during preparation of nanomaterials. $NiFe_2O_4$ nanoparticles and nanorods has been prepared by hydrothermal route using nickel dodecyl sulphate, $Ni(DS)_2$, ferric chloride and sodium hydroxide aqueous solutions at 120 °C [83]. $Ni(DS)_2$ was prepared using aqueous solutions of sodium dodecyl sulphate and nickel chloride. For typical synthesis of $NiFe_2O_4$ nanocrystals (90 % yield), $Ni(DS)_2$ (0.15 M), $FeCl_3.6H_2O$ (0.30 M) and NaOH (2.5 M) aqueous solutions after thorough mixing with stirring was autoclaved at 120 °C for 8 hours. The resultant product was a mixture of many nanoparticles (size, 10-35 nm), some quasi nanotubes (size, 45-115 nm), and few nanorods.

A decrease in concentration of $Ni(DS)_2$ and $FeCl_3.6H_2O$ (0.075 M and 0.15 M respectively) or increase (0.26 M and 0.51 respectively) while keeping other parameters constant, resulted in a product consisted of $NiFe_2O_4$ with spinel structure nanoparticles and nanorods with no obvious change in size, however, increasing concentration of nickel and iron increased the amount of nanorods. It has been speculated that the existence of two morphologies may be related to functional activity of surfactant $Ni(DS)_2$ as a soft structure directing reagent. At surfactant concentration greater than the critical micellar concentration (cmc) of $Ni(DS)_2$ ($1.2x10^{-3}M$), spherical micelles gradually transform into rod-like micelles. $NiFe_2O_4$ nanocrystals have Saturation magnetization ($M_s$) 30.4 emu/g which is lower than the bulk material (about 55 emu/g) [84] and coercivity ($H_c$) 87.4 Oe, higher than the bulk (nearly zero Oe) which should be related to their nanomorphology [85].

In reverse emulsion method, adjustment of water surfactant ratio controls the size and its distribution and morphology of the product. Reverse emulsion assisted hydrothermal method has added benefits of both techniques and it has been used for the preparation of nickel spinel ferrite nanospheres with only one size spread [86]. The composition of reverse emulsion includes water, cetyltrimethyl ammonium bromide (CTAB), iso-amyl alcohol, hexane and polyoxyethylene(10)nonyl phenyl ether(NP10). In a typical synthesis, aqueous solutions of nitrates of nickel (II) and iron (III) in 1:2 mole ratio and appropriate volumes of components of reverse emulsion were mixed with constant stirring. To the resulting mixture urea was dissolved with stirring to obtain transparent reverse emulsion. An aliquot of resulting emulsion was autoclaved and heated in an oven maintained 120 °C for 2-30 hours hold time. After natural cooling to normal temperature, washing thoroughly with distilled water, the product was kept for air drying at 90 °C. Phases of the sample changed with increasing hold time. Initial phases identified as β-FeO(OH), nanorods of hydrated iron nickel carbonate hydroxide (INCHH), $Ni_{0.75}Fe_{0.25}(CO_3)_{0.125}(OH)_2$. 0.38 $H_2O$, changed to nickel spinel ferrite nanospheres with increasing hold time up to 30 hours maximum. The role of CTAB and NP10 was dual; it not only helped to form nano- or micro-reactor, but also affected the thermodynamics of $NiFe_2O_4$ solid nuclei through the process of reverse emulsion. Out of these two roles, the former controls the size and morphology of the product and the latter is a deciding parameter in nucleation growth that results in different products in hydrothermal processes with and without surfactant. The reverse emulsion facilitated hydrothermal treatment is a potential strategy for the preparation of spherical nanoparticles having similar structures by varying the reaction conditions. Room temperature saturation magnetization increases with increasing reaction/hold time at 120 °C, *i.e.*, for 24 hours it is 51.78 emu/g and for 30 hours 55.01 emu/g which is almost near to saturation magnetization of bulk $NiFe_2O_4$(55 emu/g). Coercivity also increased with increasing time, *i.e.*, from 35.66 Oe for 24 hours and 42.77 Oe for 30 hours.

Size controlled nickel nanoferrite particles have been prepared by solvothermal treatment in ethylene glycol solvent and magnetic characteristics are discussed [87(a)]. Almost near mono-dispersed superparamagnetic spinel-type nickel nanoferrite particles are prepared by interface reaction of Ni(II)Fe(III)-oleate complex with alkali (such as urea), under hydrothermal crystallization process [87(b)]. Precursor is prepared by co-precipitation of oleate anions with aqueous solutions of Ni(II) and Fe(III) cations taken in 1:2 mole ratio. Sodium oleate is dissolved in 1:3v/v water-ethanol solvent. It resulted in Ni(II)Fe(III)-oleate complex, after thorough washing with solvent and drying in vacuum at 100 °C. For hydrothermal crystallization, known amounts of precursor and alkali solution (urea) were autoclaved and kept at a certain temperature for many hours. At the interface of precursor and solution, the Ni(II)Fe(III)-oleate complex reacted with $OH^-$ ions to give a hydroxide precursor, $NiFe_2(OH)_x$, that transformed into nickel ferrite, $NiFe_2O_4$, on dehydration in hydrothermal conditions at low temperature (150-180 °C). It is suggested that *in situ* generated oleate is adsorbed on the surface of $NiFe_2O_4$ nanoparticles and inhibited disordered accumulation of the particles, restricting further growth in size of particles. Produced $NiFe_2O_4$ nanoparticles are dispersed into the aqueous medium under the influence of Brownian movement (a key factor to affect uniformity and dispersity of the product) that allowed the reaction to occur in reverse micelle method. With a fine control of reaction conditions, almost mono-dispersed superparamagnetic spinel-type nickel nanoferrite particles with size range of 5-12 nm could be easily formed. Hydrothermal conditions of temperature, time, solvents, and nature of alkali control the size and shape of nanoparticles. Water-ethanol solvent and urea as alkali produced better size and uniform morphology of $NiFe_2O_4$.

Three-dimensional-ordered macro-porous (3DOM) $NiFe_2O_4$ as anode fabrication material for lithium ion batteries has shown superior cycling performance [88(a)]. It is composed of unvarying spherical macro-pores of nearly 140 nm diameter with 20 nm size walls. At 0.2 mA $cm^{-2}$current density, it exhibits high initial discharging capacity up to 1370 mAh $g^{-1}$ and reversible capacity of 670 mAh $g^{-1}$. These values propose that 3DOM $NiFe_2O_4$ nanomaterial is a prospective contender for anode material in Li ion batteries.

Continuous hydrothermal synthesis of mixed metal oxides, *i.e.*, nickel ferrite nanoparticles, has been achieved with the help of a flow-through apparatus connected to central collision-type micro-mixer (CCM) [88(b)]. This innovation is a production technology for producing sufficient amounts of nanoferrite particles with desirable characteristics within very short residence times in the range of several milliseconds to several tens of seconds on the concept of green chemistry [88(b)]. The term desirable characteristics implies mean particle size and its spread, crystal structure/morphology, stoichiometry of constituents, crystallinity and these are significant parameters for the development of industry, to broaden utilization fields and to minimize product development time. This innovation works without use of organic solvents, growth restraining agents and use of high base concentration. Furthermore no additional treatments of calcination or comminution are needed. The micro-mixer is designed for rapid heating of precursor solution (containing nitrates of $Ni^{2+}$ and $Fe^{3+}$ in 1:2 mole ratios) to the reaction temperature and evenly distributed nucleation. Temperature (range 573-673 K), residence time/hold time (0.02-2.00 s), total molality of solution (0.05-0.50 mol/kg) were varied. The molal concentration of NaOH solution (0.53-2.6 mol/kg), corresponding to the molality of $HNO_3$ (due to metal nitrates in the reaction mixture), was set to control pH of the system. In hydrothermal process, the nucleation of $NiFe_2O_4$ is supposed to continue *via* two-step reaction of hydrolysis and dehydration condensation. Temperature (673 K), and pressure (30 MPa) conditions, which are above critical point of water, result in formation of stable $NiFe_2O_4$ nanoferrite particles with monomodal spread even at higher molal concentration of 0.5 mol/kg. With central collision-type micro-mixer production system, synthesis of ferrite nanoparticles (5.0-7.0 nm) with monomodal spread at a preparation rate of nearly 50 g/h is manageable.

Nanocrystalline $NiFe_2O_4$, has been synthesized using reaction of metal chlorides in ethylacetate solvent with and without surfactant facilitated treatments [89]. Appropriate amounts of chlorides of Fe(III) and Ni(II) in 2:1 mole ratio dissolved in double distilled deionized water at pH 10, attained with 5M triethyl amine solution in ethylacetate, are used for hydrothermal process at a given temperature in a Teflon-lined autoclave reactor. Surfactant assisted $NiFe_2O_4$ nanoparticles were also prepared hydrothermally at 150 °C using surface active agents (Glycerol or Sodium dodecyl sulfate in deionized water) keeping rest of reagents and experimental conditions same. At 150 °C well crystalline $NiFe_2O_4$ nanoparticles are obtained, however, crystallinity decreased when surfactant was added. The average size of $NiFe_2O_4$ nanoferrite particles increased from 100 °C to 150 °C. With and without surfactant the average sizes of nanoparticles are approximately 10-15 nm and 60 nm respectively. Without surfactant the morphology of $NiFe_2O_4$ nanoparticles is non-uniform cubic-like, and with it sphere-like uniform in morphology and particle size.

With and without surfactant, all $NiFe_2O_4$ nanoparticles exhibited superparamagnetic characteristics at normal temperature. Lower saturation magnetization and coercivity values could be related to crystallinity and particle size.

Magnetic ferrite nanoparticles are used as components in recording tapes, flexible disc recording media, magnetic fluids and biomedical materials. Knowledge of properties of nanostructure systems and how modification in bulk properties takes place with decrease in crystal size to nano-metric range is required for these applications. Among various spinel ferrites, $CuFe_2O_4$ is a focus of great interest fundamentally and technologically for a variety of applications. Being a Jahn-Teller ion, it exhibits anomalously favorable properties and phase transition from tetragonal to cubic depending on reaction temperature. To study temperature dependent finite-size and surface effects in spinel $CuFe_2O_4$ nanoparticles in ferrofluids, it is synthesized by hydrothermal co-precipitation in alkaline pH by taking chlorides of Cu(II) and Fe(III) in 1:2 mole ratio [90]. The particles are easily peptized in aqueous solution to obtain a stable sol of good quality. It is observed that finite size and surface effects severely influence the thermal variation of magnetization and that the structure of nanoparticles is composed of a single-domain ordered core and a surface shell of disordered spins leading to further magnetic contribution.

The type of spinel structure of nanophase has been identified for nanoferrite particles in a continuous synthetic process in near critical and supercritical water [65]. Rapid and continuous preparation of $M^{II}Fe_2O_4$ (M = Ni, Cu, Zn) has been achieved by hydrothermal synthesis using a continuous flow reactor at 673 K and 30 MPa with variation in residence time 0.002-4 seconds [66]. The reaction resulted in solid solution of $MFe_2O_4$ and $\gamma$-$Fe_2O_3$ nanoparticles having non stoichiometric M/Fe molar ratio, with cubic spinel structure and mean particle size below 10 nm (slightly increased with increase in residence time).

Low temperature isothermal reduction behavior of $CuFe_2O_4/MgFe_2O_4$ core/shell nanocrystallites has been reported [91]. $CuFe_2O_4$ nanoparticles were coated with $MgFe_2O_4$ nanoferrite particles to obtain core/shell magnetic structures by hydrothermal precipitation method. The drying and calcining of as-synthesized nanoparticles were carried out at 343 K for 24 hours and 900 °C for 2 hours respectively to yield $CuFe_2O_4/MgFe_2O_4$ core/shell structure. Pellets were prepared from the fired product and sintered at 500 °C for 2 hours. Isothermal reduction of $CuFe_2O_4/MgFe_2O_4$ compacts was carried out with $H_2$ at 400-700 °C. Reaction kinetics suggests that in the initial stages rate is governed by diffusion of gas and in final stages by joint effect of gas diffusion and chemical reaction mechanism at the interface. A decrease in the coercivity and increase in saturation magnetization was observed with an increase in reduction temperature; however, remnant magnetization remained unaffected by reduction temperature.

$CuFe_2O_4$ nanoparticles have very active catalytic chemistry [92]. It has been used as catalyst in thermal decomposition of RDX [93], thermal decomposition of ammonium perchlorate [94], hydrogen production from dimethyl ether [95], catalytic decomposition/reduction of $CO_2$ [96, 97], and hydrocarbon (HC) oxidation [98]. Nano-$CuFe_2O_4$ has been prepared by hydrothermal method to study its catalytic behavior on $CuSO_4$ [99]. Aqueous solutions of nitrates of Fe(III) and Cu(II) were taken in appropriate molar ratio and NaOH (4 mol/L) was added drop wise with constant stirring to maintain 9-10 pH. The mixture was continuously stirred for 30 minutes, separated by filtration and washed with distilled water. The resultant precursor was stirred with polyvinyl alcohol in high pressure reactor at 320 °C for 3 hours to obtain nano-$Cu_nFe_mO_{(n+1.5m)}$ powders. Catalytic studied indicated that catalytic activity of $Cu_nFe_mO_{(n+1.5m)}$ is related with its Cu/Fe molar ratio and its percentage in the reaction mixture. With Cu/Fe 1:1 mole ratio and 4% (in mass), $Cu_nFe_mO_{(n+1.5m)}$ showed the superior activity. It has a weak influence on the dehydration step of $CuSO_4$. 5 $H_2O$, however, it showed strong effect on the thermal decomposition of $CuSO_4$.

Technological importance of electrode materials for Li ion batteries is a focus of interest because rechargeable batteries for conveniently movable electronic devices are in great demand. Furthermore electric vehicles and hybrid vehicles also require batteries with high energy and power density for their working. Traditionally carbon materials are fabricated into anodes for use in Li ion batteries. Electrochemical characteristics of iron-containing materials with different morphologies and particle/crystal structures are much studied in order to find an alternate with relatively low cost and environmental friendly electrode material. A variety of ferrite compounds have shown improved electrochemical features than those of traditional compounds used as anodes in Li ion batteries. $MFe_2O_4$ (Cu, Zn) nanoparticles have been prepared by hydrothermal method [100]. The synthesized $CuFe_2O_4$ and $ZnFe_2O_4$ are cubic spinel nanoferrites with sphere-like morphology and average respective sizes 60-70 nm and 30-40 nm. Application of current density of 0.2 $mAcm^{-2}$ at potential a 0.0-3.0 V, the initial discharging powers of $CuFe_2O_4$ and $ZnFe_2O_4$ electrodes were 1412.3 $mAhg^{-1}$ and 1287.5 $mAhg^{-1}$ respectively. These values indicate that Cu is a superior counter ion than that of Zn. The synthesized $MFe_2O_4$ nanoferrite particles are hoped to be future contenders for anode materials in Li ion batteries.

$CuFe_2O_4$ nanoferrite particles are synthesized by assisted ultrasonic hydrothermal treatment taking nitrates of Fe (III) and Cu (II) in 2:1 mole ratio with sodium hydroxide as pH regulator [101]. The precursors were mixed with continuous stirring for 15 minutes for a better homogeneity and pH 11 was adjusted with 1M NaOH. The contents were autoclaved in a Teflon lined steel autoclave fitted with an immersed sonotrode (tip diameter 7 mm, made up of titanium alloy) and kept at 200 °C for 2 hours for hydrothermal treatment under continuous sonication at 100W and 40 KHz. For this temperature the auto-generate pressure in hydrothermal process was about 40 Barr. Ultrasonation played an important role in this assisted hydrothermal synthesis of $CuFe_2O_4$ nanostructures by accelerating the reaction and found to be useful to produce smaller and unvarying nanoparticles. The hydrothermal treatment in similar conditions but without ultrasonation resulted in a big amount of unreacted precursors present in the produced ferrite. In addition to homogeneous cavitation in pure liquid, the cavitation creates mechanochemical effects in liquid–solid system and increase mass transport. Crystalline spherical $CuFe_2O_4$ nanoparticles are superparamagnetic at room temperature and have cubic spinel structure with 16 nm diameter and narrow size distribution in the range of 15-18 nm with the highest saturation magnetization value 29 emu/g.

Zinc ferrite and its solid solutions with other ferrites exhibit specific magnetic properties and show high performance for application in electronic industry. A novel magnetic substance, named as ferrite/metal composite thin film has been prepared [102]. Thin films with spinel phase $MFe_2O_4/Fe$ (M = Zn, Mg) with restrained grain sizes and morphology are prepared by direct fabrication on an iron surface by a soft solution processing route at temperatures lower than 200 °C. Ferrite/Fe composite thin films exhibit superparamagnetism and have shown much improved magnetic performance than that of single-phase ferrite materials. $ZnFe_2O_4$ ultrafine particles with high magnetization have been synthesized using solid-liquid interfacial reaction between Zn metal sheet and $Fe^{2+}$ in ammonia under hydrothermal conditions at 180 °C for 6-12 hours [103]. As-synthesized $ZnFe_2O_4$ particles exhibited excellent magnetic characteristics with highest possible saturation magnetization and coercivity ($H_c$) 61.2 emg/g and 110 Oe at 80 K and 54.6 emg/g and 52 Oe at 300 K respectively. The enhanced magnetic performance is related to ultrafine particles with special micro-structure produced by hydrothermal route.

Various amines are used as templates to prepare microporous zeolite structures. Well crystalline and well dispersed with little conglomeration, nanometer-microporous spinel $ZnFe_2O_4$ with grain size about 7.3 nm, is prepared by using triethylamine as a template under hydrothermal treatment with optimum conditions of pH 11, reaction temperature 448 K and reaction time 6 hours [104]. Precursor pH, reaction temperature and time influence the morphology. Adsorption studies showed that pores in $ZnFe_2O_4$ are perforated and adsorb 24.8 mg/g of water.

A successful modified solvothermal reduction process for the synthesis of monodisperse magnetic single-crystal $Fe_3O_4$ microspheres has been reported [5]. It is also a generalized simple, one step and low cost method for the production of monodisperse microspheres ferrite series *i.e.*, $M^{II}Fe_2O_4$ ($M^{II}$ = Fe, Mn, Co, Zn) with a controllable diameter in the range of ~200-800 nm. For microspheres with diameter of nearly 200nm, the saturation magnetization ($M_s$) values are reported as; $Fe_3O_4$, 81.9 emu/g; $MnFe_2O_4$, 53.2 emu/g; $CoFe_2O_4$, 61.6 emu/g; $ZnFe_2O_4$, 60.0 emu/g at 300 K. The prepared magnetic ferrite microspheres exhibit good colloidal stability.

Zinc ferrite is a normal spinel showing antiferromagnetic behavior with a Neel temperature of 10.5 K. Magnetic properties of nanoferrites depend upon their shape and size. Different synthetic methods produce zinc ferrite nanoparticles with different ferrimagnetic properties exhibiting a variety of Neel temperatures, coercive fields and saturation moments depending on the particle size. The magnetic behavior of nanostructured $ZnFe_2O_4$ overall is in good agreement with the enhanced magnetization and ferromagnetic behavior reported for nanostructured, ultrafine and thin films of $ZnFe_2O_4$ [105]. Zinc and nickel nanoferrite with uniform size are synthesized using low temperature hydrothermal route [106]. Aqueous solutions of nitrates of iron and zinc or nickel were taken in required stoichiometric amounts and pH 11 was attained with the addition of KOH (2 M) solution. Mixture containing precipitated hydroxides of iron and zinc or nickel was autoclaved and reacted under hydrothermal conditions for 3 to 15 hours at 160 °C for $ZnFe_2O_4$ and at 180 °C for $NiFe_2O_4$. The resultant ferrites were single phase, well crystalline with size 6-10 nm for $ZnFe_2O_4$ and 50-60 nm for $NiFe_2O_4$ with well-formed octahedron. Zinc ferrite exhibited superparamagnetic behavior with blocking temperature of 20 K, whereas, that of nickel ferrite to be above 300 K. At low temperature both ferrites showed ferrimagnetic behavior arising from total inversion of ground state in $NiFe_2O_4$ and the partially inverse in case of $ZnFe_2O_4$.

In spinel $ZnFe_2O_4$, $Zn^{2+}$ cation was substituted into the lattice of the original structure of ($Fe^{2+}Fe_2^{3+}O_4$), it proved to be an effective future material in sensing reducing gases like carbon mono oxide and methane [107]. $ZnFe_2O_4$ magnetic nanoparticles have been synthesized *via* PEG-assisted route [108]. A known volume of Fe(III) and Zn(II) salt aqueous solutions in 2:1 mole ratio was introduced drop wise into Teflon lined stainless steel autoclave, then molten and heated polyethylene glycol-400 (PEG 400) was injected in the autoclave and pH 11 was adjusted by addition of NaOH. The contents were continuously stirred at 400 rpm for 60 minutes to obtain a homogeneous solution. The reaction vessel was maintained at 130 °C for 12 hours. After cooling to normal temperature, thorough washing with water and absolute ethanol, the drying of resultant material was done in vacuum oven at 25 °C for 12 hours. $ZnFe_2O_4$ nanoparticles showed superparamagnetism and had spherical morphology with an average size 12±3 nm and narrow size distribution. Measured saturation magnetization ($M_s$) decreased with increase in temperature and $M_s$ values in the temperature range 305K to 10 K are 5, 15, 26, 32 36 emu/g respectively. The nanoparticles also showed spin glass-like behavior.

Zinc ferrite nanoparticles exhibit interesting magnetic and electromagnetic properties and fulfill specific requirements such as an absorbent material for hot gas desulfurization [109], and a microwave absorbing material [110]. High density of $ZnFe_2O_4$ makes it heavy and prohibits its application for light weight devices. Polypyrrole (ppy) has a unique flexible structure, enhanced eco-stability, highly conductive, relatively less dense and ease of preparation and find use in practical applications including as core-shell outer material [111]. Core-shell structured materials have an outer coating material and an inner component and thus exhibit unusual properties. Such materials may find application in a number of fields [112, 113]. $ZnFe_2O_4$/ppy core-shell nanocomposites have relatively low density and improved compatibility as compared to $ZnFe_2O_4$ nanoparticle and thus more effective as shielding material for different electromagnetic sources. Synthesis of $ZnFe_2O_4$ nanoparticle is carried out using an easy, low cost, nontoxic and reproducible solvothermal treatment [114]. $ZnFe_2O_4$/ppy core-shell nanoparticles are consequently prepared *via in situ* chemical oxidation polymerization of pyrrole monomers on the surface of solvothermally produced $ZnFe_2O_4$ nanoparticle with ammonium persulphate as oxidizing agent at a lower temperature (80 °C). Amount of pyrrole monomers may be adjusted to control shell thickness of core-shell nanoparticles.

To prepare $ZnFe_2O_4$ [114] a known amounts of chlorides of $Fe^{3+}$ and $Zn^{2+}$(in 2:1 mole ratio) were dissolved in ethylene glycol solvent and a required amount of

KAC(potassium citrate) was introduced with continuous stirring. The contents were autoclaved and solvothermally treated at 180 °C for 24 hours and cooled to room temperature. The product was separated by centrifugation, after washing with distilled water and ethanol, air drying was done at 50 °C for 6 hours. $ZnFe_2O_4$/ppy core-shell nanoparticles were prepared by dispersing a known amount of $ZnFe_2O_4$ in a known volume of deionized water under continuous sonication for 30 minutes followed by injection of pyrrole monomer and stirring for 30 minutes. Then aqueous solution of ammonium persulphate was slowly added into the mixture and polymerization was carried out at 80 °C for 8 hours under continuous mechanical stirring. The washing of resultant core-shell composite product was done with distilled water and ethanol and air dried at 50 °C for 6 hours. As-synthesized $ZnFe_2O_4$ nanoparticles showed saturation magnetization ($M_s$) value 31.92 emu/g and coercivity ($H_c$) 94Oe. As prepared $ZnFe_2O_4$/ppy core-shell nanoparticles exhibited ferrimagnetism with saturation magnetization ($M_s$) value 17.8 emu/g, coercivity ($H_c$) 130 Oe. They showed excellent microwave absorption than that of $ZnFe_2O_4$ nanoparticle. The non-magnet ppy-layer on the surface of magnetic particles can reduce $M_s$ value, whereas increase in $H_c$ value of core-shell composite can be attributed to surface anisotropy due to coating and external morphology.

In general, Magnetorheological (MR) materials are suspended magnetic particles dispersed in non-magnetic liquid carriers. MR fluids are characterized by rapid transformation from a liquid- to solid-like phase; such that they exhibit their rheological behavior (for example increased shear stress). MR fluids regain their initial state in several milliseconds on withdrawing the applied magnetic field. Due to rapid and controllable *MR effect*, MR fluids extensively find use in shock absorbers, clutches, hydraulic valves, polishing devices *etc*. [115]. Water as a matrix in MR polishing is a nontoxic, inexpensive solvent, and has desired lubricating characteristics and high polishing effect. However, due to subsidence stability, highly dense classical magnetic particles settle down in water. It is expected that nanoparticles with hollow sphere morphology may enhance the stability of suspension with respect to their other similar solids [116, 117]. $ZnFe_2O_4$ magnetic hollow spheres are synthesized using an improved one-pot solvothermal method [5(a), 5(b)]. Hollow spheres are prepared *via* one-step solvothermal method without using high temperature and additional templates during reaction [118]. In a typical procedure, appropriate amounts of chlorides of Fe(III) and Zn(II) in 2:1 mole ratio were dissolved in an appropriate volume of ethylene glycol and polyethylene glycol 600 (PEG 600) was added drop wise as stabilizer. It was followed by slow addition of required amount of urea with stirring which was continued for about 20 minutes to obtain homogeneous suspension. The mixture was autoclaved and kept at 200 °C for 24 hours in an oven. On cooling and washing of the black product thoroughly with water and absolute ethanol, drying was carried out in vacuum at 50 °C for 12 hours. To obtain MR fluid as-synthesized product was dispersed directly in distilled water with particle volume fraction of about 30%. As-synthesized $ZnFe_2O_4$ hollow spheres were single crystalline with face-centered cubic spinel structure. $ZnFe_2O_4$ hollow spheres were ferromagnetic with saturation magnetization ($M_s$) value 83.4 emu/g, coercivity ($H_c$) 62.06 Oe and remnant magnetization ($M_r$) 8.78 emu/g. $ZnFe_2O_4$ hollow spheres exhibited low density and well dispersion in water. MR fluid showed excellent stability and MR effect in water and suggesting that it may have possible uses in polishing and biomedical areas.

Synthesis of Fluorescent compound containing $ZnFe_2O_4$ with porous nanorod assembly was achieved under moderate hydrothermal and calcination treatments in the presence of cetyltrimethylammonium bromide as template directing agent [119]. Fluorescent compound containing $ZnFe_2O_4$ nanostructures were made up of porous nanorods having mean length 122 nm and diameter 29 nm with a band-gap approximately 1.94 eV. As-prepared $ZnFe_2O_4$ are used as a photocatalyst driven by visible light for hydrogen production and show exceptional photo-stability in an aqueous colloidal phase by using methanol as sacrificial reagent.

Quasi-cube morphology $ZnFe_2O_4$ nanocrystals have been prepared in typically small sizes of 5–15 nm by the reaction of ferric chloride hexahydrate and zinc chloride in the presence of sodium dodecyl sulphate under hydrothermal conditions and its nanocomposite with $TiO_2$ (Degussa P25) has been studied for photocatalytic property in the mineralization of Rhodamine B [120].To prepare $ZnFe_2O_4$/P25 nanocomposite, the $ZnFe_2O_4$ nanocrystals and $TiO_2$ (commercial Degussa P25) are physically ground at ambient temperature. It was found to be a powerful photo-catalyst in the mineral formation of Rhodamine B.

$ZnFe_2O_4$/C hollow spheres have been prepared *via* an easy one-pot solvothermal method using low cost starting materials [121]. The synthesized composite exhibit good rate compatibility and are suggested as superior material to fabricate anodes for Li ion Batteries.

Preparation of crystalline ultrafine nanoparticles of $ZnFe_2O_4$ with octahedral morphology has been reported *via* a hydrothermal route [103]. The characteristics of nanomaterials are very much dependent on the morphology of their internal structure [6(c)]. With a morphological control of $ZnFe_2O_4$, synthesis of octahedral

$ZnFe_2O_4$ nanocrystals using nitrates of zinc (II) and Fe (III) as staring materials in the presence of polyvinyl pyrrolidone (PVP) *via* simple hydrothermal procedure has been achieved [122]. In a typical procedure, nitrates of zinc (II) and Fe (III) in 0.5:1.0 mmol ratio, optimal amount of polyvinyl pyrrolidone (PVP) 0.3g and urea 4.5 mmol were added to 20 ml distilled water and stirred for 30 min. Drop wise addition of ammonia and hydrazine (3 ml each) into the above mixture solution was followed by addition of 9 mmol of sodium hydroxide to raise the solution pH to 13. The contents were autoclaved and kept at 180 °C for 8 hours. After natural cooling to normal temperature, a black solid powder was collected by centrifugation, on washing with alcohol and distilled water three times; its drying at 50 °C for 5 hours was accomplished. In the preparation of octahedral $ZnFe_2O_4$, PVP should selectively adsorb onto the crystallographic planes for the formation of the final octahedral particles. Magnetization measurements showed that octahedral $ZnFe_2O_4$ nanocrystals have good magnetic properties with a saturation magnetization value 44.30 emu/g and a high coercivity of 299.2 Oe at normal temperature; all of which are higher than those (0.33 emu/g, 27 Oe) reported for $ZnFe_2O_4$ powders [123]. In fact, with the reduction of particles size to nano-scale, the surface/volume ratio of the particles can be increased notably, such that up to 50% of the particle atoms can be located on the surface, directly influencing the magnetic properties of the particles [124]. Magnetic properties of product let these octahedral $ZnFe_2O_4$ nanaocrystals to have a widespread potential application in tumor therapy, semiconductor devices, magnetic recording and drug carriers.

### *4.2.3. Mixed Metal Ferrites and composites*

The spinel ferrites with formula, $M^{II}Fe_2O_4$ (M = Mn, Fe, Co, Ni, Zn, *etc.*) show unusual magnetic, magneto-resistive, and magneto-optical characteristics. Systematic alterations in magnetic characteristics of these ferrites may be achieved by total or partial substitution of the divalent $M^{2+}$ cation while retaining the fundamental spinal crystalline phase. It is an established fact that the composition and microstructure (particle size and shape) of ferrite nanoparticles are sensitive to the preparation methodologies and effectively modify the chemical, structural and magnetic properties.

Monodisperse magnetic compounds with more than three components are focus of interest and not many synthetic methods are reported for magnetically stable ferrite nanoparticles. In solvothermal synthetic methods ethylene glycol is commonly used solvent; however, in some cases *n*-octanol is also reported. A liquid-phase redox reaction with *n*-octanol as solvent and sodium dodecylbenzene sulfonate (SDBS) as an additive has been used to prepare high-quality $Co_3O_4$

nanocrystals from cobalt nitrate as a precursor [125]. This methodology has been extended for high quality monodisperse ferrite Co-Ni-ferrite with composition ($Co_xNi_yFe_{3-x-y}O_4$, x = 0.6, 0.3, 0.2, 0, $y$ = 0, 0.6, 0.7, 0.8) nanoparticles composed of many constituents with no more size-selectivity [91,126]. The sizes and compositions are controllable by suitable choice of treatment time and the initial stoichiometry of inorganic precursors respectively.

Cobalt nitrate, ferric chloride and nickel acetate were made anhydrous. In a typical procedure [126], a calculated amount of anhydrous $Co(NO_3)_2$ and/or anhydrous $Ni(Ac)_2$ (total 1.147 mmol), anhydrous $FeCl_3$ ( 0.372 g, 2.294 mmol) and SDBS (3.30 g, 8.58 mmol) were dissolved in 20.0 mL of *n*-octanol under magnetic stirring. After complete dissolution of precursors, the resultant solution was autoclaved in a Teflon lined autoclave and kept at 90 °C for 6 hours and then under solvothermal conditions at 240 °C for different lengths of time. The reaction vessel was quickly brought to normal temperature by dipping it into water. To precipitate the product, anhydrous ethanol was added to resultant black-brown solution. The product composed of nanoparticles was separated by centrifugation (8500 rpm, 10min). It could be monodispersed in nonpolar and weakly polar solvents. To produce water-dispersible nanoparticles, the product was washed with ethanol twice, then with water 5 times and finally with acetic acid (10 wt %) to change the surface of nanoparticles. The size of $Co_xNi_yFe_{3-x-y}O_4$ nanoparticles increased with reaction time and thus could be controlled by changing the reaction time, however, no obvious linear correlation was observed. $Co_{0.6}\,Fe_{2.4}O_4$ nanoparticles with uniform sizes of 9.5±0.2, 10.5±0.3, and 13.5±0.3 nm could be obtained with reaction times of 6, 8, and 12 hours, respectively. As prepared nanoparticles samples exhibited superparamagnetic behavior, except $Co_{0.6}\,Fe_{2.4}O_4$ sample, that behaved as ferromagnetic in nature. The as-synthesized nanoparticles were well-dispersed in nonpolar cyclohexane and, after simple surface modification, also showed good dispersion in water.

Hydrothermal treatment facilitated by surfactant cetyltrimethylammonium bromide has been applied to prepare $Ni_xCo_{1-x}Fe_2O_4$ (x = 0.0, 0.6, 1.0) nanoparticles in the presence of conc. $NH_3$ and NaOH as hydrolyzing reagents [127]. An appropriate amount of CTAB, stoichiometric amounts of chlorides of Fe(III), Co(II) and Ni(II) were dissolved with vigorous stirring in deionized water to make solution up to a known volume. To adjust the 11.0 pH, 2M NaOH and conc. $NH_3$ solution were added to the mixture, which was then subjected to pretreatment in ultrasonically operated water bath for 30–40 min and further hydrothermally treated at 130 °C for 15 hours in an oven. After cooling and separating of the resultant dark product, washing with distilled water and then annealing was done at 500 °C for 5 hours. For x = 0.0 and 1.0, pure $CoFe_2O_4$ and $NiFe_2O_4$ nanoparticles were obtained respectively, whereas for x = 0.6, mixed spinel ferrite was the product. With $NH_3$, a decrease in particle size, and no significant change in morphology of the samples was observed with increase in the amount of Ni (x) in the ferrite phase. Whereas increase in particle size and change in structure from nearly spherical to more edgy crystallites (polygons) was shown in the presence of NaOH with increase in the amount of Ni. Sodium hydroxide treated ferrites show larger size than those in the presence of $NH_3$, it is evident for x = 1.0. Replacement of $Co^{2+}$ by $Ni^{2+}$ ions in $CoFe_2O_4$, grains decreases the antiferromagnetic properties and increases the superexchange phenomenon leading to lowering in coercivity and decreased magnetization. When all of $Co^{2+}$ cations are substituted by $Ni^{2+}$ cations, both coercivity and magnetization further reduce.

Development of gas sensors is very important for the detection of volatile organic and semi-volatile nerve agent compounds [128]. Semiconducting metal oxides such as ZnO and $WO_3$ nanostructures are used as chemical vapour detectors [129, 130]; however, problem of low selectivity exists. New gas selective compounds like ferrites are fast developing because they are chemically more stable with enhanced sensitivity for a specific gas as compared to metallic oxide semiconductors. Large surface area is a requirement for gas sensing materials. In addition to spinel type single metal ferrites ($CoFe_2O_4$ and $NiFe_2O_4$) their mixed ferrites are also significant for application as gas sensor [56(b)]. $CoFe_2O_4$ has been used as ethanol vapour detecting material [58(a)]. Gas sensitivity of ferrite can be effectively improved by replacing A- or B-site with a number of lanthanides and Ni, Co, and Mn ions [128]. Surface area of $CoFe_2O_4$ increases when it is substituted with Mn [58(b)]. A larger surface area has been observed for as-synthesized nanoparticles of $MFe_2O_4$ (M = Mn, Co, Ni) prepared by a simple solvothermal method [54(a)]. When $NiFe_2O_4$ is doped with small quantity of Co and Mn ions its gas sensing ability is improved [78(a)].

With ferrous sulphate as original iron source, Fe-Co/$CoFe_2O_4$ nanocomposite, bulk composite and $CoFe_2O_4$ nanopowder are prepared under hydrothermal conditions in the presence of cetyltrimethylammonium bromide (CTAB) [68]. Nanocrystalline $Ni_{0.8}Coi_{0.2}Fe_2O_4$ has been synthesized under refluxing conditions in ethylene glycol as solvent [131].

Nickel-substituted cobalt ferrite ($Co_{1-x}Ni_xFe_2O_4$, x=0-0.8) magnetic nanoparticles are prepared *via* a solvothermal treatment in ethylene glycol as the solvent [132].

For $Co_{1-x}Ni_xFe_2O_4$ (x = 0-0.8) NPs, the initial molar ratio (1:2) for $Me^{2+}/Fe^{3+}$ was maintained, $Me^{2+}$ being ( $Co^{2+}$ + $Ni^{2+}$), and 5 mmol of $Fe^{3+}$ was taken in every ferrite. All the ferrite samples were single phase spinel type spherical NPs with mean particle size in the range of 40-90 nm and composed of crystallites having average size of about 5.8–9.8 nm. In $Co_{0.8}Ni_{0.2}Fe_2O_4$ NPs, the particles were generally spherical along with a few particles in disordered form. As-prepared samples show superparamagnetic behavior. It may be inferred that size of these spherical nanoparticles is smaller than the critical size required for superparamagnetic behavior [131,133]. $M_s$ of $Co_{1-x}Ni_xFe_2O_4$ decreases linearly with increase in $Ni^{2+}$ contents. $Co_{1-x}Ni_xFe_2O_4$ samples showed high selectivity for ammonia, when examined for selectivity of ammonia, ethanol, toluene, benzene, methanol and methanal vapors with $Co_{0.8}Ni_{0.2}Fe_2O_4$ showing the best response.

Mn–Zn ferrite ($Mn_{1-x}Zn_xFe_2O_4$) is such a member of ferrite group, which is a focus of meaningful research potential due to its unique magnetic and electromagnetic characteristics. It is chemically more stable and corrosion resistive. A simple, facile solvothermal method, which is general, efficient, and environmental friendly, has been designed for preparing $Mn_{1-x}Zn_xFe_2O_4$ nanospheres which are buildup of tiny nanocrystallites [134]. The unusual structure of $Mn_{1-x}Zn_xFe_2O_4$ ferrites may have a great effect on their characteristic properties and also be significantly useful for basic studies and function-based uses.

A series of $Mn_{1-x}Zn_xFe_2O_4$ nanospheres have been prepared with stoichiometry $0 < x < 1$ [134]. A specimen synthetic procedure for $Mn_{0.6}Zn_{0.4}Fe_2O_4$ nanospheres is described: dissolving of salts of $Fe^{3+}$ (5 mmol), $Zn^{2+}$ (1 mmol), and $Mn^{2+}$ (1.5 mmol) in ethylene glycol (50ml) was followed by addition of $NH_4Ac$ (0.02 mol) and polyethylene glycol (0.8 ml) and subjected to solvotheramal treatment at 200 $^o$ C for 12 hours. It is observed that morphology of as-prepared $Mn_{0.6}Zn_{0.4}Fe_2O_4$ is 100–200 nm nanospheres, which are constructed from tiny nanocrystallites of mean grain size of about 5–10 nm diameter indicating that the tiny nanocrystallites aggregate to form nanospheres. The maximum room temperature saturation magnetization value is 52.4 emu/g for $Mn_{0.6}Zn_{0.4}Fe_2O_4$. The nanospheres are magnetically very sensitive and may have a great future for its use in biomedical areas.

Three basic steps are required for the magnetic fluid synthesis: (i) preparation of ferrite nanomaterial, (ii) surface preparation/curing and (iii) homogenization of particles in a stable colloidal medium. Hydrothermal co-precipitation method has been used to prepare $Mn_{0.5}Zn_{0.5}Fe_2O_4$ nanoferrite particles with tunable Curie temperature and saturation magnetization. [135]. In zinc-doped manganese ferrite, magnetic properties of $Mn_{0.5}Zn_{0.5}Fe_2O_4$ composition can be improved by control of size or variation in site possession. Incubation time and pH of the reaction is the controlling factors for particle size in the range of 54 to 135 °A. All the particles show superparamagnetic properties at room temperature. The surface of above-synthesized nanoparticles is improved by treatment with oleic acid. Two types of stable Ferro-fluids, kerosene-based magnetic fluid, diester-based magnetic fluid, are prepared from $Mn_{0.5}Zn_{0.5}Fe_2O_4$ ferrite nanoparticles after surface treatment.

Nanoparticles of Cobalt-doped Zn-ferrite, $Co_xZn_{1-x}Fe_2O_4$ (x = 0.0–1.0), have been prepared by a polyethylene glycol (PEG) facilitated hydrothermal method [136]. Zn-ferrite nanoparticles with Co-doping are spherical with mean particle sizes in the range 7.3 and 13.4 nm.

In the class of various spinel ferrites, Ni–Zn ferrites are the most common magnetic materials. They are commonly used in microwave devices, power transformers in electronics, broad band transformers, filters, rod antennas, read/write heads for high speed digital tape, *etc*., because of their high electrical resistivity, mechanical hardness, chemical stability and large permeability at high frequency. In numerous technological uses, magnetic nanoparticles need to be firmly held in a non-magnetic matrix such as aniline polymer, and diblock copolymers. Furthermore, use of surfactant such as polyethylene glycol (PEG) overcomes the problem of agglomeration of nanoparticles in wet processes. For spinel ferrites, the magnitude of magnetization and Curie temperature can be easily controlled by change in stoichiometry. In various magnetic materials, spinel-ferrites with Zn-substitution are important, as Zn replacement can change the magnetic properties to a great extent.

Well-defined spinel phases of Co-, Ni-Zn-, and Co-Zn nanoferrites are synthesized by coprecipitation followed by microwave-hydrothermal (M-H) treatment [137]. The elevation in reaction temperature and time promotes crystal growth in the spinel ferrites. The average sizes observed for Co-ferrite, Co-Zn ferrite and Ni-Zn ferrite are 10.3, 9.1–10.4, and 10.2–10.5 nm respectively.

$Ni_{1-x}Zn_xFe_2O_4$ (x=0, 0.5, 1.0) nanophase ferrites are produced by precipitation of hydroxides chemically and proceeded by hydrothermal treatment [138]. Ferrite formation is affected by the methods used for the synthesis of hydroxide slurry before hydrothermal process.

$Ni_{0.5}Zn_{0.5}Fe_2O_4$ nanoparticle powders having well-defined spinel phases have been produced by hydrothermal treatment [139]. For $Ni_{0.5}Zn_{0.5}Fe_2O_4$, stoichiometric amounts of nitrates of $Fe^{3+}$, $Zn^{2+}$ and $Ni^{2+}$ are dissolved in deionized water followed by addition of aqueous ammonia solution at room temperature until the suspension attains desired pH value. The resulting suspension is hydrothermally treated using reported experimental conditions. It is observed that in spinel Ni–Zn ferrite powder, only one crystal phase is obtained at pH value 9.0 or 9.8 with a mean particle size in the range of 10–20 nm. The elevation in temperature and time in the hydrothermal treatment promotes the crystals formation in the spinel ferrites. The calculated energy of activation for this treatment is 56.2 kJ/mole.

Co-precipitated amorphous Ni, Zn and Fe hydroxides has been converted into nanocrystalline ferrite $Zn_{0.5}Ni_{0.5}Fe_2O_4$ and $NiFe_2O_4$ powders in addition of $ZrO_2$, $TiO_2$ with a narrow particle size distribution by using ultrasonically facilitated hydrothermal processing [80].

For the preparation of pure $Zn_{0.6}Ni_{0.4}Fe_2O_4$ magnetic nanoparticles in high yield, a simple PEG-facilitated hydrothermal treatment has been reported [140]. $Zn_{0.6}Ni_{0.4}Fe_2O_4$ nanoparticles have cubic spinel structure with average particle size 15±1 nm. The nanoparticles are superparamagnetic in nature with room temperature saturation magnetization of 40.3emu/g and at 10K 65.35emu/g. PEG being a surface active compound prevents growth of particle with decrease in interactions in them and thus controls the morphology.

The PEG-assisted hydrothermal procedure [140] has been adopted for the synthesis of high purity and high yield Zn doped $NiFe_2O_4$ nanoparticles, $Zn_xNi_{1-x}Fe_2O_4$, where x is equal to 0, 0.2, 0.4, 0.6, 0.8 and 1, using appropriate stoichiometric amounts of metal salts of ($Fe^{3+}/Zn^{2+}/Ni^{2+}$ [141]. Their structure shows the presence of magnetic cores and non-magnetic shells. The average particle size of the resultant nanoparticles varies from 13–47 nm and magnetic core size from 9–20 nm. With increasing concentration of Zn, morphology of as-synthesized nanoparticles transformed from unsymmetrical polygon nanoparticles to symmetrical nanospheres. As-synthesized samples show almost inverse-spinel structure. Lattice constant increases linearly from 8.38 to 8.45 A on increasing Zn substitution in $Zn_xNi_{1-x}Fe_2O_4$. All particles exhibited superparamagnetic behavior and saturation magnetization ($M_S$) slightly decreased from 100 to 50 emu/g when Zn contents increased, because magnetic moment of Zn is zero. The magnitude of Remanence magnetization ($M_r$) values are less than 25 emu/g with the exception for Zn ferrite with x =0.

An easy hydrothermal method, using cetyltrimethlammonium bromide and urea, for preparation of single-crystalline Ni-Zn-Co ferrite nanoparticles at low temperature has been reported [142]. The effect of the synthetic method and the doping of Co and Ni ions in structure on the magneto-dielectric properties of the ferrite have been studied. Nanoparticles of $(Ni_xZn_{1-x})Co_yFe_{2-y}O_4$ where x = 0, 0.4, 0.5, 0.6; a d y = 0, 0.01 are prepared by the hydrothermal method using urea as precipitating agent and a pretreatment in an ultrasonic water bath[142]. The resultant ferrite is a mono-phase cubic spinel structure with mean crystallite size of 10–40 nm. The $Ni^{2+}$ or $Co^{2+}$ doped zinc ferrite has crystallite size bigger than 8.6 nm of zinc ferrite, showing that the particles size increases on substitution of $Ni^{2+}$ for $Zn^{2+}$. Increase in dielectric constant from 7 to 11 for Ni-Zn ferrite is observed when stoichiometric ratio for cobalt is 0.01. The $M_s$ value of Ni-Zn ferrite increases from 0.041 to 0.074 A·m2/g for composition $Ni_{0.49}Zn_{0.5}Co_{0.01}Fe_{1.98}O_4$.

$CoFe_2O_4$ is a potential photo- magnetic material. There is a focus in research in magneto-optical memory of nanomaterials [143]. In order to restrain from accumulation of particles in disorderly pattern and to control particle size in nano-crystals, nanocomposites are produced. Furthermore, nanocomposites may have more useful technological features such as improved catalytic, magnetic, magneto-optic and mechanical behavior. Formation of magnetic crystals with nucleation sites can be obtained by formation of titanium dioxide matrix due to its non-magnetic porosity nature. Such composites can reduce the disordered accumulation of particles and produce refined and evenly packed magnetic crystals. The nanocomposites of $xTiO_2$ + (1-x) $CoFe_2O_4$ ($0 \leq x \leq 1$) powders are synthesized using microwave-assisted hydrothermal process at a temperature of 165 °C/45min [144]. The $CoFe_2O_4$ is prepared from nitrates of Co(II) and Fe(III) and NaOH whereas $TiO_2$ from the reaction of $TiCl_4$ and NaOH. As-synthesized $CoFe_2O_4$ and $TiO_2$ nanopowders were homogenized using ball mill to prepare $xTiO_2$ + (1-x) $CoFe_2O_4$ nano-composites ($0 \leq x \leq 1$). The particle size for all the nanopwders varies from 18nm to 34nm. No transformation from anatase to rutile phase has been observed. The resultant nanopowders are sintered using a microwave method at 500 °C/30min. After sintering; composites show evidence for $TiO_2$ and $CoFe_2O_4$ only and have grain sizes in between 54 to 78 nm. An increase of $TiO_2$ content in $CoFe_2O_4$ results in decrease in $M_s$ value of ferrite-composites.

Mg-Cu-Zn-based ferrites have better electro-magnetic characteristics, with higher magnitude of resistivity and Curie temperature, and eco-stability and thus can be a potential material for a variety of application. These ferrites have reduced magnetostriction constant as compared to Ni-Cu-Zn-based ferrites and show much improved magnetic properties. It has been observed that the magnitude of initial permeability increases with a decrease in the magnetostriction constant. Mg-containing polycrystalline ferrites have problems of porosity and resistivity. PbO is one of the additives with low melting point that can solve the problem.

Mg-Cu-Zn ferrite powders containing PbO are synthesized using co-precipitation reactions in a microwave-assisted hydrothermal (M-H) set-up [145] taking appropriate stoichiometric amounts of nitrates of magnesium, copper, zinc, and iron at a proper pH for pure phase$Mg_{0.2}Cu_{0.3}Zn_{0.5}Fe_2O_4$. These ferrites with the addition of 0.6wt % PbO show increase in Bulk density, electrical resistivity, permeability and saturation magnetization, however, porosity decreases. Sintered samples have higher particle size than that of nanoferrite powder.

Importance of conducting polymers lies in their prospective use in electromagnetic interference (EMI) shielding, rechargeable battery, chemical sensor, corrosion devices and microwave absorption. Polyaniline (PANI) is one of conducting polymers with remarkable electrochemical and physicochemical properties, environmentally more stable, and can be synthesized easily. Polyaniline- nanocomposites with inorganic oxide, sulfide, and nitride are new functional hybrids with organized structures between organic and inorganic materials. The Li–Ni ferrites are soft magnetic materials with properties such as large saturation magnetization and electrical resistivity coupled with small eddy current losses. They find application in devices such as isolators, circulators, phase shifters and gyrators that all work on microwaves. However, use of Li–Ni ferrites is limited by their heavy weight and physical rigidity.

The La-based Li–Ni ferrites with composition ($LiNi_{0.5}La_{0.02}Fe_{1.98}O_4$) are prepared in a new rheological phase-thermolysis reaction using oxalate as a starting material and then converted into the polyaniline-$LiNi_{0.5}La_{0.02}Fe_{1.98}O_4$ composite nanoferrites by polymerization in aqueous phase [146]. In the rheological phase reaction method, nanomaterials are prepared from a rheological mixture containing solid and liquid. The solid starting materials are thoroughly combined in the required stoichiometric ratio, and converted into a solid–liquid rheological body (with uniform distribution of components) by addition of a sufficient volume of water/suitable solvents in to it. The reaction is carried out using required experimental methodology to obtain the ferrite. $LiNi_{0.5}La_{0.02}Fe_{1.98}O_4$ ferrite having mono-phase spinel, spherical morphology of particles produced in this manner shows a mean particle size under 100 nm.

Similarly a series of polyaniline- $LiNi_{0.5}La_{0.02}Fe_{1.98}O_4$ composites are synthesized using aniline and ferrite particles as precursors [146]. Composites thus produced show morphology that ferrite particles are firmly enclosed into polymeric chains of aniline indicating that a core effect of ferrite particles on the polymerization of aniline is apparent. Saturation magnetization and coercivity increases with increase in magnetic ferrite contents in the polyaniline- $LiNi_{0.5}La_{0.02}Fe_{1.98}O_4$ Composites showing the inherent magnetic hysteresis of ferromagnetic materials.

Heavy metal environmental pollutant (like Cu, Ni, Cr and Zn) present in electroplating wastewater released in industrial processes can be treated by techniques as reported in literature (chemical precipitation, coagulation–flocculation, ion exchange, and membrane filtration) [147]. Electroplating wastewaters can be disposed of and utilized by an efficient ferrite-treatment and heavy metal cations can be held in the spinel lattice structure as ferrites ($AB_2O_4$), which can exist as an environmentally stable compound. In addition, its recycling as a magnetic material, *e.g.*, electric wave absorbents can also be done [148]. Modification in electrical, magnetic and structural features of ferrite materials can be achieved by the replacement of para- or dia-magnetic ions in their composition.

Effect on the properties of Ni-Zn ferrite on replacement of Cu has been studied by microwave- assisted hydrothermal treatment [149]. Nitrates of $Ni^{2+}$, $Cu^{2+}$, $Zn^{2+}$ and $Fe^{3+}$ are taken in stoichiometric amounts required for ferrite composition $Ni_{0.5-x}Zn_{0.5}Cu_xFe_2O_4$ ($0.0 \leq x \leq 0.4$). Accordingly aqueous solutions of metal salts are mixed; pH is adjusted to 9.5 and contents are poured into the microwave heated reactors with inside lining of polytetrafluoroethylene and subjected to hydrothermal process at 180°C for half an hour. After filtration, the resultant product is washed thoroughly using de-ionized water and placed in an oven for overnight drying at 80°C. It is observed that grain size is less affected by substitution of $Cu^{2+}$ for $Ni^{2+}$ because the mean size of ferrite crystallites is found in 8.5~10.3 nm range. Up to x=0.3, as-synthesized ferrites show cubic spinel phase, while at x=0.4 another phase with stoichiometry $Fe_2O_3$ forms. It is observed that substitution of $Cu^{2+}$ has negative effect on the saturation magnetization of $Ni_{0.5-x}Zn_{0.5}Cu_xFe_2O_4$ ferrite.

## CONCLUSION

Synthesis of nanoparticles of various compositions and exciting properties such as sizes and a variety of morphologies covering extensive areas of applications is a vast expanding subject of nanoscience. Preparation of good quality nanoparticles with desired size and morphology in a precisely regulated and repeatable method is an art by itself. Solvothermal/hydrothermal based synthetic methods are economical and environmental friendly because they are carried out in a close system avoiding toxic solvents and reagents.

Solvothermal/hydrothermal syntheses of magnetic nanomaterials especially iron oxide nanoparticles and other structures, metal ferrites and mixed metal ferrites and composites have been reviewed. In many cases solvothermal/hydrothermal process is a single one-pot step for the synthesis of nanoparticles. There are many instances when it is used as a step towards the formation of desired/controlled morphologies and one or more pre-solvothermal or post- solvothermal steps are involved depending on the nature of the final product.

On the technical side a number of innovations have been added to conventional solvothermal/hydrothermal process to improve its efficiency. Microwave heating in solvothermal method shortens the reaction time and also controls the morphology, size and its distribution. The autoclave systems are equipped with magnetic stirrer and/or in-vessel temperature sensors (thermocouple) to control heating rate and maintain the desired temperature of reaction mixture in the reaction vessel. Similarly microwave digestion systems are specially designed to fulfill all the requirements of a precise synthetic technique. Sometime inert atmosphere (argon or nitrogen gas) is also used to minimize undesired oxidation/reduction metal component. Magnetic field-assisted hydrothermal and sonochemical-hydrothermal treatments are the innovations that help control the morphology, size and kinetic growth of nanoparticles. The method based on continuous hydrothermal synthesis at supercritical water conditions is an effective way to produce magnetic material in bulk.

In terms of chemical/physical interactions in the synthesis of nanoparticles *via* solvothermal/hydrothertmal route, many strategies have been devised for controlling the shape, growth rate of particle size and its distribution, morphology, particle stoichiometry, prevent particle aggregation and to improve particle solubility in various solvents. Stabilization of isolated nanoparticles has been achieved by capping or passivation agents. Hydrothermal treatment of the precursors prepared by co-precipitation, results in controlled stoichiometry of nanoparticles and improved chemical stability and magnetic properties.

Biomolecule-assisted route in hydrothermal process is a new, environmental-friendly, and prospective strategy for the synthesis of a variety of nanostructures. Biomolecules such as aspartic acid, hydroxyl rich polysaccharide (such as starch) and DNA with well-defined helical structures *etc.*, serve as structure-directing agents/templates. This strategy has been useful for the synthesis of complicated nanostructures, including 2- or 3- dimensional and distinct 1-dimensional nano-assemblies.

Hydrothermal decomposition step is facile in the intergrowth of spinel ferrite and M-hexaferrite. Similarly seed-hydrothermal approach is applied to control the morphology of nanoparticles. For the formation of magnetic spinel nanoparticles nanoparticles, in solvothermal reduction method, some reducing agent (such as trisodium citrate dihydrate, hydrazine hydrate, formamide *etc.*) is added along with other required reactants before proceeding for the solvothermal step.

For solvothermal synthesis of magnetic nanoparticles, some reagents are so chosen that act bi-functionally. $KMnO_4$ is used as a metal source and an oxidant. For shape-controlled synthesis, dodecyl sulphate of metals, such as Ni, Cu, Pb, Co, can be used as a precursor and a surface active stabilizer in hydrothermal synthetic route of nanomaterials of a variety of morphologies.

To improve intrinsic properties of nanoparticles polyethylene glycol (PEG 400/600) is used as a template reagent in template assisted-solvothermal route synthesis. Reverse emulsion-assisted hydrothermal process is a way for the preparation of nanomaterials of different kinds. Dielectric property of a solvent exerts strong influence on the formation of particle under hydrothermal conditions. Addition of low dielectric medium (such as alcohols: isopropyl alcohol) to the aqueous solutions (particle forming medium) of inorganic salts to speed up particle formation kinetics is an important strategy.

For use in the interface reaction between a solid precursor and alkali solution followed by hydrothermal crystallization for the controlled synthesis of magnetic ferrites, the solid precursor is synthesized by co-precipitation. The interface chemical reaction occurs at the junction of solid and liquid phase and released nanoparticles are distributed in aqueous solution *via* Brownian movement. Sometimes, solid is a soft metal like Zn metal sheet and is brought in contact with solution of $Fe^{2+}$ in ammonia solution during hydrothermal treatment for the production of zinc ferrite.

For bulk synthesis of monodisperse $M^{II}Fe_2O_4$ (M = Ni, Co, Mn) and $\gamma$-$Fe_2O_3$ nanoparticles, water-non polar organic solvent such as toluene interface and oleic acid as dispersing agent is used in conventional and microwave hydrothermal process. A simple separating funnel is required to collect as synthesized nanoparticles in a non-polar organic solvent from reaction mixture. The organic solvent can be reused after concentrating it under reduced pressure. This method is suitable for large scale and economical production of monodisperse ferrites under conventional hydrothermal and microwave hydrothermal conditions.

Studies on composite formation and functionalization of surface of nanoparticles and to add more functional activity opens the door for their novel and exciting properties and hence their application. Furthermore to improve the intrinsic stability of monodisperse nanoparticles and other structures especially for biological applications, exploration of new ways for effective surface coating needs to be more focused. Since the application of ferrites and other magnetic nanoparticles in the remediation and treatment of pollutants and removal of these particles from the environments is gaining more momentum, more solvothermal/hydrothermal based synthethetic studies need to be carried out. All such studies require collaborative and multi-disciplinary efforts around the globe.

## ACKNOWLEDGEMENT

Professor Dr. Shabbar Attique, Director Institute of Advanced Materials is acknowledged for providing printing facilities for reading materials and Mr. Mohammad Adnan Asghar (M. Phil Student) for help in collection and arrangement of material.

## CONFLICT OF INTEREST

The authors confirm that this chapter contents have no conflict of interest.

## REFERENCES

[1] C. N. R. Rao, A. Muller, A. K. Cheetham, Eds., *The Chemistry of Nanomaterials: Synthesis, Properties and Applications*. Vol. 1 & 2; Wiley-VCH Verlag GmbH & Co. KGaA, 2005.

[2] Geoffrey A. Ozine and Andre C. Arsenault, *Nanochemistry: A Chemical Approach to Nanomaterials*. The Royal Society of Chemistry, 2005.

[3] E. Roduner, “Size matters: Why nanomaterials are different,” *Chem. Soc. Rev.*, vol.35, pp.583-592, 2006.

[4] (a) A. H. Lu, E. L. Salabas, and F. Schuth, “Magnetic Nanoparticles: Synthesis, Protection, Functionalization, and Application,” *Angew. Chem. Int. Ed.*, vol. 46, pp. 1222 – 1244, 2007. (b) M. Faraji, Y. Yamini and M. Rezaee, “Magnetic Nanoparticles: Synthesis, Stabilization, Functionalization, Characterization, and Applications,” *J. Iran. Chem. Soc.*, vol. 7, no. 1, pp.1-37. 2010.

[5] (a) H. Deng, X. Li, Q. Peng, X. Wang, J. Chen, and Y. Li, “Monodisperse Magnetite Single-Crystal Ferrite Microspheres,” *Angew. Chem. Int. Ed.*, vol .44, pp. 2782 –2785, 2005. (b) X. Chen, Z. Zhang, X. Li and C. Shi, “Hollow magnetite spheres: Synthesis, characterization, and magnetic properties,” *Chem. Phy. Lett.*, vol.422, pp.294–298, 2006.

[6] (a) H. L. Niu, Q. W. Chen, M. Ning, Y. S. Jia, X. J. Wang, “Synthesis and One-Dimensional Self-Assembly of Acicular Nickel Nanocrystallites under Magnetic Field,” *J. Phys.Chem. B*, vol. 108, pp.

3996-3999, 2004. (b) J. Wang, Q. W. Chen, C. Zeng, B. Y. Hou, "Magnetic-Field-Induced Growth of Single-Crystalline $Fe_3O_4$ Nanowires," *Adv. Mater.*, vol .16(2), pp. 137-140, 2004. (c) M. Wu, Y. Xiong, Y. Jia, H. Niu, H. Qi, J. Ye, Q. Chen, "Magnetic field-assisted hydrothermal growth of chain-like nanostructure of magnetite," *Chem. Phy.*, vol. 401, pp.374–379, 2005.

[7] (a) X. M. Liu, S.Y Fu and H.M. Xiao, "Synthesis of maghemite sub-microspheres by simple solvothermal reduction method, " *J. Sol. Stat. Chem*, vol 179, pp.1554–1558, 2006. (b) V. Sreeja and P.A. Joy, "Microwave–hydrothermal synthesis of $\gamma$-$Fe_2O_3$ nanoparticles and their magnetic properties," *Mat. Res. Bull.*, vol 42, pp.1570–1576, 2007.

[8] O. Horner, S. Neveu, S. De Montredon, J. M. Siaugue and V. Cabuil, "Hydrothermal synthesis of large maghemite nanoparticles: influence of the pH on the particle size," *J. Nanopart. Res.*, vol 11, pp. 1247-1250, 2009.

[9] K. Sue, M. Suzuki, K. Arai, T. Ohashi, H. Ura, K. Matsui, Y. Hakuta, H. Hayashi, M. Watanabe and T. Hiaki, "Size-controlled synthesis of metal oxide nanoparticles with a flow-through supercritical water method," *Green. Chem.*, vol. 8, no. 7, pp. 634-638, 2006.

[10] T. Taniguchi, K. Nakagawa, T. Watanabe, N. Matsushita, and M. Yoshimura, "Hydrothermal Growth of Fatty Acid Stabilized Iron Oxide Nanocrystals," *J. Phys. Chem. C*, vol 113, pp. 839–843, 2009.

[11] Z. Jing and S. Wu, "Synthesis and characterization of monodisperse hematite nanoparticles modified by surfactants *via* hydrothermal approach," *Mat. Lett.*, vol 58, pp. 3637– 3640, 2004.

[12] L. Vayssieres, "An Aqueous Solution Approach to Advanced Metal Oxide Arrays on Substrates," *Appl. Phys. A: Mater. Sci. Process.*, vol 89, pp. 1-8, 2007.

[13] (a) Debajeet K. Bora, Artur Braun, Rolf Erni, Giuseppino Fortunato, Thomas Graule, and Edwin C. Constble, "Hydrothermal Treatment of a Hematite Film Leads to Highly Oriented Faceted Nanostructures with Enhanced Photocurrents," *Chem. Mater.*, vol 23, pp. 2051–2061, 2011. (b) J. R. Huang, M. Yang, C. P. Gu, *et al.*, "Hematite solid and hollow spindles: Selective synthesis and application in gas sensor and photocatalysis," *Mater. Res. Bull.*, vol. 4, pp. 1211–1218, 2011.

[14] Q. Sun, X. Lu and G. Liang, "Controlled template-free hydrothermal synthesis of hematite nanoplates," *Mater. Lett.*, vol. 64, no.18, pp. 2006-2008, 2010.

[15] W. Li, J. Guan, W. Wang, G. Tong and X. Fan, "Synthesis and formation mechanism of hematite hollow microspheres by a one-pot templateless surfactant-free hydrothermal process," *Mater. Chem. Phys.*, vol. 118, pp. 496-500, 2009.

[16] H. Chen, Y. Zhao, M. Yang, J. He, P. K. Chu, J. Zhang and S. Wu, "Glycine-assisted hydrothermal synthesis of peculiar porous $\alpha$-$Fe_2O_3$ nanospheres with excellent gas-sensing properties," *Anal. Chim. Acta*, vol. 659, pp. 266-273, 2010.

[17] T. J. Daou, G. Pourroy, S. Begin-Colin, J. M. Greneche, C. Ulhaq-Bouillet, P. Legare, P. Bernhardt, C. Leuvrey and G. Rogez, "Hydrothermal Synthesis of Monodisperse Magnetite Nanoparticles," *Chem. Mater.*, vol. 18, pp. 4399-4404, 2006.

[18] (a) X. Sun, C. Zheng, F. Zhang, Y. Yang, G. Wu, A. Yu, and N. Guan, "Size-Controlled Synthesis of Magnetite ($Fe_3O_4$) Nanoparticles Coated with Glucose and Gluconic Acid from a Single Fe(III) Precursor by a Sucrose Bifunctional Hydrothermal Method," *J. Phys. Chem. C*, vol. 113, pp.16002–16008, 2009. (b) G.Y. Li, Y.R Jiang, K.- Huang, P. Ding and J. Chen, "Preparation and properties of magnetic $Fe_3O_4$–chitosan nanoparticles," *Journal of Alloys and Compounds*," vol. 466, pp. 451–456, 2008. (c) Q. Y. Lu, F. Gao, S. Komarneni, "A Green Chemical Approach to the Synthesis of Tellurium Nanowires," *Langmuir*, vol. 21, no.13, pp. 6002-6005, 2005.

[19] (a) H. Xin, A. T. Woolley, "DNA-templated nanotube localization," *J. Am. Chem. Soc.*, vol. 125, pp. 8710-8711, 2003. (b) Xiao-Fei Qu, Gen-Tao Zhou, Qi-Zhi Yao, and Sheng-Quan Fu, "Aspartic-Acid-Assisted Hydrothermal Growth and Properties of Magnetite Octahedrons," *J. Phys. Chem. C*, vol. 114, pp. 284–289, 2010. (c) B. Zhang, X. Ye, C. Wang, Y. Xie, "A facile solution-phase deposition approach to porous selenium materials," *J. Mater. Chem.*, vol. 17, pp. 2706-2712, 2007.

[20] S. Laurent, D. Forge, M. Port, A. Roch, C. Robic, L. V. Elst and R. N. Muller, "Magnetic Iron Oxide Nanoparticles: Synthesis, Stabilization, Vectorization, Physicochemical Characterizations, and Biological Applications," *Chem. Rev.*, vol. 108, pp. 2064–2110, 2008.

[21] S. Komarneni, E. Fregeau, E. Breval and R. Roy, "Hydrothermal preparation of ultrafine ferrites and their sintering," *J. Am. Ceram. Soc.*, vol. 71, C-26-C-28, 1988.

[22] K. Li, H. Gu and Q. Wei, "A Novel Hydrothermal Synthesis Method for Barium Ferrite," *China Particuology*, vol. 2, no.1, pp. 41-43, 2004.

[23] J. Wang, Y. Wu, Y. Zhu and P. Wang, "Formation of rod-shaped $BaFe_{12}O_{19}$ nanoparticles with well magnetic properties," *Mater. Lett.*, vol. 61, no.7, pp. 1522-1525, 2007.

[24] S. G. Kim, W. N. Wang, T. Iwaki, A. Yabuki, and K. Okuyama, "Low-Temperature Crystallization of Barium Ferrite Nanoparticles by a Sodium Citrate-Aided Synthetic Process," *J. Phys. Chem. C.*, vol. 111, pp. 10175-10180, 2007.

[25] V. B. Bregar, "Advantages of Ferromagnetic Nanoparticle Composites in Microwave Absorbers," *IEEE Trans. Magn.*, vol. 40, pp. 1679-1684, 2004.

[26] M. Drofenik, I. Ban, G. Ferk, D. Makovec, Andrej Znidarsic, D, Hanzel and D. Lisjak, "The concept of low temperature synthesis of super-paramagnetic barium hexaferrite particles," *J. Am. Ceram. Soc.*, vol. 93, no.6, pp. 1602-1607, 2010.

[27] M. Drofenik, I.Ban, D. Makovec, A. Znidarsic, Z. Jaglicic, D. Hanzel and D. Lisjak, "The hydrothermal synthesis of super-paramagnetic barium hexaferrite particles," *Materials Chemistry and Physics*, vol. 127, pp. 415–419, 2011.

[28] D. Makovec and M. Drofenik, "Synthesis of Plate-like Spinel Particles and Spinel-Hexaferrite Intergrowth Nanocomposite Particles Using the Hydrothermal Decomposition of Ba-Hexaferrite," *Crystal Growth & Design*, vol. 8, no. 7, pp. 2182-2186, 2008.

[29] P. Veverka, K. Kanizek, E. Pollert. J. Bohazeek, S. Vasseur, E. Duguet and J. Porttier, "Strontium ferrite nanoparticles synthesized in presence of polyvinyl alcohol: Phase composition, microstructural and magnetic properties," *J. Magnet. Magnet. Mater.*, vol. 309, pp.106-112, 2007.

[30] N. Chen, G. Mu, X. Pen, K. Gan and M. Gu, "Microwave absorption properties of $SrFe_{12}O_{19}/ZnFe_2O_4$ composite powders," *Mater. Sci. Eng. B*, vol. 139, pp. 256-260, 2007.

[31] Dong-Lin Zhao, Qiang Lv and Zeng-Min Shen, "Fabrication and microwave absorbing properties of Ni-Zn spinel ferrites," *J. Alloys. Comp.*, vol. 480, pp. 634–638, 2009.

[32] Dao-lai Fang, Chu-sheng Chen, A. J. A. Winnubst, "Preparation and electrical properties of $Fe_xCu_{0.10}Ni_{0.66}Mn_{2.24-x}O_4$ ($0 \leq x \leq 0.90$) NTC ceramics," *J. Alloys. Comp.*, vol. 454, pp. 286–291, 2008.

[33] S. Lorentzou, A. Zygogianni, K. Tousimi, C. Agrafiotis and A.G. Konstandopoulos, "Advanced Synthesis of Nanostructured Materials for Environmental Applications," *J. Alloys. Comp.*, vol. 483, pp. 302–305, 2009.

[34] R. Wu and J. Qu, "Removal of Water Soluble Azo Dyes by Magnetic Material $MnFe_2O_4$" *J. Chem. Technol. Biotechnol.*, vol. 80, pp. 20-27, 2005.

[35] Y. Ao, J. Xu, X. Shen, D. Fua, C. Yuan, "Magnetically Separable Composite Photocatalyst with Enhanced Photocatalytic Activity," *J. Hazard. Mater*, vol. 160, pp. 295–300, 2008.

[36] L. Zhang, M.Su and X. Guo, "Studies on the treatment of brilliant green solution by combination microwave induced oxidation with $CoFe_2O_4$," *Sep. Purif. Technol*, vol. 62, pp. 458–463, 2008.

[37] M. A. Ahmad, N. Okasha and S.I. El-Dek, "Preparation and characterization of nanometric Mn ferrites *via* different methods," *Nanotechnol*, vol. 19, pp. 1-6, 2008.

[38] S. Lv, X. Chen, Y.Ye, S.Yin, J. Cheng and M. Xia, "Rice hull/$MnFe_2O_4$ composite: Preparation, characterization and its rapid microwave-assisted COD removal for organic wastewater," *J. Hazard. Mater.*, vol. 171, pp. 634–639, 2009.

[39] T. Valdes-Solıs, P. Valle-Vigon, S. Alvarez, G. Marban and A. B. Fuertes, "Manganese ferrite nanoparticles synthesized through a nanocasting route as a highly active Fenton catalyst," *Catal. Commun.*, vol. 8, pp. 2037–2042, 2007.

[40] O. A. Fouad, K. S. Abdel Halim and M. M. Rashad, "Catalytic Oxidation of CO over Synthesized Nickel Ferrite Nanoparticles from Fly Ash," *Top. Catal.*, vol. 47, pp. 61–65, 2008.

[41] (a) S.Komarneni, M.C.D'Arrigo, C. Leonelli, G.C.Pellacani, H. Katsuki, "Microwave-Hydrothermal Synthesis of Nanophase Ferrites," *J. Am. Ceram. Soc.*, vol. 81, no. 11, pp. 3041–3043, 1998. (b) J. Wang, Q. Chen, B. Hou and Z. Peng, "Synthesis and Magnetic Properties of Single-Crystals of $MnFe_2O_4$ Nanorods," *Eur. J. Inorg. Chem.*, pp. 1165-1168, 2004.

[42] L. Zhen, K He, C. Y. Xu and W. Z. Shao, "Synthesis and characterization of single-crystalline $MnFe_2O_4$ nanorods *via* a surfactant-free hydrothermal route," *J. Magnet. Magnet. Mater.*, vol. 320, no. 21, pp. 2672-2675, 2008.

[43] Jinting Jiu, Keiichi Murai, Doseop Kim, Keunsoo Kim and Katsuaki Suganuma, "Preparation of Ag Nanorods with High Yield by Polyol process," *Mater. Chem. Phys*, vol. 114, pp. 333–338, 2009.

[44] S. Kundu and H. Liang, "Polyelectrolyte-mediated Non-micellar Synthesis of Monodispersed 'Aggregates'," *Colloids and Surfaces A: Physicochem. Eng. Aspects*, vol. 330, pp. 143–150, 2008.

[45] Woo-Jin Lee and William H. Smyrl, "Oxide Nanotube Arrays Fabricated by Anodizing Processes for Advanced Material application," *Curr. Appl. Phys.*, vol. 8, pp. 818–821, 2008.

[46] C. Pana, D. Zhanga and L. Shia, "CTAB Assisted Hydrothermal Synthesis, Controlled Conversion and CO Oxidation Properties of $CeO_2$ Nanoplates, Nanotubes and Nanorods," *J. Solid. State. Chem*, vol. 181, pp. 1298–1306, 2008.

[47] L. Cui, H. Gu, H. Xu and D. Shi, "Synthesis and Characterization of Superparamagnetic Composite Nanorings," *Mater. Lett.*, vol. 60, pp. 2929–2932, 2006.

[48] X. Hou, J. Feng, X. Xu and M. Zhang, "Synthesis and characterizations of spinel $MnFe_2O_4$ nanorod by seed–hydrothermal route," *J. Alloy. Comp.*, vol. 491, pp. 258–263, 2010.

[49] Y. Wang, R. Cheng, Z. Wen and L. Zhao, "Synthesis and Characterization of Single-crystalline $MnFe_2O_4$ Ferrite Nanocrystals and their Possible Application in Water Treatment," *Eur. J. Ing. Chem.*, pp. 2942-2947, 2011.

[50] S. Zhang, H. Niu, Y. Cai, X. Zhao and Y. Shi, "Arsinite and Arsenate Adsorption on Coprecipitated Bimetal Oxide Nanomaterials: $MnFe_2O_4$ and $CoFe_2O_4$," *Chem. Eng. J.*, vol. 158, pp. 599–607, 2010.

[51] J. Hu, I.M.C. Lo and G. Chen, "Comparative Study of Various Magnetic Nanoparticles for Cr(VI) Removal," *Sep. Purific. Technol.*, vol. 56, pp. 249–256, 2007.

[52] B. Aslibeiki, P. Kameli, H. Salamati, M. Eshraghi and T. Tahmasebi, "Superspin Glass State in $MnFe_2O_4$ Nanoparticles," *J. Magn. Magn. Mater.*, vol. 322, no. 19, pp. 2929-2934, 2010.

[53] X. Hou, J. Feng, X. Liu and M. Zhang, "Comparable Studies of Adsorption and Magnetic Properties of Ferrite $MnFe_2O_4$ Nanoparticles, Bulk and nanowires," *Chem. Res. Chinese Universities*, vol. 27, no. 4, pp. 543-546, 2011.

[54] (a) S. Yanez-Vilar, M. Sanchez-Andujar, C. Gomez-Aguirre, *et al.*, "A simple sovothermal synthesis of $MFe_2O_4$ (M= Mn, Co, Ni) nanoparticles," *J. Solid State Chem.*, vol. 182, pp. 2685–2690, 2009. (b) Gui-hua Ren and Zhi-song Yu, "Synthesis of Monodisperse $Fe_3O_4$ and $MnFe_2O_4$ Nanospheres by using a Solvothermal Reduction Method," *Sol. Stat. Phenom.*, 181-182, pp. 393-396, 2012.

[55] B.E. Kashevsky, V.E. Agabekov, S. B Kashevsk, K.A. Kekalo, E.Y. Manina, I.V. Prokhorov and V. S. Ulashchik, "Study of cobalt ferrite nanosuspensions for low-frequency ferromagnetic hyperthermia," *Particuology*, vol. 6, no. 5, pp. 322-333, 2008.

[56] (a) G. Baldi, D. Bonacchi and C. Innocent, "Cobalt ferrite nanoparticles: The control of particle size and surface state and their effects on magnetic properties," *J. Magnet. Magnet. Mater.*, vol. 311, no. 1, pp. 10-16, 2007. (b) C. H. Kim, Y. Myung, Y. J. Cho, *et al.*, "Electronic Structure of Vertically Aligned Mn-Doped $CoFe_2O_4$ Nanowires and their Application as Humidity Sensors and Photodetectors," *J. Phys. Chem. C.*, vol. 113, no.17, pp. 7085–7090, 2009.

[57] M. Pita, J. M. Abad, C. Vaz-Dominguez, C. Briones, E. Mateo-Marti, J. A. Martin-Gago, M. D. P. Morales and V. M. Fernandez, "Synthesis of cobalt ferrite core/metallic shell nanoparticles for the development of specific PNA/DNA biosensors," *J. Colloid. Interface Sci.*, vol. 321, no. 2, pp. 484-492, 2008.

[58] (a) X. F. Chu, D. L. Jiang, Y. Guo, C. M. Zheng, "Ethanol gas sensor based on $CoFe_2O_4$ nano-crystallines prepared by hydrothermal method," *Sens. Actuators. B.*, vol. 120, pp. 177-181, 2006. (b) I. Sandu, L. Presmanes, P. Alphonse, *et al.*, "Nanostructured cobalt manganese ferrite thin films for gas sensor application," *Thin. Sol. Films.*, vol. 495, pp. 130–133, 2006.

[59] G. B. Ji, S. L. Tang, S. K. Ren, F. M. Zhang, B. X. Gu and Y. W. Du, "Simplified synthesis of single-crystalline magnetic $CoFe_2O_4$ nanorods by surfactant–assisted hydrothermal process," *J. Cry. Growth.*, vol. 270, pp. 156-161, 2004.

[60] Z. Y. Huo, C. Chen, Y. D. Li, "Self-assembly of uniform hexagonal yttrium phosphate nanocrystals," *Chem. Commun.*, pp. 3522-3524, 2006.

[61] Y. Hakuta, T. Haganuma, K. Sue, T. Adchiri and K. Arai, "Continuous production of phosphor YAG:Tb nanoparticles by hydrothermal synthesis in supercritical water," *Mater. Res. Bull.*, vol. 38, pp. 1257–1265, 2003.

[62] D. Zhao, X. Wu, H. Guan and E. Han, "Study on supercritical hydrothermal synthesis of $CoFe_2O_4$ nanoparticles," *J. Supercritical. Fluids.*, vol. 42, pp. 226–233, 2007.

[63] Y. Hao and A. S. Teja, "Continuous hydrothermal crystallization of $\alpha$-$Fe_2O_3$ and $Co_3O_4$ nanoparticles,' *J. Mat. Res.*, vol. 18, pp. 415–422, 2003.

[64] L. J. Cote, A. S. Teja, A. P. Wilkinson and Z. Zhang, "Continuous Hydrothermal Synthesis of $CoFe_2O_4$ Nanoparticles," *Fluid. Phase. Equilib.*, vol. 210, pp. 307-317, 2006.

[65] M. H. Nilsen, N. Nordhei, A. L. Ramstad, D. G. Nicholson, A. Kabanas and M. Poliakoff, "XAS (XANES and EXAFS) Investigations of Nanoparticulate Ferrites Synthesized Continuously in Near Critical and Supercritical Water," *J. Phys. Chem. C.*, vol. 111, no. 17, pp. 6252-6562, 2007.

[66] T. Sato, K. Sue, W. Suzuki, M. Suzuki, K. Matsui, Y. Hakuta, H. Hayashi, K. Arai, S. Kawasaki, A. Kawai-Nakamura and T. Hiaki, "Rapid and Continuous Production of Ferrite Nanoparticles by Hydrothermal Synthesis at 673 K and 30 MPa," *Ind. Eng. Chem. Res.*, vol. 47, pp.1855-1860, 2008.

[67] B. Baruwati, M. N. Nadagouda, and R. S. Varma, "Bulk Synthesis of Monodisperse Ferrite Nanoparticles at Water-Organic Interfaces under Conventional and Microwave Hydrothermal Treatment and Their Surface Functionalization," *J. Phys. Chem. C*, vol. 112, pp. 18399–18404, 2008.

[68] L. Zhao, H. Zhang, Y. Xing, S. Song, S. Yu, W. Shi, X. Guo, J. Yang, Y. Lei and F. Cao, "Studies on the magnetism of cobalt ferrite nanocrystals synthesized by hydrothermal method," *J. Sol. Stat. Chem.*, vol. 181, pp. 245–252, 2008.

[69] Q. Liu, J. Sun, H. Long, X. Sun, X. Zhong and Z. Xu, "Hydrothermal Synthesis of $CoFe_2O_4$ Nanoplates and Nanoparticles," *Mater. Chem. Phy.*, vol. 108, pp. 269-273, 2008.

[70] S. H. Xiao, W. F. Jiang, L. Y. Li and X. J. Li, "Low-temperature auto-combustion synthesis and magnetic properties of cobalt ferrite nanopowders," *Mater. Chem. Phy.*, vol. 106, no. 1, 82-87, 2007.

[71] K. Maaz, A. Mumtaz and S. k. Hasanain, "Synthesis and Magnetic Properties of Cobalt Ferrite ($CoFe_2O_4$) Nanoparticles Prepared by Wet Chemical Route," *J. Magnet. Magnet. Mater.*, vol. 308, 289-295, 2007.

[72] S.C. Goh, C.H. Chia, S. Zakaria, M. Yusoff, C.Y. Haw, Sh. Ahmadi, N.M. Huang and H.N. Lim, "Hydrothermal preparation of high saturation magnetization and coercivity cobalt ferrite nanocrystals without subsequent calcination," *Mater. Chem. Phys.*, vol. 120, 31–35, 2010.

[73] J. Gao, H. Gu and B. Xu, "Multifunctional Magnetic Nanoparticles: Design, Synthesis and biomedical Applications," *Acc. Chem. Res.*, vol. 42, no. 8, 1097-1107, 2009.

[74] Y. Zhang, Y. Liu, C. Fei, Z. Yang, Z. Lu, R. Xiong, D. Yin, J. Shi, "The temperature dependence of magnetic properties of cobalt ferrite nanoparticles by hydrothermal method," *J. Appl. Phys.*, vol. 108, no. 8, 084312-6 pages, 2010.

[75] V. Cabuil, V. Dupuis, D. Talbot and S. Neveu, "Ionic magnetic fluid based on cobalt ferrite nanoparticles: Influence of hydrothermal treatment on the nanoparticle size," *J. Magnet. Magnet. Mater.*, vol. 323, pp. 1238–1241, 2011.

[76] D. H. Chen, D. R. Chen, X. Jiao, Y. Zhao and M. He, "Hydrothermal Synthesis and Characterization of Octahedral Nickel Ferrite Particles," *Powder. Technol.*, vol. 133, pp. 247-250, 2003.

[77] V. K. Sankaranarayanan and C. Sreekumar, "Precursor Synthesis and Microwave Processing of Nickel Ferrite Nanoparticles," *Curr. Appl. Phys.*, vol. 3, pp. 205-208, 2003.

[78] (a) K. M. Reddy, L. Satyanarayana, S. V. Manorama, *et al.*, "A comparative study of the gas sensing behavior of nanostructured nickel ferrite synthesized by hydrothermal and reverse micelle techniques," *Mater. Res. Bull.*, vol. 39, no. 10, pp.1491–1498, 2004. (b) J. Zhou, J. Ma, C. Sun, L. Xie, Z. Zhao, H. Tian, Y. Wang, J. Tao and X. Zhu, "Low-Temperature Synthesis of $NiFe_2O_4$ by a Hydrothermal Method," *J. Am. Ceram. Soc.*, vol. 88, no. 12, pp. 3535–3537, 2005.

[79] D. E. Zhang, X. J. Zhang, X. M. Ni, H. G. Zheng and D. D. Yang, "Synthesis and characterization of $NiFe_2O_4$ magnetic nanorods *via* a PEG-assisted route," *J. Magnet. Magnet. Mater.*, vol. 292, pp. 79–82, 2005.

[80] P. E. Meskin, V. K. Ivanov, A. E. Barantchikov, B. R. Churagulov, Y. D. Tretyakov, "Ultrasonically assisted hydrothermal synthesis of nanocrystalline $ZrO_2$, $TiO_2$, $NiFe_2O_4$ and $Ni_{0.5}Zn_{0.5}Fe_2O_4$ powders," *Ultrasonics. Sonochem.*, vol. 13, pp. 47–53, 2006.

[81] J. Huo and M. Wei, "Characterization and magnetic properties of nanocrystalline nickel ferrite synthesized by hydrothermal method," *Mater. Lett.*, vol. 63, pp. 1183–1184, 2009.

[82] M. Srivastava, S. Chaubey and A. K. Ojha, "Investigation on size dependent structural and magnetic behavior of nickel ferrite nanoparticles prepared by sol–gel and hydrothermal methods," *Mater. Chem. Phys.*, vol. 118, pp. 174–180, 2009.

[83] Q. Liu, H. Huang, L. Lai, J. Sun, T. Shang, Q. Zhou and Z. Xu, "Hydrothermal synthesis and magnetic properties of $NiFe_2O_4$ nanparticles and nanorods," *J. Mater. Sci.*, vol. 44, pp. 1187–1191, 2009.

[84] J. Jiang and Y-M. Yang, "Facile synthesis of nanocrystalline spinel $NiFe_2O_4$ *via* novel soft chemistry route," *Mater. Lett.*, vol. 61, pp. 4276-4279, 2007.

[85] J-H. Liu, L-F. Zhang, G-F. Tian, J-C Li and F-S Li, "Structure and magnetic properties of $NiFe_2O_4$ nanoparticles prepared by low-temperature solid-state reaction," *Acta. Phys. Sin.*, vol. 56, pp. 6050-6055, 2007.

[86] J. Zhang, J. Shi and M. Gong, "Synthesis of magnetic nickel spinel ferrite nanospheres by a reverse emulsion assisted hydrothermal process," *J. Sol. Stat. Chem.*, vol. 182, pp. 2135–2140, 2009.

[87] (a) J. Y. Wang, R. Yi, F. L. Ren, *et al.*, "Solvothermal synthesis and magnetic properties of size-controlled nickel ferrite nanoparticles," *J. Alloys. Comp.*, vol. 479(1-2), pp. 791–796, 2009. (b) L. Wang, J. Ren, Y. Wang, X. Liu and Y. Wang, "Controlled synthesis of magnetic spinel-type nickel ferrite nanoparticles by the interface reaction and hydrothermal crystallization," *J. Alloys. Comp.*, vol. 490, pp. 656–660, 2010.

[88] (a) T. Zhang, H. Cao, J. Peng, Q. Xiao, Z. Li, G. Lei, "Excellent cycling performance of three-dimensional-ordered macroporous $NiFe_2O_4$ as anode material for lithium ion batteries," *Funct. Mater. Lett.*, vol. 4(4), pp. 327-331, 2011. (b) K. Sue, M. Aoki, T. Sato, D. N. Hamane, S. Kawasaki, Y. Hakuta, Y. Takebayashi, S. Yoda, T. Furuya, T. Sato, and T. Hiaki, "Continuous Hydrothermal Synthesis of Nickel Ferrite Nanoparticles Using a Central Collision-Type Micromixer: Effects of Temperature, Residence Time, Metal Salt Molality, and NaOH Addition on Conversion, Particle Size, and Crystal Phase," *Ind. Eng. Chem. Res.*, vol. 50, pp. 9625–9631, 2011.

[89] K. Nejati and R. Zabihi, "Preparation and magnetic properties of nano size nickel ferrite particles using hydrothermal method," *Chem. Cen. J.*, vol. 6, no. 23, 6 pages, 2012.

[90] C. R. Alves, R. Aquino, M. H. Sousa, H. R. Rechenberg, G. F. Goya, F. A. Tourinho, J. Depeyrot, "Low temperature experimental investigation of finite-size and surface effects in $CuFe_2O_4$ nanoparticles of ferrofluids," *J. Metastable. Nanocryst. Mater.*, vol. 20-21, pp. 694-699, 2004.

[91] M. Bahgat, M. H. Khedr, W. M. A. El Rouby, "Low temperature isothermal reduction behaviour of hydrothermally precipitated $CuFe_2 O_4/MgFe_2O_4$ core/shell nanocrystallites," *Mater. Technol: Adv. Performance. Materials*, vol. 23, no. 4, pp. 224-230, 2008.

[92] S. Kameoka, T. Tanabe, A.P. Tsai, "Spinel $CuFe_2O_4$: A precursor of copper catalyst with high thermal stability and activity," *Catal. Lett.*, vol. 100, no. 1-2, pp. 89–93, 2005.

[93] W.L. Hong, J. H Liu, D.Y Tian *et al.*, "Synthesis of nanometer-$CuFe_2O_4$ and its catalysis on thermal decomposition of RDX," *J. Propulsion. Technol.*, vol. 24(6), pp. 560-562, 2003.

[94] T. Liu, L. Wang, P. Yang, B. Hu, "Preparation of nanometer-$CuFe_2O_4$ by auto-combustion and its catalytic activity on thermal decomposition of ammonium perchlorate," *Mater. Lett.* Vol. 62(24), pp. 4056–4058, 2008.

[95] K. Faungnawakij, R. Kikuchi, T. Fukunaga, K. Eguchi, "Catalytic hydrogen production from dimethyl ether over $CuFe_2O_4$ spinel-based composites: hydrogen reduction and metal dopant effects," *Catal. Today.*, vol. pp.138,157–161, 2008.

[96] M. H. Khedr, A. A. Farghali, "Microstructure, kinetics and mechanisms of $CO_2$ catalytic decomposition over freshly reduced nano-crystallite $CuFe_2O_4$ at 400–600 °C," *Appl. Catal. B.*, vol. 61, pp. 219-226, 2005.

[97] A. A. Farghali, M. H. Khedr, A. A. A. Khalek, "Catalytic decomposition of carbondioxide over freshly reduced activated $CuFe_2O_4$ nano-crystals" *J. Mater. Processing. Technol.*, vol. 181, no. (1-3), pp. 81-87, 2007.

[98] J. E. Tasca, C. E. Quincoces, A. Lavat, A. M. Alvarez, M. G. Gonzalez, "Preparation and characterization of $CuFe_2O_4$ bulk catalysts," *Ceram. Internat.*, vol. 37, pp. 803–812, 2011.

[99] H. W. Liao, L. Mu, M. Zhen, Y. Y. Meng, "Study on Catalysis of Cu-Fe Metal Composite Oxides to Thermal Decomposition Behavior of Bluestone," *Iran. J. Chem. Chem. Eng.*, vol. 30, no. 2, pp. 33-36, 2011.

[100] Hongxiao Zhao, Huimin Jia, Shumin Wang, Dengqi Xue, Zhi Zheng, "Fabrication and Application of $MFe_2O_4$ (M = Zn, Cu) nanoparticles as Anodes for Li Ion Batteries," *J. Experimental. Nanosci*, vol. 6, no. 1, pp. 75-83, 2011.

[101] S. F. Rus, P. Vlazan, S. Novaconi, P. Sfirloaga, I. Grozescu, "Synthesis and characterization $CuFe_2O_4$ nanoparticles prepared by the hydrothermal ultrasonic assisted method," *J. Optoelectronics. Advan. Mater*, vol. 14, no. 3-4, pp. 293–297, 2012.

[102] S.H.Yu and M. Yoshimura, "Ferrite/Metal Composites Fabricated by Soft Solution Processing," *Adv. Funct. Mater.*, vol. 12, no. 1, pp. 9-15, 2002.

[103] S. H. Yu, T. Fujino, M. Yoshimura, "Hydrothermal synthesis of $ZnFe_2O_4$ ultrafine particles with high magnetization," *J. Magn. Magn. Mater.*, vol. 256, pp. 420-424, 2003.
[104] X. Hu, P. Guan and X. Yan, "Hydrothermal Synthesis of Nano-meter Micrporous Zinc Ferrite," *China Particuology.*, vol. 2, no. 3, pp.135-137, 2004.
[105] M. Hofmann, S. J. Campbell, H. Ehrhardt, R. Feyerherm, "The magnetic behavior of nanostructured zinc ferrite," *J. Mater. Sci.*, vol. 39, pp. 5057-5065, 2004.
[106] R. Saez Puche, M. J. T. Fernandez, V. B. Gutierrez, R. Gomez, V. Marquina, M. L. Marquina, J. L. Perezmazariego, R. Ridaura, "Ferrites nanoparticles $MFe_2O_4$ (M = Ni and Zn): hydrothermal synthesis and magnetic properties," *Bol. Soc. Esp. Ceram. V.,* vol. 47, no. 3, pp. 133-137, 2008.
[107] L. Satyaranaya, K. M. Reddy, S. V. Manorama, "Synthesis of nanocrystalline $Ni_{1-x}Co_xMn_xFe_{2-x}O_4$: A material for liquefied petroleum gas sensing," *Sens. Actuators B*, vol. 89, no. 1-2, pp. 62-67, 2003.
[108] Y. Koseoglu, A. Baykal, M. S. Toprak, F. Gozuak, A. C. Basaran, B. Aktas, "Synthesis and characterization of $ZnFe_2O_4$ magnetic nanoparticles *via* a PEG-assisted route," *J. Alloys. Comp.*, vol. 462, no. pp. 209–213, 2008.
[109] R. Zhang, J. Huang, J. Zhao, Z. Sun, and Y. Wang, "Sol-Gel Auto-Combustion Synthesis of Zinc Ferrite for Moderate Temperature Desulfurization," *Energ. Fuels.*, vol. 21, pp. 2682-2687, 2007.
[110] N. Chen, G. Mu, X. Pan, K. Gan, M. Gu, "Microwave absorption properties of $SrFe_{12}O_{19}/ZnFe_2O_4$ composite powders," *Mater. Sci. Eng. B.*, vol. 139, pp. 256–260, 2007.
[111] Y. Lu, A. Pich, H. Adler, "Synthesis and Characterization of Nanometer-Sized Polypyrrole Composites," *Synth. Met.*, vol. 135–136, pp. 37-38, 2003.
[112] P. Xu, X. Han, C. Wang, H. Zhao, J. Wang, X. Wang, and B. Zhang, "Synthesis of Electromagnetic Functionalized Barium Ferrite Nanoparticles Embedded in Polypyrrole," *J. Phys. Chem. B.*, vol. 112, no. 10, pp. 2775-2781, 2008.
[113] X. Feng, H. Huang, Q.Ye, J.J. Zhu, and W. Hou, "Ag/Polypyrrole Core-Shell Nanostructures: Interface polymerization, Characterization, and Modification by Gold Nanoparticles," *J. Phys. Chem. C.*, vol. 111, no. 24, pp. 8463-8468, 2007.
[114] Y. Li, R. Yi, A. Yan, L. Deng, K. Zhou, X. Liu, "Facile synthesis and properties of $ZnFe_2O_4$ and $ZnFe_2O_4$/polypyrrole core-shell nanoparticles," *Sol. Stat. Sci.*, vol. 11, pp. 1319–1324, 2009.
[115] H. J. Choi, B. J. Park, M. S. Cho and J. L. You, "Core-shell structured poly(methyl methacrylate)coated carbonyl iron particles and their magneto-rheological characteristics," *J. Magn. Magn. Mater.*, vol. 310, no. 2, pp. 2835- 2837, 2007.
[116] Y. Zhang, Z. Huang, F. Tang, J. Ren, "Ferrite hollow spheres with tunable magnetic properties," *Thin. Solid. Films.*, vol. 515, no. 4, pp. 2555–2561, 2006.
[117] J. Zhao, L. Mi, H. Hou, X. Shi, Y. Fan, "The preparation of zinc ferrite nanorods by using single ferrocenyl complex as precursor" *Mater. Lett.*, vol. 61, pp. 4196–4198, 2007.
[118] W. Jiang, Z. Cao, R. Gu, X. Ye, C. Jiang and X. Gong, "A simple route to synthesize $ZnFe_2O_4$ hollow spheres and their magnetorheological characteristics," *Smart. Mater. Struct.*, vol. 18, 125013 (4pp), 2009.
[119] H. Lv, L. Ma, P. Zeng, D. Ke, and T. Peng, "Synthesis of floriated $ZnFe_2O_4$ with porous nanorod structures and its photo-catalytic hydrogen production under visible light," *J. Mater. Chem.*, vol. 20, pp. 3665-3672, 2010.
[120] G. Y. Zhang, Y.Q Sun, D. Z. Gao, Y. Y Xu, "Quasi-cube $ZnFe_2O_4$ nanocrystals: Hydrothermal synthesis and photocatalytic activity with $TiO_2$ (Degussa P25) as nanocomposite," *Mater. Res. Bull.*, vol. 45, pp. 755–760, 2010.
[121] Y. Deng, Q. Zhang, S. Tang, L. Zhang, S. Deng, Z. Shi, and G. Chen, "One-pot synthesis of $ZnFe_2O_4$/C hollow spheres as superior anode material for lithium ion batteries," *Chem. Commun.*, vol. 47, no. 24, pp. 6828-6830, 2011.
[122] Z. Pang, M. Fang, M. Wu, Y. Liu, Y. Ma, X. Liu, L. Zhang, "Polyvinyl pyrrolidone-assisted hydrothermal synthesis of octahedral $ZnFe_2O_4$ nanoparticles," *Micro. Nano. Lett*, vol. **6,** no. 12, pp. 1012–1015, 2011.
[123] M. Mozaffari, M. E. Arani, J. Amighian, "The effect of cation distribution on magnetization of $ZnFe_2O_4$ nanoparticles," *J. Magnet. Magnet. Mater.*, vol. 322, pp. 3240–3244, 2010.
[124] C. Yao, Q. Zeng, G. F. Goya, T. Torres, J. Liu, H. Wu, M. Ge, Y. Zeng, Y. Wang, and J. Z. Jiang, "$ZnFe_2O_4$ Nanocrystals: Synthesis and Magnetic Properties," *J. Phys. Chem. C.*, vol. 111, pp. 12274-12278, 2007.

**[125]** T. He, D. Chen, X. Jiao, Y. Wang and Y. Duan, "Solubility-Controlled Synthesis of High-Quality $Co_3O_4$ Nanocrystals," *Chem. Mater.*, vol. 17, no. 15, pp.4023-4030, 2005.

[126] X. Jia, D. Chen, X. Jiao, T. He, H. Wang and W. Jiang, "Monodispersed Co, Ni-Ferrite Nanoparticles with Tunable Sizes: Controlled Synthesis, Magnetic Properties, and Surface Modification," *J. Physic. Chem. C.*, vol. 112, pp. 911–917, 2008.

[127] A. Baykal, N. Kasapoglu, Z. Durmus, H. Kavas, M. S. Toprak and Y. Koseoglu, "CTAB-Assisted Hydrothermal Synthesis and Magnetic Characterization of $Ni_xCo_{1-x}Fe_2O_4$ Nanoparticles (x = 0.0, 0.6, 1.0)," *Turk. J. Chem.*, vol. 33, pp. 33-45, 2009.

[128] A. B. Gadkari, T. J. Shinde, P. N. Vasambekar, "Ferrite Gas Sensors," *IEEE. Sens. J.*, vol. 11, no. 4, pp. 849–861, 2011.

[129] R. Q. Chen, C. W. Zou and X. D. Yan, *et al.*, "Zinc oxide nanostructures and porous films produced by oxidation of zinc precursors in wet-oxygen atmosphere," *Progress in Natural Science: Mater. Int.*, vol. 21, no. 2, pp. 81–96, 2011.

[130] X. T. Su, Y. N. Li and J. K. Jian, Jide Wang, "*In situ* etching $WO_3$ Nanoplates: Hydrothermal synthesis, photoluminescence and gas sensor properties," *Mater. Res. Bull.*, vol. 459, no. (12), pp. 1960–1963, 2010.

[131] J. Jiang, "A facile method to the $Ni_{0.8}Co_{0.2}Fe_2O_4$ nanocrystalline *via* a refluxing route in ethylene glycol," *Mater. Lett.*, vol. 61, pp. 3239–3242, 2007.

[132] Y. Tang, X. Wang, Q. Zhang, Y. Li and H. Wang, "Solvothermal synthesis of $Co_{1-x}Ni_xFe_2O_4$ nanoparticles and its application in ammonia vapors detection," *Progr. Nat. Sci: Mater. Inter.*, vol. 22, no. 1, pp. 53–58, 2012.

[133] H. T. Zhang, J. Ding and G. M. Chow, "Synthesis and characterization of Ni-Fe@spinel oxide core-shell nanoparticles," *Mater. Res. Bull.*, vol. 44, no. 5, pp. 1195–1199, 2009.

[134] Q. Zhang, M. Zhu, Q. Zhang, Y. Li, H. Wang, "Fabrication and magnetic property analysis of monodisperse manganese–zinc ferrite nanospheres," *J. Magnet. Magnet. Mater.*, vol. 321, pp. 3203–3206, 2009.

[135] R. Desai, V. Davariya, K. Parekh and R. V. Upadhyay, "Structural and magnetic properties of size-controlled $Mn_{0.5}Zn_{0.5}Fe_2O_4$ nanoparticles and magnetic fluids," *J. Phys.*, vol. 73, no. 4, pp. 765-780, 2009.

[136] F. Gozuak, Y. Koseoglu, A. Baykal, H. Kavas, "Synthesis and characterization of $Co_xZn_{1-x}Fe_2O_4$ magnetic nanoparticles *via* a PEG-assisted route," *J. Magnet. Magnet. Mater.*, vol. 321, pp. 2170–2177, 2009.

[137] C. K. Kim, J. H. Lee, S. Katoh, R. Murakami and M. Yoshimura, "Synthesis of Co-, Co-Zn and Ni-Zn ferrite powders by the microwave-hydrothermal method," *Mater. Res. Bull.*, vol. 36, pp. 2241–2250, 2001.

[138] C. Upadhyay, D. Mishra, H.C. Verma, S. Anand and R.P. Das, "Effect of preparation conditions on formation of nanophase Ni–Zn ferrites through hydrothermal technique," *J. Magnet. Magnet. Mater.*, vol. 260, pp. 188–194, 2003.

[139] H. W. Wang and S.C. Kung, "Crystallization of nanosized Ni–Zn ferrite powders prepared by hydrothermal method," *J. Magnet. Magnet. Mater.*, vol. 270, pp. 230–236, 2004.

[140] M. Sertkol, Y. Koseoglu, A. Baykal, H. Kavas and A. C. Basaran, "Synthesis and magnetic characterization of $Zn_{0.6}Ni_{0.4}Fe_2O_4$ nanoparticles *via* a polyethylene glycol-assisted hydrothermal route", *J. Magnet. Magnet. Mater.*, vol. 321, pp. 157-162, 2009.

[141] Huseyin Kavas, Abdulhadi Baykal, Muhammet S. Toprak, Yuksel Koseoglu, Murat Sertkol and Bekir Aktas, "Cation distribution and magnetic properties of Zn doped $NiFe_2O_4$ nanoparticles synthesized by PEG-assisted hydrothermal route," *J. Alloys. Comp*, vol. 479, pp. 49–55, 2009.

[142] X. Shen, Y.X.Wang, X. Yang, Y. Xia, J. F. Zhuang and P. D. Tang, "Megahertz magneto-dielectric properties of nanosized NiZnCo ferrite from CTAB-assisted hydrothermal process," *Trans. Nonferrous. Met. Soc. China*, vol. 19, pp. 1588-1592, 2009.

[143] Y. Chen, M. Ruan, Y.F. Jiang and S.G. Cheng, W. Li, "The synthesis and thermal effect of $CoFe_2O_4$ nanoparticles," *J. Alloys Compd.*, vol. 493, no.1-2, L36–38, 2010.

[144] P. Raju and S. R. Murthy, "Microwave-Hydrothermal synthesis of $CoFe_2O_4TiO_2$ nanocomposites," *Adv. Mater. Lett.*, Accepted 2012: DOI: 10.5185/amlett. 2012. icnano. 130, Copyright © 2012 VBRI Press.

[145] V. S. R. Raju, S. R. Murthy, F. Gao and Q. Lu, S. Komarneni, "Microwave hydrothermal synthesis of nanosize PbO added Mg-Cu-Zn ferrites," *J. Mater. Sci.*, vol. 41, pp. 1475–1479, 2006.

[146] L. Li, J. Jiang and F. Xu, "Novel polyaniline-$LiNi_{0.5}La_{0.02}Fe_{1.98}O_4$ nanocomposites prepared *via* an *in situ* polymerization," *Europ. Polym. J.*, vol. 42, pp. 2221–2227, 2006.

[147] T. A. Kurniawan, G. Y.S. Chan, W.H.Lo and S. Babel, "Physico-chemical treatment techniques for wastewater laden with heavy metals," *Chem. Eng. J.*, vol. 118, no.1-2, pp. 83-98, 2006.

[148] J.C. Lou and C.K.Chang, "Completely treating heavy metal laboratory waste liquid by an improved ferrite process," *Sep. Purif. Technol.*, vol. 57, pp. 513-518, 2007.

[149] D. Chen, Y. Guo, C.Y.Mei, Y. Li, G.R.Qian, L.H. Yao and H.M. Jin, "Effect of Cu substitution on Ni-Zn ferrite by microwave hydrothermal method," *Bioinformatics & Biomedical Engineering, 5th International Conference,* DOI:10.1109/icbbe.5781173, 4 pages, 2011.

CHAPTER 6

# DFT Study of II-VI Semiconducting Nano-Clusters: An Overview

**Abdul Majid***

*Physics Department, University of Gujrat, Pakistan*

**Abstract:** Semiconductors have been potential materials for device grade applications due to their exceptional properties. In range of other compound semiconductors, II-VI's are such compound semiconducting materials having cations from the Group II and anions from Group VI of the periodic table. Keeping research and technological importance of these materials, density functional theory (DFT) based study of II-VI semiconducting nano-clusters was carried out and is described in this chapter.

**Keywords:** II-VI semiconductors, Density Functional Theory, Raman scattering.

## 1. INTRODUCTION

Semiconductors are prosperous materials for applications in devices and have decades long history of use in appliances due to exceptional properties. Though they exhibit electrical behavior similar to that of insulators at absolute zero but show conductivity between those of insulators and conductors at finite temperatures. The semiconductors in their pure form are known as intrinsic semiconductors. However, if a suitable impurity is added, for technological applications and making devices, the resulting semiconductor is known as extrinsic semiconductor [1]. The introduction of impurities in semiconductors modify their properties in the favor of devices for applications point of view. The electronic and optoelectronic devices are made usually from semiconducting materials containing intentionally introduced impurities. The functioning of all such devices and products depends upon physical and chemical properties of the semiconducting material used. The structural properties, for instance, pointed out the special arrangement of atoms which is of prime importance since worthless coals is converted into precious diamond only upon changing atomic order. The sophisticated computers are made from chips which are made from well organized silicon atoms extracted from sand. Though silicon has been prime research and industrial focus in the past but now compound semiconductors are replacing it for applications. Therefore it can be said that majority of scientific and technological

***Corresponding author Abdul Majid:** Physics Department, University of Gujrat, Pakistan;
E-mail: abdulmajid40@yahoo.com

**Sher Bahadar Khan, Abdullah M. Asiri & Kalsoom Akhtar (Eds)**

advances made in the past depend on the properties of materials at atomic level. One of such technological advancement, rather scientific miracle, is known as "nanotechnology". The prefix nano is derived from the Greek word nanos which bears meaning of dwarf which is explained as something much smaller than normal size. Nanotechnology is aimed at fabrication of structures and functional materials on nanometer scale for applications in future appliances and uses.

Since, one nanometer is one billionth of a meter *i.e.* 1 nm = $10^{-9}$ m, the following examples may be helpful to understand the smallness of a nanometer. The diameter of a red blood cell is 6,000 nm, that of a human hair is 80,000 nm and that of a virus is around 100 nm. In the field of nanotechnology, the size of 100 nm is important because at this dimensional range, one can observe new and astonishing features of the matter. Owing to extensive research efforts of outstanding human brains, we are now in position to organize the matter on the atomic scale. This has made possible realization of numerous products as a reward of our rapidly increasing ability to fabricate the objects of size less than 100 nm. Miniaturization is a prime intention of nanotechnology which has become famous to produce smaller, faster, lighter, efficient and cheaper devices with less energy consumption. Therefore, ongoing research efforts on fabrication of nanomaterials are mainly focused on miniaturization which is expected to contribute significantly towards usage of raw material and energy. Historically, fabrication of nanomaterials begins from fourth century of A.D when glassmakers in Rome fabricated glasses comprising nanosized metals [1,23]. An artifact fabricated during this period named as Lycurgus cup has been placed in British Museum of London. The color of this cup changes from green to dark red on illuminating in light.

In eighteenth century, the British scientists Thomas Wedgewood and Sir Humprey Davy produced images using silver nitrate and silver chloride but their images were not permanent. Several famous researchers such names as Daguerre, Niecpce, Talbot, Archer and Kennet worked on the problem in the nineteenth century. James Clark Maxwell, in his famous work related to electromagnetism, produced the first colored photograph in 1861. Around 1883, an American inventor George Eastman, who later founded the famous Kodak Corporation, developed a film consisting of a long ribbon paper coated with an emulsion of silver halides. This was later modified into form of a delicate film for use in modern photography. In 1959, Richard Feynman in a well known presentation "There's Plenty of Room at the Bottom" raised a question that, is it not possible to write all 24 volumes of Encyclopedia Britannica on pin-head? That question mark put by Feynman had set the pace for modern nanotechnology at the time when viewing the atom using microscope was just a dream. A short historical description of nanomaterials research, with emphasis on semiconducting materials, is presented here to understand the widespread interest in the nanotechnology.

As it was described earlier that semiconductors are having conductivity in between the conductor and insulator materials. They are found in both forms; bulk and nano sizes. The structural properties of the semiconductor remain nearly same in both forms of the material but their physical properties change drastically. At nanoscale the effectiveness of any property of the semiconducting material is enhanced which attracted majority of research groups worldwide to work on. Semiconductors can be classified into elemental and compound semiconductors. Elementary semiconductors, Si and Ge for example, are natural occurring semiconductors whereas compound semiconductors are fabricated by artificial means in laboratory. Compound semiconductors can be differentiated further into sub groups known as binary, ternary and even quaternary semiconductors. Binary semiconductors consists of elements of two different groups of the periodic table, GaN, CdS, InP for example, and ternary semiconductors are made by combining three elements e.g GaInN, AlInN *etc*. Famous compound semiconductors are formed by mixing elements of III-V groups (GaN, GaP, AlN *etc*) or by combining elements of II-VI groups (ZnO, CdS *etc*). In 1947, researchers from Bell's laboratory invented first transistor based on semiconductor. This happened to be a milestone after which research and industrial interests in field of semiconductors grown abruptly. Consequent upon it, fabrication of faster, smaller in size, economical, powerful and efficient devices became possible. Though, first transistor was of Germanium (Ge) but later silicon gained more research interests due to its exception properties. Silicon then became an industrial trade mark and dominated economy of several countries due to progressive development in its fabrication technology. It has been used as discrete component of devices and integrated circuits for switching, data storage, computing and communications.

By the end of last century, besides elemental semiconductors, several groups started research activities in compound semiconductors. At that time GaAs emerged as a best choice for high speed and optoelectronic devices, such as laser diodes and high speed integrated circuits. It is III-V compound semiconductor having high effective carrier velocity and mobility in comparison to that of Si. Moreover, GaAs due to its direct band gap nature unlike Si, which is indirect band gap, attracted extensive scientific interests for use optoelectronic applications. Despite the promising physical properties, some other characteristics which are necessary for high power and high temperature electronics and in the applications of blue/UV emitter are missing in the GaAs and Si. This led the community to search for the alternate materials. The focal point then was to explore new wide band gap semiconductors possessing additional inherent properties like high break down field strength and carrier mobility. Several top research groups were involved in growing compound semiconductors suitable for the short wavelength, high power and high temperature optoelectronic devices.

II-VI semiconductors are such compound semiconducting materials having cations from second and anions from fourth group of periodic table. Renowned II-VI semiconductor materials are cadmium selenide (CdSe), cadmium telluride (CdTe), zinc sulphide (ZnS), zinc oxide (ZnO), and mercuric selenide (HgSe).

Among II-VI semiconductors family, CdS is a potential candidate for photonic devices. Bulk CdS in hexagonal wurtzite type structure has lattice parameters a = 4.136 Å, c = 6.714 Å [2], density 4.82 g/cm$^3$[3], melting point 1600°C [4] and band gap 2.42 eV [5] at room temperature and pressure of 1 atm. Due to its band gap it is most important contestant among other II-VI semiconductors for harvesting visible radiation of solar spectrum. It is direct and wide band gap, so it can be implemented in the form of window material for applications in hetero-junction solar cells for avoiding the recombination process of photo generated electrons and holes [6]. The harvest of sunlight using CdS quantum dots to produce sufficiently high number of electrons and to enhance solar cell efficiency have been reported [7-9]. CdS NCs are reported to have applications in light emitting diodes [10], photo resistors [11], and the most recent in biosensors to enhance the sensitivity in order to detect DNA hybridization [12, 13].

Nanostructures of ZnO have attracted an accelerated research interest in the recent past due to their potential properties for electrical and optoelectronic appliances. ZnO is also a wide band gap semiconductor with room temperature energy gap of 3.4 eV. It is also found in stable hexagonal wurtzite structure having lattice constants of a=3.24992 Å, c=5.20658 Å, internal parameter u 0.375 and the c/a ratio has value of 1.626. It is a direct band gap II-VI compound semiconductor having requisite properties for use in blue/UV optoelectronics, spintronics and sensor applications. Usually ZnO is in the form polycrystalline material and extensively used in different fields e.g ointment, facial powder, sunscreen, catalyst, lubricant additives, paint, pigment, piezoelectric transducer, varistors *etc.*

Among II-VI semiconductors ZnSe is another important material having direct band gap of 2. 67 eV at room temperature. It has very efficient luminescence in the blur parts of spectral region. It is widely used in the fabrications of blue diode lasers. ZnSe NCs have been studied due to their fascinating band structure and applications for applications in future electronic and optoelectronics devices [14-20]. Due to broad range of expected potential applications [21, 22] this material, existing in hexagonal (wurtzite) or the cubic (sphalerite, zinc blende) phases is hopefully a future focus of material scientific community.

Since nano-clusters (NCs) are renowned building blocks of nanomaterials, this study involves study of semiconducting NCs using density functional theory based computations. The calculated results on NCs will be helpful in realizing the corresponding nanomaterials. A detailed description on vibrational dynamics of NCs of compound semiconductors zinc oxide (ZnO), cadmium sulfide (CdS) and zinc selenide (ZnSe) is given in this chapter.

## 2. COMPUTATIONAL DETAILS

ADF (Amsterdam Density functional) is a computational package based on DFT (Density Functional Theory) used for the calculations of different periodic and non-periodic systems *i.e.* band diagram and density of states for crystal structures and spectroscopic properties for different molecules and NCs *etc.* It has two main categories; ADF-Molecule and ADF-Band. The ADF-Band program is used for the calculations of periodic systems [24] but this work reports the results calculated using ADF- Molecule for Raman spectroscopic properties of different sized NCs of II-VI series.

### 2.1. Introduction to ADF-Molecule

ADF-Molecule is a program for calculating properties of molecules and NCs. It works within Kohn-Sham approach and is very similar to ADF- Band. Like ADF-Band, ADF- Molecule employs atomic orbitals in such a way that it is able to handle all elements in the periodic table to investigate spectroscopic, thermodynamic and electronic properties *etc.* of the materials. It is conventional to write the ADF-Band as BAND only and ADF-Molecule as ADF only.

### 2.2. Computing Spectroscopic Properties with ADF

Raman spectroscopy (RS) is a well established and resourceful tool for chemical and structural identification of materials in any form. Its analytical capabilities, on both macro and micro level without perturbing a sample, have made this technique a versatile one. Additionally, it is being used in condensed matter physics, chemistry, biochemistry, medical sciences and allied disciplines for study of the vibrational modes of the system of atoms or molecules. RS works on the basis of Raman effect which illustrates the effects of inelastic scattering of photons by structures under investigation and interaction of photons with molecules to induce spectral transitions. The scattering changes rotational and electronic energy of a molecule and the energy difference of incident photons and the scattered photons is equal to vibrational energy of the scattering molecule [25]. Thus Raman spectrum exhibits the plot showing intensity of scattered light as a function of energy difference. Using the ADF-Molecule one can calculate the variety of spectroscopic properties like spectra as Raman, IR and UV/Vis, NMR with appropriate quantum mechanical method. These spectra can be displayed and analyzed in the ADF graphical user interface [26]. The IR and Raman spectra can be calculated in the single run in which the run type should be of 'Frequencies' selected by the key of frequencies. These calculations become very costly if we increase the size of NCs and are therefore more favorable for small NCs (*i.e.* these calculation takes less time for small size NCs). This chapter deals with modes of vibrations calculated using Raman spectrum for the NCs of different sizes using run type of Frequencies in ADF-Molecule interface. All the calculations were carried out within Local Density Approximation (LDA), double zeta (DZ) basis set and 'Large' core type. The Raman frequency in full ranging from (0-1000) 1/cm was calculated in all cases. The comparison of Raman spectra of molecules of different number of atoms shows that some modes of vibrations are shifted towards higher frequencies and some towards lower frequencies with the increase in size.

## 3. RESULTS AND DISCUSSION

The calculations were carried out using density functional theory to obtain the Raman spectra of cadmium sulfide (CdS), zinc oxide (ZnO) and zinc selenide (ZnSe) NCs containing 2, 4, 8 and 12 number of atoms. All the modes of vibrations are discussed in detail with their NCs size dependent variation in the Raman shift and in Raman intensity.

### 3.1. Study of CdS NCs

Raman Modes of CdS NCs, modifications in frequency and intensity of different modes of vibration of CdS as a function of number of atoms in NCs were studied and interpreted to explore the underlying mechanism [27]. The Raman spectrum of CdS NCs of two atoms is shown in Fig. (**1**).

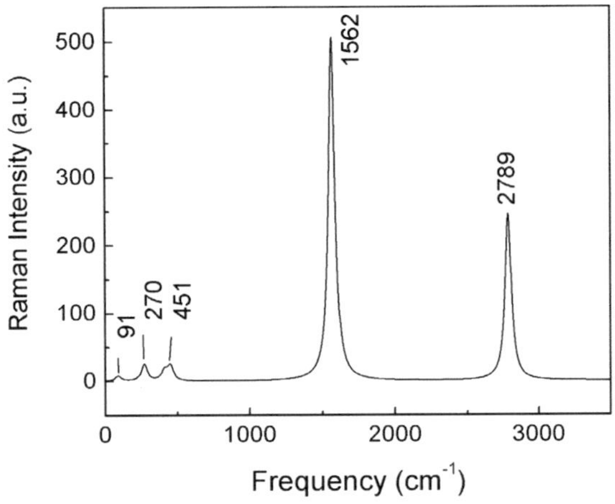


**Figure 1:** Raman spectrum calculated for two atoms CdS NCs [27].

Different peaks are observed in the spectrum which point to different Raman modes for two atoms CdS NCs. The modes were observed at 91 $cm^{-1}$, 270 $cm^{-1}$, 451 $cm^{-1}$, 1562 $cm^{-1}$, and 2789 $cm^{-1}$ whereas identification of these modes will be given in the coming sections. Similar procedure of calculations was used for the CdS NCs of 4, 8, and 12 atoms and corresponding Raman spectra were recorded. Fig. (**2**) gives the Raman spectra calculated for the WZ CdS NCs of 2, 4, 8 and 12 atoms.

The spectra are distributed into low frequency (<600 $cm^{-1}$) and high frequency regions (>600 $cm^{-1}$) for sake of explanation. The modes in low frequency region are present in extreme left whereas high frequency modes are shown the form of prominent peaks in high energy region of the spectra. The modes in high frequency region are discussed first. The low frequency Raman modes need de-convolution and will be discussed in later section. For the case of two atoms CdS NCs, high frequency modes are observed at frequencies of 1562 $cm^{-1}$ and 2789 $cm^{-1}$. The mode at the frequency of 1562 $cm^{-1}$ is Cd-H anti-symmetric stretching and at 2789 $cm^{-1}$ is S-H anti-symmetric stretching mode in accordance with the ADF user interface and the literature. Hydrogen was used to saturate the CdS NCs in order to complete the outer most orbital. These Raman modes of vibrations were also observed for the CdS NCs of 4, 8 and 12 atoms as given elsewhere [27]. It indicates the Raman spectra for hydrogenated CdS NCs of 2, 4, 8 and 12 atoms in the frequency range of 1000 to 3000 $cm^{-1}$.

In the following session, each of the Raman modes will be individually discussed to shed light on vibrational properties of the NCs with the increase in size by increasing the number of atoms.

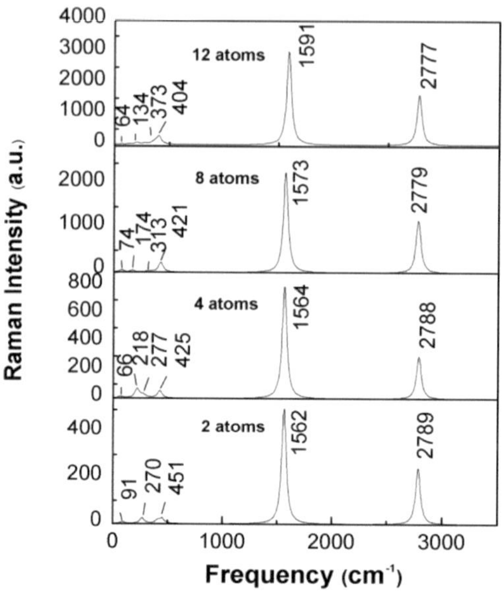


**Figure 2:** Raman spectra calculated on NCs of CdS containing 2, 4, 8, and 12 atoms [27].

**Cd-H Anti-symmetric stretching (strong):** The respective values of frequencies observed for Cd-H anti-stretching Raman mode are 1562 $cm^{-1}$, 1564 $cm^{-1}$, 1573 $cm^{-1}$ and 1591 $cm^{-1}$ for the CdS NCs of 2, 4, 8 and 12 number of atoms. The literature indicates values of characteristic frequency for CdH stretching mode is 1430 $cm^{-1}$, 1775 $cm^{-1}$ and 1836 $cm^{-1}$ [28]. There is an increase in frequency with number of atoms pointing towards blue shift.

**S-H Anti-symmetric stretching (strong):** The values of frequency observed for S-H anti-symmetric stretching Raman modes are 2789 $cm^{-1}$, 278 $cm^{-1}$, 2779 $cm^{-1}$ and 2777 $cm^{-1}$ for the CdS NCs of 2, 4, 8, and 12 atoms respectively. The reported value of vibrational frequency for SH stretching is about 2675 $cm^{-1}$ [28]. The trend of frequency for this Raman mode is a decrease in value with the increase in number of atoms pointing towards the red shift.

The direct dependence of frequency with number of atoms is interpreted in terms of coherent anti-stokes Raman scattering (CARS) mechanism [30]. When an incident photon of certain frequency is scattered by promoting the vibrational state to one of the excited states the frequency of the scattered photons decreases. The scattered photon again undergoes through the scattering process by an initially excited state thereby leading this excited state to some other lower energy state. When atom returns from initially excited state to less energy state the energy is emitted by the sample and the frequency increases. As the number of atoms increases the above discussed process (CARS) goes on multiplying because the number of initially excited states show direct dependence with number of atoms, hence the frequency is increased by the process of CARS. This process will rarely happen if numbers of molecules or NCs in excited state is small. The observed decrease in frequency with number of atoms is interpreted *via* stimulated Raman scattering mechanism. When incident photon of certain frequency is allowed to fall on the hydrogenated CdS NCs of two atoms, containing bonds Cd-S, S-H and Cd-H, the incident photons scattered by promoting the atoms from lower to higher energy states. It should be noted that no further vibrational state will be available for the scattering of photons in case of 2- atom NCs. On increasing the number of atoms there are more bonds and hence more atoms in available vibrational states [30]. When the incident photon is allowed to fall on the CdS NCs of more number of atoms then it is scattered from one of the allowed vibrational states. Some of its energy is absorbed by the NCs in promoting it to the higher energy vibrational state and therefore the frequency of scattered photon is reduced. This scattered photon act as an incident photon for the next atoms present in a specific vibrational state because for bigger NCs there exist more atoms in the same vibrational state. This scattered photon is further scattered by the next atoms importing some of its energy in promoting these atoms from their initial to some excited vibrational state.

The frequency of finally scattered photon is further reduced as in case of Raman stokes scattering therefore this process is called Raman stimulated scattering. Figures given in ref. 27 shows variation of the frequencies of the Raman shift Cd-S symmetric stretching The respective values of Cd-S symmetric stretching mode for the NCs of 2, 4, 8, and 12 atoms are 260 $cm^{-1}$, 213 $cm^{-1}$, 174 $cm^{-1}$ and 134 $cm^{-1}$. A decreasing trend in Raman shift is observed as the number of atoms increases which points to the red shift which is assigned to Raman stimulated scattering. The reported value of this characteristic frequency for CdS stretching mode is however 228 $cm^{-1}$ [29].

**Cd-S anti-symmetric stretching mode:** The values of frequency for Cd-S anti-symmetric stretching mode are 270 $cm^{-1}$, 273 $cm^{-1}$, 313 $cm^{-1}$ and 373 $cm^{-1}$ for CdS NCs of 2, 4, 8 and 12 numbers of atoms. This Raman mode show the blue shift as the frequency shifts towards higher frequency with the increase in number of atoms and is assigned to CARS mechanism. The frequency of CdS anti - symmetric stretching mode increases because the value of frequency for CdS symmetric stretching decreases.

**S-H rock (strong):** The respective values of SH strong rocking mode are 308 $cm^{-1}$, 304 $cm^{-1}$, 295 $cm^{-1}$ and 275 $cm^{-1}$ in the CdS of 2, 4, 8 and 12 atoms NCs. The observed decreasing trend in the frequency with increase in number of atoms indicates a red shift which is likely due to stimulated Raman scattering.

**Cd-H rock (medium):** The respective values of frequency for Cd-H rocking mode are 451 $cm^{-1}$, 430 $cm^{-1}$, 426 $cm^{-1}$ and 404 $cm^{-1}$ for the CdS NCs of 2, 4, 8 and 12 atoms. The frequency for Cd-H rocking mode of vibration is observed shifted towards smaller frequency which indicate the red shift and is designated to stimulated Raman scattering.

**S-H twisting (strong):** The values of Raman shift for S -H twisting mode are calculated as 413 $cm^{-1}$, 419 $cm^{-1}$, 433 $cm^{-1}$ and 442 $cm^{-1}$ for the CdS NCs of 2, 4, 8 and 12 atoms. A blue shift is observed as a function of number of atoms. The reported value of vibrational frequency, calculated using DFT method, for SH twisting mode is 439 $cm^{-1}$ [29].

### 3.2. Study of ZnO NCs

Raman scattering studies were also carried out for hydrogenated WZ ZnO NCs using similar procedure. Modifications in RS and in relevant intensity for several modes in ZnO NCs were calculated and interpreted to explore the underlying mechanism. The spectrum of ZnO NCs of two atoms is shown in the Fig. (**3**).

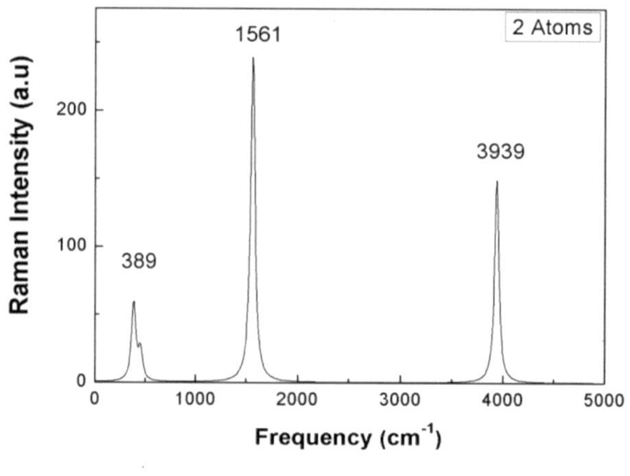


**Figure 3:** Raman spectrum of ZnO NCs consisting of two atoms.

Different peaks were observed in the spectrum which point to different Raman modes for two atoms ZnO NCs. The modes were observed at the frequencies of

389 $cm^{-1}$, 1561$cm^{-1}$ and 3939 $cm^{-1}$ as labeled in Fig. (**3**). Identification of these modes is given in coming sections. Similar procedure of calculations was performed for the ZnO NCs of 4, 8 and 12 atoms and calculations of corresponding Raman spectra were carried out. Fig. (**4**) shows the Raman spectra for the WZ ZnO NCs of 2, 4, 8 and 12 atoms.

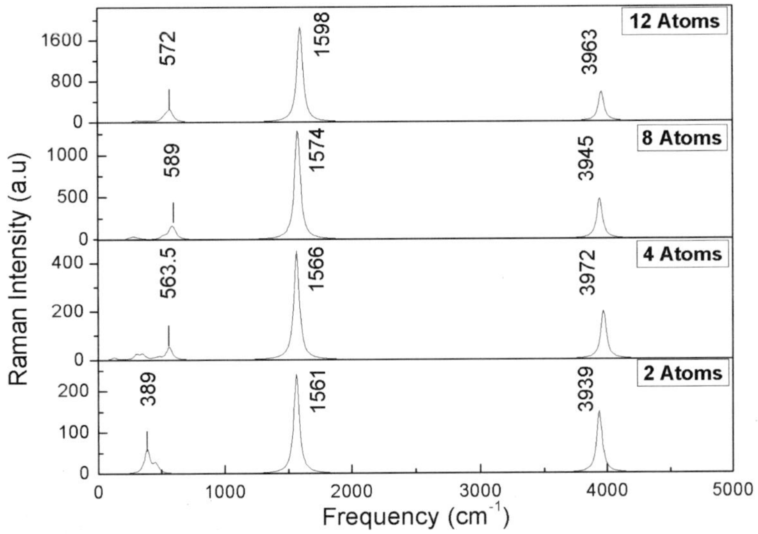


**Figure 4:** Raman spectra for ZnO NCs of 2, 4, 8, and 12 atoms.

For sake of simplicity and explanation, the above mentioned spectrum is divided into two regions, low frequency (< 600 $cm^{-1}$) and high frequency (> 600 $cm^{-1}$). The small peaks in the extreme left indicates the modes at low frequency whereas the dominant peaks in rest of the spectrum indicates high frequency Raman modes of the materials studied. Firstly, high frequency Raman modes are discussed whereas low frequency modes will be interpreted afterwards.

For the case of two atoms ZnO NCs the high frequency modes are observed at a frequency of 1561 $cm^{-1}$ and 3939 $cm^{-1}$. The Raman modes at the frequency of 1561 $cm^{-1}$ is Zn-H anti-symmetric stretching and at 3939 $cm^{-1}$ is O-H anti symmetric stretching as observed in the ADF user interface and also reviewed in the literature. These modes of vibrations are also observed for the ZnO NCs of 4, 8, and 12 atoms respectively as shown in Fig. **5**.

In the following section Raman modes are individually elaborated with the increase in size of the ZnO NCs by increasing the number of atoms.

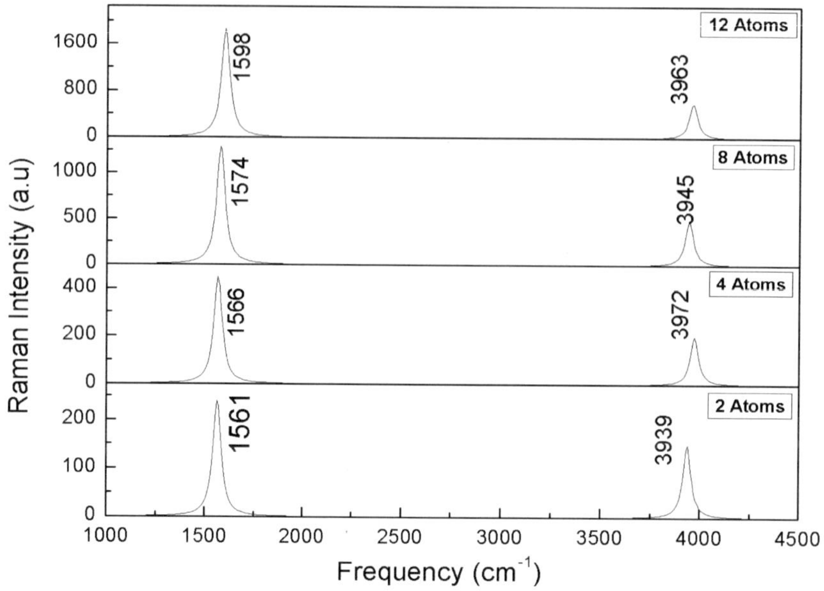


**Figure 5:** High frequency Raman modes of ZnO NCs.

**Zn-H Anti Symmetric Stretching mode:** The Zn-H anti-symmetric Raman mode is observed at frequencies of 1561 cm$^{-1}$, 1566 cm$^{-1}$, 1575 cm$^{-1}$ and 1598 cm$^{-1}$ for ZnO NCs of the 2, 4, 8 and 12 number of atoms as shown in Fig. (**5**). The reported value of this characteristic frequency of the Zn-H stretching mode is 1475 cm$^{-1}$ [32, 35] and 1705 cm$^{-1}$ [31-35]. The value of frequency may change upon using different computational and experimental conditions. An increase in the value of frequency is observed as the number of atoms increases which points to the blue shift.

**O-H Anti-symmetric stretching mode:** The O-H anti Symmetric stretching Raman mode per calculations is observed at frequencies of 3939 cm$^{-1}$,3972 cm$^{-1}$, 3945 cm$^{-1}$ and 3962 cm$^{-1}$ for the WZ ZnO NCs of 2, 4, 8 and 12 atoms respectively. In literature the vibrational frequencies of O-H is 3498 cm$^{-1}$ [31-35], 3687 cm$^{-1}$, and 3670 cm$^{-1}$ [32, 35]. The trend of frequency for this Raman mode is different for different atoms. The value of frequency for this Raman mode is maximum for 4 atoms ZnO NCs whereas 2 atoms NCs has minimum frequency for this specific Raman mode. Overall behavior of the frequency is that it increases in direct proportion with number of atoms which is indicating blue shift however this increment is nonlinear. All the frequencies of high Raman modes are listed in the Table **1** indicating their strength of vibrations.

| Sr. | Modes of vibrations | Raman shift (1/cm) | | | | |
|---|---|---|---|---|---|---|
| | | Strength | 2-atoms | 4-atoms | 8-atoms | 12-atoms |
| 1 | Zn-H Stretching (A) | v *st.* | 1561 | 1566 | 1575 | 1598 |
| 2 | O-H stretching (A) | v *st.* | 3939 | 3972 | 3945 | 3963 |

Note: (A) = anti-symmetric, (S) = Symmetric, *vs* = very slow, v st. = very strong, m= medium.

Fig. **6** shows a plot of frequency (Raman shift) against number of atoms for Zn-H and O-H stretching which clearly shows the blue shift as the number of atoms are increased in the WZ ZnO NCs of 2, 4, 8 and 12 atoms respectively.

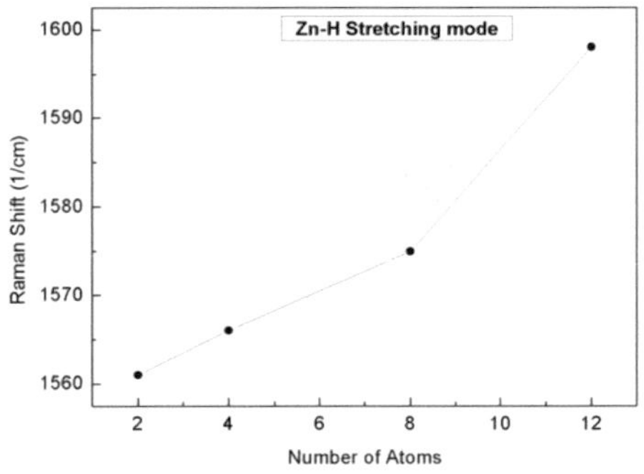


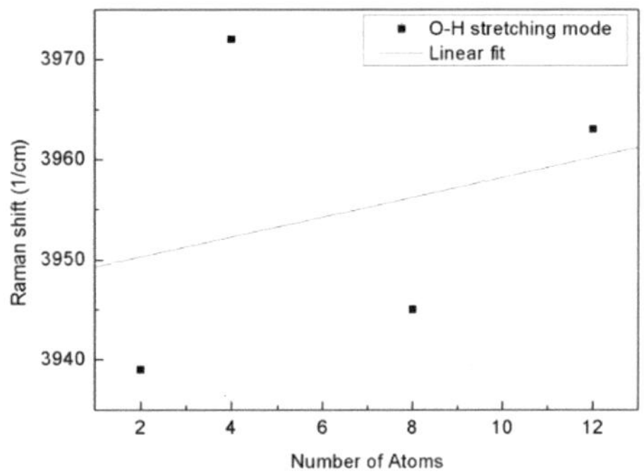


**Figure 6:** Variations in frequency of the Zn-H (left) and O-H (right).

Variation in frequency of the Zn-H and O-H stretching modes for ZnO NCs of different sizes given in Fig. (**6**) indicates that the frequency of O-H stretching mode and that of Zn-H mode increases in nonlinear way with increase in number of atoms.

**Low Frequency Raman modes:** Low frequency Raman modes of vibrations were calculated for different sized ZnO NCs. The trend of variation of frequency of low frequency Raman modes with change in number of atoms is shown in the Fig. (**7**). Just like the high frequency Raman modes, there is also increase in the frequency with change in the number of atoms. This rise in the frequency indicates the blue shift.

### 3.3. Study of ZnSe NCs

NCs Using similar computational details and approach Raman spectra of hydrogenated ZnSe NCs of different sizes were also calculated. Raman spectra of ZnSe NCs of 2, 4, 8 and 12-atoms are shown in the Fig. (**8**). The Raman Spectrum of ZnSe NCs of 2-atoms contains two peaks at 965 $cm^{-1}$ and 2412 $cm^{-1}$ which are assigned to scissoring of H-Se and strong anti-stretching of H-Se respectively. Similarly there are two main peaks in the Raman spectrum of 4-atoms first at

1569 $cm^{-1}$ and second is at 2395 $cm^{-1}$. First peak describes the strong stretching of Zn-H [31-35] whereas the peak at 2395 $cm^{-1}$ is due to strong stretching of H-Se bond. The spectrum of 8-atoms NCs also presents two main peaks. The peak at 1620 $cm^{-1}$ indicates the stretching of Zn-H in the medium range whereas the peak at 2421 $cm^{-1}$ represents the same mode of H-Se. The calculated spectrum of 12-atoms has two peaks one is at 1629 $cm^{-1}$ and other is at 2425 $cm^{-1}$. First peak tells us symmetric stretching of Zn-H and second peak describes the anti symmetric stretching mode in the medium range.

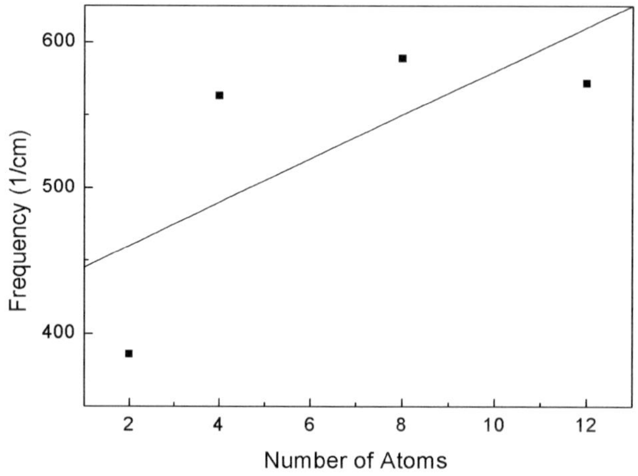


**Figure 7:** Variation in frequency of low frequency Raman mode for different sized ZnO NCs.

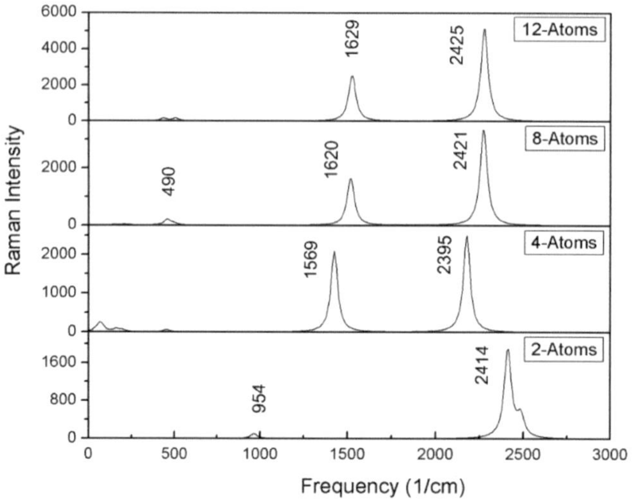


**Figure 8:** Raman spectra of ZnSe NCs of different sizes.

The Raman Spectra of high frequency modes indicate the blue shift in the frequency as was studied in the case of ZnO and CdS NCs.

## ACKNOWLEDGEMENTS

Declard none.

## CONFLICT OF INTEREST

The author confirms that this book contents have no conflict of interest.

## REFERENCES

[1] D.K. Avasthi, G.K. Mehta, *Swift Heavy Ions for Materials Engineering and Nanostructuring.* Springer series in material science, 2011.
[2] Villars P. and L. D. Calvert, *Pearson's Handbook of Crystallographic Data for Intermetallic Phases.* Metals Park, OH: ASM international, 1985.
[3] S.S. Li, *Semiconductor physical electronics.* (2nd ed.). Springer (USA), 2006.
[4] A.N. Goldstein, C.M. Echerand, A.P. Alivisatos, "Melting in semiconductor nanocrystals," *Science.*, vol. 256, pp.1425-1427, 1992.
[5] N. R. De Tacconi, C. R. Chenthamarakshan, K. Rajeshwar, "Semiconductor-Based Composite Materials: Preparation, Properties, and Performance," *Chem. Mater.,* vol. 13, pp. 2765-2782, 2001.
[6] A.M Acevedo, "Can we improve the record efficiency of CdS/CdTe solar cells?," *Solar. Energy. Materials. Solar. Cells.,* vol. 90, pp. 2213-2220, 2006.
[7] B. Kochman, A.D.S Roberts, S. Chakrabarti, J.D Phillips, S. Krishna, J. Singh, P. Bhattacharya, "Absorption, carrier lifetime, and gain in InAs-GaAs quantum-dot infrared photodetectors," *IEEE. J. Quantum. Elect.*, vol.39, pp.459-467, 2003.
[8] M. Green, A. Wang, and S. Wenham, "Very high efficiency silicon solar cells science and technology," *IEEE. Transactions. Elect. Devices.*, vol. 46, pp. 1940-1947, 1999.
[9] D. W. Parent. (2015) Solar Cell Fundamentals, SJSU Short Course
[10] M.C. Schlamp, X. Peng, and A.P. Alivisato, "Improved efficiencies in light emitting diodes made with CdSe(CdS) core/shell type nanocrystals and a semiconducting polymer," *J. Appl. Phys.* vol.82, pp. 5837-5842, 1997.
[11] J. Wang, S. Ramanathan, Q. Fan, F.Yun, H. Morkoe, S. Nandyopadhyay, J. Nanoscience. "Electric field modulation of infrared absorption at room temperature in electrochemically self assembled quantum dots". *Nanotechnology.*, vol. 6, pp. 2077-2080, 2006.
[12] Y. Xu, H. Cai, P.G. He, Y.Z. Fang, "Probing DNA hybridization by impedance measurement based on CdS-oligonucleotide nanoconjugates"., *Electroanal.*, vol. 16, pp. 150-155, 2004.
[13] Q. Xia, X. Chen, K. Zhao, J.H. Liu, "Synthesis and characterizations of polycrystalline walnut-like CdS nanoparticle by solvothermal method with PVP as stabilizer," *Mater. Chem. Phys.*, vol. 111, pp. 98-105, 2008.
[14] P. C. Ohara, D. V. Leff, J. R. Heath, W. M. Gelbart, "Crystallization of Opals from Polydisperse Nanoparticles," *Phys. Rev. Lett.*, vol. 75, pp. 3466-3469, 1995.
[15] T. Trindade, P.O. Brien, N. L. Pickett, "Nanocrystalline Semiconductors: Synthesis, Properties, and Perspectives," *Chem. Mater.* vol. 13, pp. 3843-3858, 2001.
[16] M. A. Malik, N. Revaprasadu, P.O. Brien, "Air-stable single-source precursors for the synthesis of chalcogenide semiconductor nanoparticles, " *Chem. Mater.*, vol. 13, pp. 913-920, 2001.
[17] M. A. Malik, N. Revaprasadu, P.O. Brien, "A simple route to the synthesis of core/shell nanoparticles of chalcogenides, " *Chem. Mater.,* vol. 14, pp. 2004-2010, 2002.
[18] W.Z. Wang, Y. Geng, P. Yan, F.Y. Liu, Y. Xie, Y.T. Qian, "A novel mild route to nanocrys- talline selenides at room temperature," *J. Am. Chem. Soc.*, vol. 121, pp. 4062-4063, 1999.

[19] Y. Li, Y. Ding, Y. T. Qian, Y. Zhang, L. Yang, "A solvothermal elemental reaction to produce nanocrystalline ZnSe," *Inorg. Chem.*, vol. 37, pp.2844-2845, 1998.

[20] M. Hasse, J. Qui, J. Depuydt, H. Cheng, Blue-green laser diodes. *Appl. Phys. Lett.,* vol. 59, pp.1272-1274, 1991.

[21] Y. Jiang, X. M. Meng, W. C. Yiu, J. Liu, J. X. Ding, C. S. Lee, S. T. Lee, "Zinc selenide nanoribbons and nanowires," *J. Phys. Chem. B.,* vol. 108, pp. 2784-2787, 2004.

[22] Q. Li, X. Gong, C. Wang, J. Wang, K. Ip, S. Hark, "Size-dependent periodically twinned ZnSe nanowires," *Adv. Mater.,* vol. 16, pp. 1436-1440, 2004.

[23] R. Rathi, "Core Concept of Nanotechnology with Application Spectrum", SBS publisher, 2007.

[24] R. M. Rodriguez, Synthesis, structural characterization and spectroscopic study of nanocrystalline and microcrystalline materials. PhD Thesis, 2010.

[25] B. Schrader, Infrared and Raman Spectroscopy. VCH (NY) Publishers Inc., 1995.

[26] Available: www.scm.com

[27] A. Majid, R. Ahmad, A. Nabi, A. Shakoor and N. Hassan, "A Density Functional Theory Study of Raman Modes of Hydrogenated Cadmium Sulphide Nanoparticles," *Nanomater. Nanotechnol.* vol. 2, pp.1-8, 2012.

[28] J. Polit, A. Kisiel, A. Mycielski, A. Marcelli, E. Sheregii, J. Cebulski, M. Piccinini, M. Cestelli Guidi, B. V. Robouch, and A. Nucara, "Vibrational spectra of hydrogenated CdTe," *Phys. stat. sol. (c).,* vol. 2, pp.1147-1154, 2005.

[29] S. Joshi, "Preparation and characterization of CdS NPs", PhD thesis, 2004.

[30] J. M. Holls, (1998). *High resolution spectroscopy.* (2nd ed.). Wiley and sons

[31] R. P. Eischens, W. A. Pliskin and M. J. D. Low, "The infrared spectrum of hydrogen chemisorbed on zinc oxide," *J. Catal.*, vol. 1, pp. 180-1991, 1962.

[32] F. Boccuzzi, E. Borello, A. Zecchina, A. Bossi and M. Camia, "Infrared study of Zn surface properties. I. H2 and D2 chemisorption at room temperature," *J. Catal.* vol. 51, pp. 150-159, 1978.

[33] F. Boccuzzi, C. Morterra, R. Scala and A. Zecchina, (1981). Infrared spectrum of microcrystalline zinc oxide. "Electronic and vibrational contributions under different temperature and environmental conditions," *J. Chem. Soc., Faraday Trans.,* vol. 77, pp. 2059-2066, 1981.

[34] G. L. Griffin and J. T. Yates, "Adsorption of $H_2$ Isotopes on ZnO: Coverage Induced IR Frequency Shifts and Adsorbate Geometry," *J. Chem. Phys.*, vol. 77, pp. 3744-3751, 1982.

[35] G. Ghiotti, A. Chiorino and F. Boccuzzi, "Surface chemistry and electronic effects of $H_2$ ($D_2$) on two different microcrystalline ZnO powders," *Surf. Sci.* vol. 287/288, pp. 228-234, 1993.

# Subject Index



Sher Bahadar Khan, Abdullah M. Asiri & Kalsoom Akhtar (Eds)

## D



## E

## *F*



## *G*



## *H*



## *I*

### *K*



### *L*



### *M*

***Q***



***R***



***S***

## *T*

## U



## V



## W



## X



## Z

Made in the USA
Las Vegas, NV
03 November 2026

57424576R00162